Springer Series in Optical Sciences

Volume 211

Springer Series in Optical Sciences is led by Editor-in-Chief William T. Rhodes, Georgia Institute of Technology, USA, and provides an expanding selection of research monographs in all major areas of optics:

- lasers and quantum optics
- ultrafast phenomena
- optical spectroscopy techniques
- optoelectronics
- information optics
- applied laser technology
- industrial applications and
- other topics of contemporary interest.

With this broad coverage of topics the series is useful to research scientists and engineers who need up-to-date reference books.

More information about this series at http://www.springer.com/series/624

Arkadi Chipouline · Franko Küppers

Optical Metamaterials: Qualitative Models

Introduction to Nano-Optics and Optical Metamaterials

Springer

Arkadi Chipouline
Center for Photonics and Quantum Materials
Skolkovo Institute of Science
and Technology
Moscow, Russia

Institute for Microwave Engineering
and Photonics
Darmstadt University of Technology
Darmstadt, Germany

Franko Küppers
Center for Photonics and Quantum Materials
Skolkovo Institute of Science
and Technology
Moscow, Russia

Institute for Microwave Engineering
and Photonics
Darmstadt University of Technology
Darmstadt, Germany

ISSN 0342-4111 ISSN 1556-1534 (electronic)
Springer Series in Optical Sciences
ISBN 978-3-319-77518-0 ISBN 978-3-319-77520-3 (eBook)
https://doi.org/10.1007/978-3-319-77520-3

Library of Congress Control Number: 2018938640

This Springer imprint is published by the registered company Springer Nature Switzerland AG
The registered company address is: Gewerbestrasse 11, 6330 Cham, Switzerland

Preface

In modern photonics, there is the evident inclination in favor of numerical methods in order to describe, for example, the optical properties of metamaterials (MMs) at the expense of physical intuition. There is no doubt that modern numerical algorithms and available computer facilities provide the main way to investigate more or less complicated problems. Nevertheless, the qualitative approximate type models can provide a deeper understanding of the basic physical processes, stimulate discussion of new effects, and even provide new paradigm for optimization of a particular design. The qualitative models are complementary to the numerical ones, taking advantages of careful comparison with the results of rigorous numerical calculations, but at the same time remaining analytically treatable. In order to create this type of model, accurate approximations have to be made in order to simplify the respective consideration, at the same time keeping the main physical effects and interplay between them in the model. In turn, it requires specific mathematical apparatus and specific methods, which give especially useful results if a wide range of the problems is considered in the frame of the same, unified, and self-consistent approach. The qualitative models allow us to show mutual self-consistency of the physical theory (in this particular case in application to the MM), which otherwise could be seen as a huge leap from independent experimental to theoretical facts. Elaboration of the qualitative models in the frame of the unifying paradigm appears to be useful not only for the consideration of the specific problems but also for the teaching of the respective courses as well.

The developed multipole approach assumes an expansion of the charge dynamics in atom/molecules, which in the case of the MM has to be replaced by the charge dynamics in the metaatoms (MAs). This gives us a unique avenue for the creation of a unified approach to all possible types of MMs: charge dynamics in the MAs can be described in the frame of classic or quantum models, but the algorithm for the calculation remains the same. It motivates us, in turn, to try to create a unified approach, which would unify classic, quantum, or semi classic MM.

Authors are extremely thankful for the long-term support of the colleagues from Friedrich-Schiller University of Jena. Authors also specially acknowledge inestimable editing support received from Dr. Alex Brown.

Moscow, Russia Arkadi Chipouline
Darmstadt, Germany Franko Küppers

The original version of the book was revised: Author corrections have been incorporated. The correction to the book is available at
https://doi.org/10.1007/978-3-319-77520-3_15

Contents

Chapter 1
Introduction to Optical Metamaterials: Motivation and Goals

> *It is necessary to reconstruct the textbooks and change some pedagogical methods in lecturing of electrodynamics... It turned out to be necessarily to elaborate the electrodynamics from fundamental principles taking into account possible magnetic effects.*
>
> Prof. V. G. Veselago

1.1 Appearance and Development of Metamaterials

It is now commonly accepted that the era of metamaterials (MM) was ushered in with the publication of the widely known paper by Prof. Veselago [1] who first suggested that the basic principles of electrodynamics do not forbid the possibility of materials with negative values of the real parts of both the permittivity and permeability. One consequence of this suggestion is the existence of the so called left-handed materials (i.e. the materials with simultaneously negative real parts of their permittivity and permeability), where the phase and group velocities are (in the simplest cases) opposite to each other. In fact, the idea of opposite directions of the phase and group velocities was first mentioned much earlier—credit has to be given to [2, 3] back to 1904 year. In 1944, left-handed optical materials were mentioned in a series of lectures by Prof. Mandelshtam in Moscow State University [4], which unfortunately have not been published (as has happened many times before, for instance with Leibnitz and Newton, where the discoverer did not publish the respective results in time). It took another quarter of century before Prof. Veselago formulated the fundamentals of electrodynamics with negative permittivity and permeability. It is interesting to mention, that in the middle of 80th year of the last century (about 20 years after the fundamental paper [1] had been published) Prof. Veselago did not mention this idea at his lectures, at least at the Moscow Physical Technical Institute, where one of the authors of this book attended courses given by Prof. Veselago.

A. Chipouline and F. Küppers, *Optical Metamaterials: Qualitative Models*, Springer Series in Optical Sciences 211, https://doi.org/10.1007/978-3-319-77520-3_1

At that time, a microscopic mechanism of achieving negative values for the permittivity and permeability was not even discussed, but one seminal work, predicting the crucial role of resonances in small particles and its influence on the anomalous values of the effective constants had already been published [5]. The modern era of MM began with the experimental verification of a negative refractive index, first in microwave [6] and latterly [7, 8] in the optical domains. From this time onwards, an explosive amount of publications appeared and continues to appear in scientific and popular publications regarding the fundamentals and applications of the MMs. However, it should be remembered that the physics of the MMs (as a branch of science) is still far from mature and there is much left to uncover. An optical MM is based on the nanophotonics (in order to provide metaatoms (MA) in optical domain the MA have to have nanosizes) and evolves in conjunction with the respective technological, experimental, and theoretical achievements. Nevertheless, the development of this branch of modern science has already passed its embryonic stage and now seems to have crossed the invariable dip "nanochasm" in interest after the initial excessive expectation associated with all new ideas. Renormalization of expectation has resulted in a steady, but realistic increase in interest and development in MMs, as presented in the Fig. 1.1.

MMs are artificial media which tailor the macroscopic properties of light propagation by a careful choice of a microscopic unit cell (called the metaatom—MA) from which they are constructed. By controlling the geometrical shape and material dispersion of the MA, novel effects such as negative refraction [9–11], optical cloaking [12–17], as well as a series of optical analogues to well-known physical phenomena from various disciplines within physics can be observed

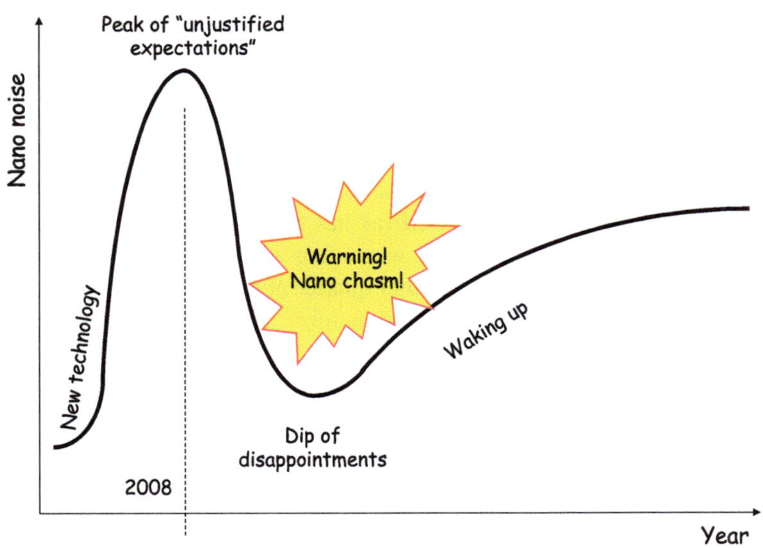

Fig. 1.1 Crossing "nanochasm"—after peak of "unjustified expectations" the development trajectory has to cross the "nanochasm" before it again reaches the trajectory of stable development

[18–22]. In addition to a bi-axial anisotropic (linear dichroism) material response [9–11, 23–25], research was recently extended towards the exploration of MAs that affect the off-diagonal elements of the material tensors (elliptical dichroism), leading to, e.g., optical activity [26–30], bidirectional and asymmetric transmission [31, 32], or chirality induced negative refraction [33–35]. However, despite the possibility of using rigorous computation for describing light propagation on the microscopic level of the MAs, an enduring problem in metamaterial research is the question of what the effective material tensor looks like for a certain MM.

MM design in the optical domain is mainly carried out using rigorous Maxwell's equation solvers like finite difference time-domain simulations [36], finite-element methods [37], and Fourier modal methods FMMs [38]. Instead of these differential methods, integral ones, such as, e.g., the boundary-element method [39] offers an alternative choice of techniques. The discrete-dipole approximation [40] and the multipole method [41] are more physical approaches, where the structure is represented by localized electric multipoles. Nevertheless, presently differential descriptions dominate in MM design.

In contrast to such numerical techniques, the analytical description of MMs is much less developed. Podolskiy et al. introduced the coupled-dipole equations, in order to approximate single and coupled metal wires [42, 43]. The direct excitation of *LC* resonances with the magnetic field of the incident plane wave in a system of two coupled rods was proposed in [44] in order to explain the observed phenomena in terms of effective parameters. Following the "effective medium" theory, an investigation of dielectric and magnetic conducting inclusions was performed for spheroids [45]. In spite of the fact that the approach is limited to the realm of quasistatics, the model was extended to describe dynamical problems. In order to simulate the current distribution in a coupled-wire structure and to calculate the permeability, the Green's function technique is applied [46]. The application of *RLC* circuit theory has been recently used to obtain the resonance frequencies and the quality factors of coupled split-ring resonators [47].

Lately, several reviews regarding the state of the art in the physics of MM have been published [48–52]. The tendency of the MM physics development was presented, for example, in [49], where the most justified and supported by main achievements prognoses are given. Among other developments, it was predicted that future activity will be largely concentrated in the following areas [49]:

1. Noble metals are replaced by structured alloys, CNT & graphene, oxides, superconductors.
2. Hybridization with functional materials (nanocarbon, organics, nanosemiconductors, phase change media).
3. NEMS structures (moving components on the scale of a few tens of nanometers).
4. Close-to-molecular level top-down fabrication, self-organization, DNA & protein, scaffolding, stereo lithography, casting around organic frameworks.

> The current trend is to think of MM as devices, where the structuring of metal and the hybridization with functional agents brings new functionality and response becomes tunable, switchable or nonlinear. In the near future, we will be able to enter the field of quantum metamaterials [49].

It is clearly seen that a new MA paradigm—hybridized with functional agent (like quantum dots, dye molecules, or bio molecules)—becomes one of the main point of the future development in this area. Taking into account that the functional agents are also supposed to be quantum systems, one can easily conclude the importance of bringing quantum concepts to bear on the physics of MM, giving raise the appearance of the new area of the quantum MM. The name quantum MM, means that the internal dynamics of the MAs is described using quantum theory tools at least in part (for example in case of spaser [53]) or in full (in case MM with superconducting MAs [54]).

Another hot topic in the physics of MM is so called structures with toroidal/anapole like MAs. The toroidal moment itself and the respective effects (including toroidal metamaterials [55]) have been deeply investigated theoretically [56, 57]. Several experimental verifications in microwave [58] and optical [59] domains for toroidal moments confirm theoretical conclusions. Anapole has been introduced in physics of elementary particles [60]. The electrodynamic analog of a stationary anapole is well known toroid with a constant poloidal surface current, which is also associated with a toroidal dipole moment. It generates no field outside, but possible nonzero potential, which might lead to a violation of the reciprocity theorem and Aharonov-Bohm like phenomena [56, 61]. Anapole mode in optical domain using a simple silicon structure has been experimentally demonstrated in [62]. In context of homogenization procedure, we have found a new form of material equations (part of macroscopic Maxwell equations) which corresponds to the material consisting of toroidal/anapole structures. It is interesting, that this new form (we named it "Toroidal" form in contrast with the previously known "Casimir" and "Landau&Lifshitz" forms) appears as an extension of so called phenomenological approach in homogenization of Maxwell equations naturally, irrespective to multipole expansion, where the toroidal moments actually appear. Presence of the elementary structures with toroidal moments (and MAs with toroidal moments) fixes a fundamental ambiguity in basics of electromagnetic theory and seems to have much deeper physical meaning than just one of the terms in the multipole expansion.

1.2 Motivation and Goals

The theoretical description of the MM was started from a priori introduced dispersive permittivity and permeability. The spectral dependence of both permittivity and permeability is supposed to follow analytical functions with resonant denominator, which assumed harmonic oscillator type model for both effective parameters [1]. This model basically assumes local media response from small MAs, possessing not only the dielectric (like any dipole-like atom) but the magnetic response as well. The assumed smallness of the hypothetical MAs allows the commonly assumed homogenization theory to be invoked. The situation changed shortly after the first experimental realization of the MM, where the typical sizes of the MAs turned out to be not much smaller than the respective wavelengths and hence usual homogenization model fails. In order to develop a homogenization procedure for MMs in general, more sophisticated models are required. Basically, the homogenization models for MM appeared to be an extension of several approaches. One of them was developed in the optics of crystals [63] with the extension to an allowed magnetic response [64, 65]; this extension follows a basic approach demonstrated by "A Course of Theoretical Physics" [Landau&Lifshits]. An extension of the exciton theory to MM [Cho K] has to be assigned to the same category. This second approach uses formalisms developed for photonic crystals with an appropriate extension over Bloch waves and the determination of the respective dispersion diagram in the form of Brillouin zones (see recent paper [66] and references therein) for a particular structure. The third approach appears to be a continuation of methods developed in the theory of compound materials [67–69] and extended to the case of a magnetic response.

In contrast a forth approach, based on the approximation of the MAs as point-like multipoles has not received much attention, but had been mentioned as a one possible way of describing the behavior of MMs in [65]. This approach has been chosen as a basis for the development of the homogenization model of the MM in this book.

A detailed comparison between the different approaches, mentioned above, is out of the scope of the material presented in this book. Nevertheless, the necessity of the taking into consideration spatial dispersion is valid for all approaches. The problem appears to be in a gap between developed approaches and the basics of the homogenization, which have to be satisfied anyway. Ignoring of the basics of the homogenization can lead in some cases to the violation of the self-evident basic assumptions like causality and passivity (the review of these found in publications violations is given in [67]). In some cases the different approaches like phenomenological [Landau&Lifshitz] and multipole [70] appeared to be mixed [64] and different representations of the macroscopic Maxwell equations (the Landau&Lifshitz "L&L" representation [64] and Casimir "C" representation [69]) sometimes appeared to be not clearly distinguished (actually, discussion about the different representations of the macroscopic Maxwell equations can rarely be found in publications at all). The connections between the different representations, the appearance of the spatial dispersion in different representations, and the mutual transformations between the different representations have not received enough attention in the literature.

The multipole approach developed originally in [70] assumes an expansion of the charge dynamics in atom/molecules, which in the case of the MM has to be replaced by the charge dynamics in the MAs (note, that in some publications under multipole expansion, an expansion of the fields over wave vector is assumed, which is obviously a different math tool entirely). The original approach [70], where averaged parameters are elaborated based on the charge dynamics, gives us a unique avenue for the creation of a unified approach to all possible types of MMs: charge dynamics in the MAs can be described in the frame of classic or quantum models, but the algorithm for the effective parameter calculation remains the same. It motivates us, in turn to try to create a unified approach, which would unify classic, quantum, or semi-classic MM. This approach can be easily extended to other forms of expansion and consider not only multipole expansion, including, for example, toroidal/anapole expansions as well [60, 71].

The last, but not least, comment is about the evident inclination in favor of numerical methods in order to describe the optical properties of MMs at the expense of physical intuition. There is no doubt that modern numerical algorithms and available computer facilities provide the main way to investigate more or less complicated problems. Nevertheless, the qualitative approximate type models can provide a deeper understanding of the basic physical processes, stimulate discussion of new effects, and even provide new paradigm for optimization of a particular design. The qualitative models are complimentary to the numerical ones, taking advantages of careful comparison with the results of rigorous numerical calculations, but at the same time remaining analytically treatable. In order to create this type of model, accurate approximations have to be made in order to simplify the respective consideration, at the same time keeping the main physical effects and interplay between them in the model. In turn, it requires specific mathematical apparatus and specific methods, which give especially useful results if wide range of the problems is considered in the frame of the same, unified, and self-consistent approach. The qualitative models allow us to show mutual self-consistency of the physical theory (in this particular case in application to the MM), which otherwise could be seen as a huge leap from independent experimental to theoretical facts.

Elaboration of the qualitative models in the frame of the same paradigm appears to be useful not only at the consideration of the specific problems, but at the teaching of the respective courses as well. It must be clearly stated, that all considerations in the presented book are performed for the bulk MM, i.e. the problems connected with the boundaries are out of the scopes.

1.3 General Concept of the Modern Education

It is worth considering the educational aspects of homogenization in connection with the creation of the model mentioned above. In a modern courses of electrodynamics the homogenization (averaging) procedure does not receive enough attention, which can partially be explained by the fact that the commonly accepted approach (local frequency dispersive permittivity and permeability) gives in most cases pretty good correspondence with the experimental data (not for MM) and the necessity of the more sophisticated approaches is pretty low. Electromagnetic theory was developed a rather long time ago, but some basics of this theory (especially elaboration of macroscopic Maxwell equations—basics of the homogenization procedure) have not been revisited since the 1970s. Due to the appearance of the MM, the averaging procedure must now be considered more fully and has to become a part of the standard courses of electrodynamics in order to teach students in a self-consistence and unified manner.

Theoretical predictions and technological achievements have made possible, the realization of a wide class of artificial materials exhibiting a hitherto unseen magnetic response at high, in particular optical, frequencies. This, in turn, forced scientists to refresh an idea about materials with both negative dielectric and magnetic constants (left-handed materials) put forward by Prof. V. G. Veselago about 45 Years ago. These attempts resulted in successful experimental demonstration of such material in RF and optical domains, which manifested start of the era of MMs. The new achievements require development of new concepts, which has been precisely formulated in [72]:

> Today we encounter a situation which reminds that of the beginning of the XX century. It was time of the crisis of classical mechanics, which became redundant with the creation of relativistic theory and quantum mechanics. Now, we go through the crisis of classical electrodynamics of condensed matter, developed by Maxwell, Heaviside and Lorentz. Already in the XX century we observed an increasing amount of experimental results which can be hardly treated with existing theories.
>
> Like in quantum mechanics we have to abandon an attempt to simultaneously measure the coordinates and velocity of a particle, now we have to abandon an attempt to connect local current with local field.

The use of the apparatus of the spatial dispersion supposes deep understanding of the basics of the homogenization theory; otherwise, inappropriate application of the respective math expressions leaded to the obviously wrong conclusions violating causality or passivity principles. The results published in [73] and presented here prove this statement:

> In terms of the coordination project "Electromagnetic Characterization of Nanostructured Materials" (ECONAM; http://econam.metamorphose-vi.org), which is financed by the Eurocomission, the author of this study, in cooperation with the group of Prof. Bilotti at the Roma Tre University (Italy), took part in surveying all of the papers devoted to metamaterials published over 8 years in one of the leading physical journals in the world, Physical Review Letters. The results proved to be as follows. The authors of 53% of papers in which the dispersion curves for the material parameters of metamaterials were presented (most frequently as results of measurements) paid no attention to the fact that their results contradict the causality principle and the principle of the passivity of the medium. That is to say, the material parameters of metamaterials that would not violate the known conditions under which the medium of particles may be replaced by a continuous medium were obtained in less than half of all publications. Nevertheless, in all these papers, the metamaterials were treated precisely as continuous media. In other words, it turned out that the authors of the papers on metamaterials published in Physical Review Letters attempted to combine the results of the dynamic analysis with the purely static notion of effective material parameters. Those 47% of papers on metamaterials published in 2000–2007 in Physical Review Letters, where the effective material parameters of homogenized metamaterials did not violate the conditions of locality, corresponded to those lucky cases in which the ordinary (static) model of homogenization turned for some reason to be applicable to metamaterials".

The problem appears to be rather significant: obvious violation of the basics cannot be tolerated. It has to be emphasised, that the problem here is not in elaboration of some "new" approaches, which would meet the forecasted resistance of the old "conservative" theories, but rather inappropriate attention of both authors from one side and reviewers from the other side to the basic principles, which could be probably partially referred to the lack of the basic education. In order to apply properly the homogenization methods which have been developed in area (in particular) of compound materials, knowledge of the basics of the homogenization procedure is the absolute and evident prerequisite. Several attempts have been performed in order to bring the discussion on a higher, more adequate, and undoubtedly more scientifically justified level: several reviews have been published [67, 73], special sessions at the conferences have been organized, but the tendency up to now seems not to be broken.

The problems facing now the education cannot be of course considered separately from the challenges facing the society at the beginning of the 21st century. The problems of the world economical model, caused by an excessive credit pumping in previous decades, will definitively affect the educational system as well, requiring new approaches in order to maximize effectiveness of the education. Analysis of these problems, of course, is far out of the scopes of the presented book; here authors would like to mention just one (may be even not mostly important, but nevertheless extremely sad) consequence of the modern scientific system, namely: disappearance of the phenomena of "scientific schools" of the first half of the 20th century, which was typical for, for example, scientific life and spirit in Germany and USSR. Science, which now has become a part of business, does not allow outstanding people working together for rather long time, which was a prerequisite for the growing of the scientific schools. Without discussion of advantages and drawback of the scientific systems, the influence of the scientific schools on the education has to be undoubtedly mentioned. The scientific schools have left the most famous courses, which are used even up to now for the education: it is enough to mention R. Feynman's course of physics or course of theoretical physics written by L. D. Landau and E. M. Lifshitz. One of the main advantages of these courses was a conceptual approach to very wide range of problems; for example, the principle of minimum action was the basis of whole course of L. D. Landau and E. M. Lifshitz. A conceptual approach, in contrast with just the teaching of huge amount of weakly correlated facts, creates solid background and, which is very important, significantly minimizes an amount of information, which has to be memorised in the frame of a course. A conceptual approach in turn was a consequence of multiple discussions about content of the course (material selection, consequence of presentation, logic links, analogies with other disciplines etc.). Unfortunately, at present the discussions about content of the courses and their mutual correlations are not a common practice, which does not allow optimizing further the structure of the educational courses.

The modern concept of high education in natural science and technology towards mainly preparation of staff for high tech industry and includes three stages: Bachelor of Science (B.S.), Master of Science (M.S.), and Philosophical Degree (Ph.D.).

Very roughly speaking, it is supposed, that after already the first stage (B.S.) the students will be capable to perform some duty in industrial sector, which do not require full understanding of the technological principles; for example, operate the technological chains with only partial understanding of the physical processes which the operation is based on. After the second stage (M.S.) the students are expected to be able to understand the industrial procedures in full and be able to suggest technological/technical modifications if necessary. At the next stage (Ph.D.) the students have to be able not only operate, understand, and modernize the technological procedures, but also develop new ones and put into the existent industrial chains; it is also supposed, that the students have basic knowledge in business and could preliminary evaluate how profitable their developments could be.

Schematically the first two stages could be summarized as it is shown in Fig. 1.2. The B.S. study (blue lines in Fig. 1.2) consists of education in areas of standard disciplines like physics, math, chemistry, biology etc. This education could be named as "traditional" and takes 3–4 years depends on the particular country.

On the next stage, M.S. (green lines in Fig. 1.2) could be designed several different ways, namely: it could repeat again B.S. structure on higher level, or it could be structured according to the other principles. The "other principles" are supposed to be dictated by the requests from the society and reflect the modern challenges which are facing the modern society. The particular set of these challenges are country (society) specific and have been identified according to an own development concept. As an example, here the two "conceptual structuring" of the requests of the societies are presented: the first one is the structure of the BMBF and the second one is the structure of the new University established in Moscow under guidance of the Massachusetts Institute of Technology—see Figs. 1.3 and 1.4.

Basically, the suggested education has to be organized according to these "clusters" (green lines in Fig. 1.2) rather than according to the separated disciplines (blue lines in Fig. 1.3). To the first approximation, any students can collect

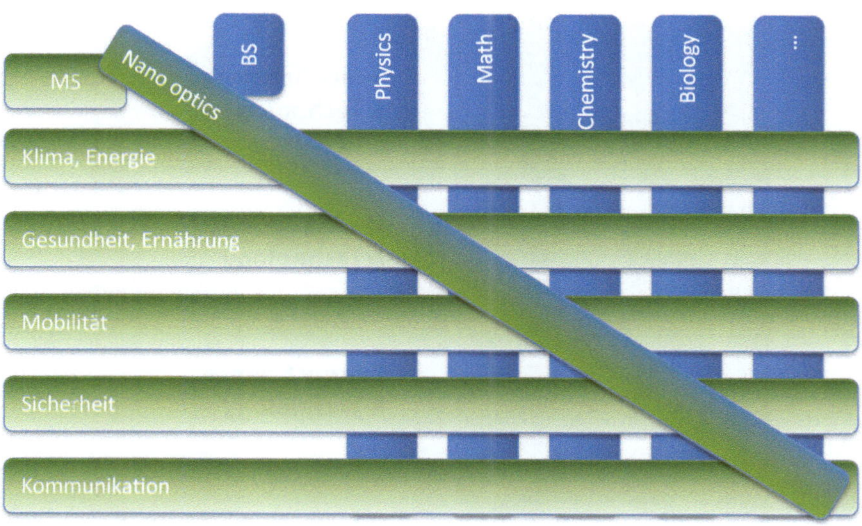

Fig. 1.2 Structure of the B.S. (blue vertical lines) and advanced modern M.S. (green horizontal lines) educational programs and position of the nanooptics in the structure of the B.S./M.S. education. Absence of an "eigen optical green line" makes the nanooptics a multidisciplinary program in terms of the both (B.S. and M.S.) presented here scheme

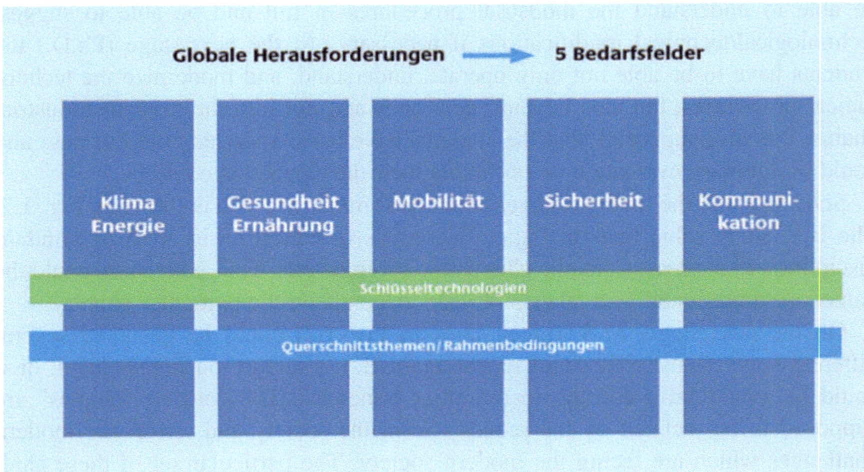

Fig. 1.3 BMBF structure of the requests from the modern society (http://www.hightech-strategie.de/de/82.php)

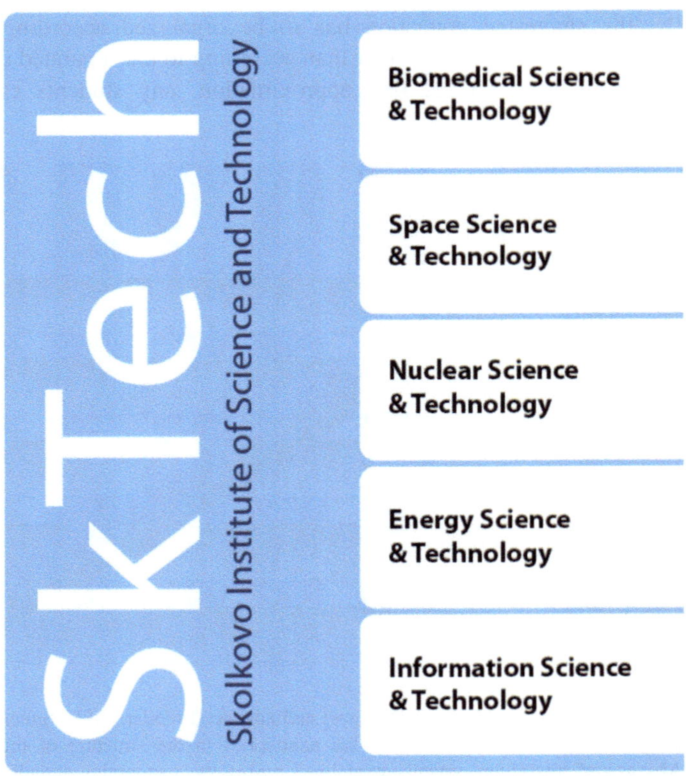

Fig. 1.4 Science & Technology clusters at the Skolkovo Institute of Technology as an example of the conceptual structure of the modern education system (http://skolkovotech.ru/)

University courses which match the chosen “cluster” under supervision of a more experienced mentor and from this prospective the educational system is universal and (provided nomenclature of the courses is reach enough) can be easily adjusted to any new “clusters”. The actual problem is in choice of the necessary amount of information: in most cases the course contains not only necessary information, but also a lot of the information, which the student does not need at the moment.

The students nevertheless have to learn “or all or nothing” in the frame of the particular course which leads to the overloading and consequently to extra problems. In the following paragraph, I would like to present some suggestions, which could, from my point of view, potentially improve the education process.

1.4 What Could Be Revisited: An Example of a Developed Course

The course “Introduction to nanophotonics” has been created as a part of the Master Program of Photonics. The course received deep revision every year in response to the student feedback and according to the acquired new experience.

The main idea here is to reduce necessary amount of information to be memorized in favour of more structured, conceptual information presentation. The course has to present a concept and its position in the “cluster” with the basic information (which is mandatory for exams), and give to the student information about sources where the more details could be found.

In response to the problem of explosive amount of knowledge, the information for course has been strongly restricted just by basic one, which would nevertheless allow student in future to consider most of the particular problems in area of nanophotonics. The information selection has been done in synergy with the structure of the course, presented in Fig. 1.5. The “construction bricks” of the course present main physical phenomena which have to be studied in a logic and self-consistent chain, and include also “retrospective” part “Phenomenological Electrodynamics of media with Negative Refractive Index”, which shows the first stage of the development of the knowledge in area of optical MMs and nanooptics. The basis of the course is the two “bricks”—“Maxwell equations for continuous media—averaging procedure” and “Different models for charge dynamics”.

The necessity of the first one (averaging procedure) is stipulated by the fact, that even elementary basic notation of plasmonics assumes the use of permittivity, and it makes full sense first to introduce this notation (and theory of the effective parameters in general) using modern and state of the art concepts. In order to consider the averaging procedure it is in turn necessary to introduce Maxwell equations itself, which also could be done different ways—see Fig. 1.6a.

The logic of a “standard” course of electrodynamics is following: after presentation of three basic laws (Faraday law, Coloumb law, and Bio-Savar law) the Maxwell equations (macroscopic Maxwell equations) are formulated as a consequence of the three laws.

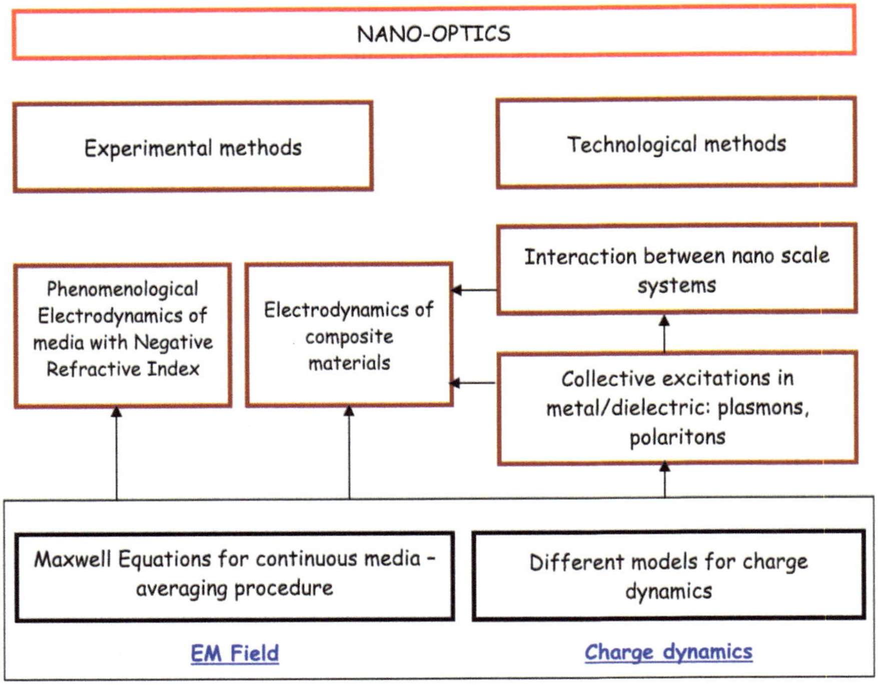

Fig. 1.5 Logic of the elaborated course of "Introduction to nano optics"

After that the formulation of the Maxwell equations in free space (microscopic Maxwell equations) is given with the statement, that the macroscopic Maxwell equations (which have been introduced based on the three laws) are direct consequences of the microscopic ones after the averaging procedure. In other words, from the postulated macroscopic Maxwell equations the microscopic Maxwell equations are elaborated, which is obviously counterintuitive. Moreover, for the Maxwell equations the relativistic invariance is proven without any explanation why Maxwell equations appeared to be relativistic invariant. In spite of the internal logic contradictions, the structure of the courses of electrodynamics is well established and well accepted by the students who study electrodynamics for the first time. There is another way to present the electrodynamics, which can be found in course of theoretical physics of L. Landau and E. Lifshitz in volumes 2 "Field theory" and volume 8 "Electrodynamics of continuous media"—see Fig. 1.6b. The logic here follows the general logic of the whole course, namely all basic equations are derived based on minimum action principle, which starts from the relativistic invariant action integral. In this case microscopic Maxwell equations are derived first, and the charge dynamic equation appears to be as one of Maxwell equations. Initially required relativistic invariance results obviously in relativistic invariant microscopic Maxwell equations, and the averaging procedure (transition from the obtained microscopic Maxwell equation to the macroscopic ones) is developed in volume 8 following the phenomenological approach. This way of teaching, in spite

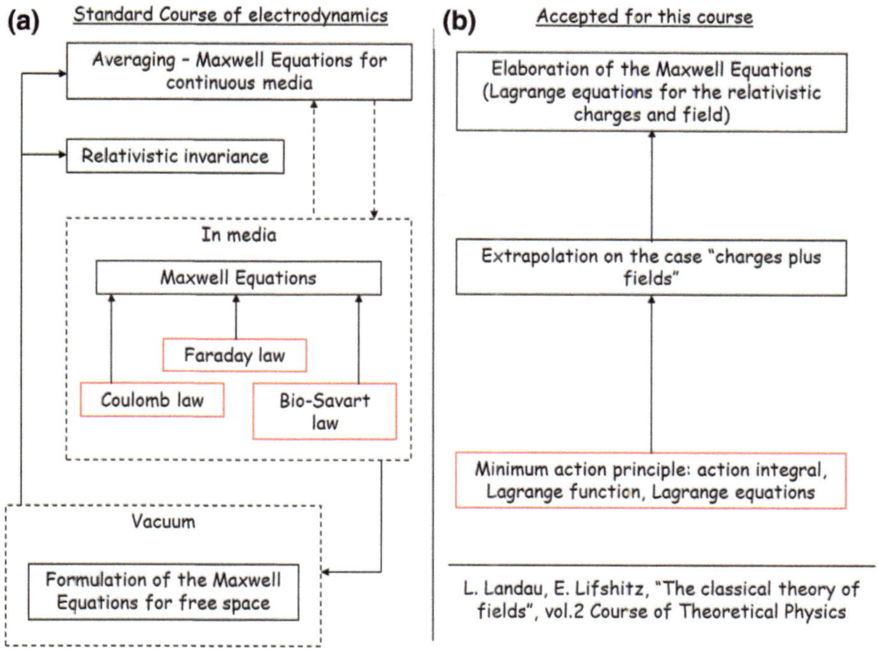

Fig. 1.6 Two ways of introducing of Maxwell equations: **a** in standard University courses, **b** in course of theoretical physics of L. Landau and E. Lifshitz

of the much better logic and self-consistency, is pretty complicated in terms of the understanding and math involved. Nevertheless, the second way is perfectly suited for the M.S. programs, where the students are assumed to have studied Maxwell equation using the first way.

The homogenization procedure follows the approach accepted in course of theoretical physics of L. Landau and E. Lifshitz. The *Serdyukov-Fedorov transformations,* which plays a crucial role in understanding of the basics of homogenization of Maxwell equations and gives a clear structured view of the different forms of representing Maxwell equations and the mutual relation between them, is given as a part of the general homogenization theory—this information is not only excluded from the standard courses, but very rarely appears in publications.

I would like briefly compare the averaging procedure presented in "standard" courses (for example, following [74]) and in course of L. Landau and M. Lifshitz. In "standard" courses the transition between micro and macroscopic Maxwell equations is given in form of postulates, without elaboration and explanation of how it could be rigorously justified. There is also usually no comparison between phenomenological and multipole approaches; the expressions (including expressions for the case of spatial dispersion) are often given without explanation how they have been elaborated. The lack of attention to the homogenization procedure could be partially explained by the fact, that up to now there is no established and commonly accepted way to perform the homogenization: there are different models for different situations, and a systematic presentation of the homogenization procedure is the rather complicated task. Nevertheless, there are several models for static case (models of Clausius-Mossotti-Lorenz-Lorenz, Maxwell-Garnet, Bruggemann etc.—see [69]) which are pretty much established and which are given in some courses. The static Clausius-Mossottii-Lorenz-Lorenz model can be extended on the dynamic case as well [75]. Anyway, even though that up to now there is no well-established and commonly accepted way of teaching of the homogenization procedure, it should not be a reason to avoid the respective discussions in the frame of a course and would help to avoid multiple mistakes connected with this topic.

The homogenization procedure requires (one or another way) the description of the charge dynamics under the action of the electro-magnetic field—the second basic "brick" (different models for charge dynamics) in Fig. 1.5. This can be done evidently either in the frame of classic or quantum mechanical approaches. Both classic and quantum mechanics are well established and self-consistent courses, which are usually given without special attention of the application of their methods to the problem of light-matter interaction. In case of classic mechanics it does not cause any problems, while in the case of the quantum mechanics a significant adaptation of the respective methods is required. One of the parts of quantum mechanics which has to be given in view of this adaptation is the density matrix formalism, which is widely used in theory of optical amplifiers and lasers. The main difference with the more traditional approach based on Schrödinger equation is in the including of the relaxation processes in the consideration; in other words, the microsystem (quantum object) is considered as a part of a big macrosystem with huge numbers of the degrees of freedom. This macrosystem is called thermostat (or thermo bath) and the respective interaction leads to the relaxation processes, which in vast majority of the experimental realizations have to be taken into account. This approach is unavoidable for the description of gain in MMs, dynamics of spasers, saturation caused nonlinearity, transition processes etc. Moreover, the same formalism is applicable to the superconductive MAs, which form the so called superconducting or quantum MMs [54]. It has to be emphasized that the required education in area of density matrix formalism has almost nothing to do with quantum optics, which deals with the situation with small number of photons per available state and which is required, for example, for adequate consideration of Purcell effect [76].

The next logic step is the elaboration of the models for plasmons/polaritons and the respective physical phenomena like extraordinary transmission etc. The next step is in integration of the elementary plasmonic nanoresonators in coupled systems, including coupling not only between classic objects (two and more plasmonic resonators, for example double wires system), but also with the quantum ingredients like Quantum Dots, which gives rise consideration of spasers, nonlinear response enhancement (including well known Stimulated Enhanced Raman Scattering—SERS), and modification of the luminescence rates (Purcell effect). The coupled systems considered here could be further used as MAs for the MMs. This basically completes all possible variants of the charge dynamics in the MAs.

Now all necessary information for the consideration of the effective properties of the MMs is prepared and can be used for the elaboration of the effective parameters. Nevertheless, before coming to the modern state of the art concepts, it makes sense to show student how the theory of the optical MMs has been first introduced by assumption of the negative permittivity and permeability without detailed explanation of the sources for their negative values. The "brick" "Phenomenological electrodynamics…" shows that this assumption about possible negative values of the effective permittivity and permeability does not contradict to the basic principles of electrodynamics and figures out necessary corrections to the known expressions (for example, transmission/reflection formulas, expression for Brewster angle, Ferma principle, causality principle, expressions for the field energy, conditions for the negative refraction etc.) which have to be done in case of both negative permittivity and permeability. Among the others, this consideration emphasizes again that there are several requirements (causality and passivity), which have to be met irrespective to any micro models for the charge dynamics.

Finally, electrodynamics of the MMs is combined from the prepared before the multipole approach (which gives constructive expressions for the effective parameters through the charge dynamics) and the charge dynamics. From this consideration the students clearly see how the multipole expansion leads to the appearance of the magnetic response in the media, correspondence between different types of eigenmodes of the charge oscillations in MA and the presence/absence of the magnetization (and, respectively, negative refractive index). Prepared before models for MAs consisting of plasmonic nanoresonators and quantum ingredients allows us to consider loss compensation (and respective influence of the loss compensation on the effective MM properties) in the frame of the same unified approach. It is worth noting the last sentence again—all physical phenomena in the course are considered in the frame of the same unified approach which gives clear and solid scheme and recipes for qualitative consideration of the most physical phenomena in area of nanooptics and optical MMs.

Review of the respective experimental and technological methods completes the course.

The course intentionally uses mostly an analytical treatment of the respective problems. The numerical methods are evidently necessary for any more or less complicated systems, where analytical methods can be applied only for qualitative estimations. It appears to be necessary to complement the powerful numerical methods by relatively simple analytical models, which nevertheless keep main physical processes inside. As an example, the multipole approach complements the direct numerical methods (like FMM or FTDT), uses the data of the respective numerical simulation to find numerical values for the parameters to be fitted, and in turn gives a tool for the fast estimation of new effects like anisotropy, nonlinearity, optical activity etc. Moreover, after the first fitting the multipole model allows us to estimate variation of the effective parameters for the random MMs, where numerical approach is rather time consuming. Using the same parameters, one can expand the model on the case of the MMs with gain or nonlinear ingredients coupled to the MAs—all of that have been demonstrated in the presented work.

From the other side, the development and understanding of the analytical models requires change of the education programs and add to the courses more qualitative math methods.

The peculiarity of the course is in the proposed grading system as well. At the exam the students are allowed to use any sources of information, but the proposed set of the problems is extremely wide and covers a lot of different aspects. Actually, the student has to demonstrate that he/she is able find the necessary information fast and can give a qualitative explanation for a wide range of the problems rather than demonstrate rigorous solutions for only a limited number of problems.

The course has to be matched with the other ones, which requires a lot of systematic work. For example, both black “bricks” in Fig. 1.5 have to correlate with the courses of electrodynamics and solid state physics. The parts about experimental and technological methods have to be matched with the respective Practicum and Internship opportunities. The careful matching of the courses and elaboration of the specifically oriented programs could be an adequate respond to the challenges facing the modern education at the moment.

One more problem in the modern education structure is in not very clear difference between the courses of electrodynamics for engineering and physical departments in area of nanophotonics and optical MMs. The problem of education which would combine good basic and applied knowledge is rather old one and roots back to the 30–50th years of the last century, when the necessity of this education has been stipulated by a technological revolution. It has also been realized that the classical University education does not respond to the specific engineering requirements from one side; at the same time, purely engineering education does not allow to its carrier to adopt knowledge to the new and fast appearing technological challenges. The required combination of the good basic education and enough initial practical experience has been achieved in new type of the Universities established in the USA (Massachusetts Institute of Technology), France (Ecole Polytechnique), and former USSR (Moscow Physical-Technical Institute). All these Universities have developed rather similar programs, which consisted of the basic education packed in three first years (now B.S.), while the rest two or three years have been reserved for the more special and applied practice (M.S.). Moreover, even during the first three years for different departments a different amount and combination of basic courses have been carefully selected. This system has functioned pretty well for the more than 50 years, but now it evidently requires again revisit and deep reformation. Comprehensive discussion of this problem is not a goal of the presented work, but the statement of the necessity of adaptation of explosive amount of new information and development of carefully structured courses in response to the modern requirement of science and technology is appropriate here in connection with the attempt to develop a course, responding to this challenge.

The identified weaknesses and the respective responses are summarised in the Table 1.1.

The course has been given to the groups of the first year master students in 2008–2011 years, has been evaluated and corrected according to the produced feedback. The course is seen as a first step in creating a more general course of

Table 1.1 Identified weakness of the education programs in area of nanophotonics (left) and proposed respective responds (right) in the developed course "Introduction to nanooptics"

Lack of qualitative models with appropriate math tools	The course is based on the qualitative models; moreover, all used models are developed in the frame of the same approach
Absence of well-established and commonly accepted way of teaching of the homogenization procedure	The homogenization procedure is given at the beginning in form of the phenomenological (L. Landau and E. Lifshitz) and multipole models
Lack of education in area of light-matter interaction based on quantum mechanical approach	Basics of the light-matter interaction based on the density matrix formalism are given
Necessity of adaptation of explosive amount of new information and development of carefully structured courses in response to the modern requirement of science and technology	The course is self-consistent and is strongly restricted by fundamental phenomena. Any particular applications are supposed to be elaborated on demand. The course does not require memorize all information, but rather helps to navigate students to the appropriate theoretical constructions. At the exam the students are allowed to use any sources of information
Lack of clear difference between engineer and academically oriented education programs	The course is clearly identified as an academic one with appropriate set of problems for seminars

nanophotonics, and it was very important to see how well the students understood and memorized the information. The conclusions are:

1. In spite of higher level of complexity, the information is rather well received due to pedagogical consistency of the course.
2. The teaching has to be accompanied by seminars with examples explaining using the new ideas; otherwise the relatively complex constructions like multipoles will not be fully understood.
3. The knowledge of the students, in the area of quantum mechanics is not satisfactory. The basics of quantum mechanics formalism based on density matrix approach have to be included in the course and also accompanied by appropriate and detailed examples.

1.5 Structure of the Book

To conclude, the goals of this book are formulated as follows:

1. Develop solid basis for homogenization procedure, which fills the gap between microscopic Maxwell equations and various forms of averaging procedures, suggested for MMs.

2. Develop homogenization model for MMs using multipole expansion approach and find relations between this approach and the elaborated basics of the homogenization.
3. Apply the aforementioned model to the optical MM regime and consider linear and nonlinear effects in MMs in the frame of the developed unified approach.
4. Using the fact that the created model uses the charge dynamics in MAs as a theoretical basics, extend the developed approach to the case of quantum MMs.
5. Apply the developed model for quantum MM's in the particular cases of non-linear MM, dynamics of spaser, and MMs with gain.

The book consists of 14 chapters including "Introduction" (this chapter) and Chap. 14 "Conclusion". In order to better visualize the logic and structure of the book, Fig. 1.7 has been drawn.

Chapters 2 and 3 respond to the first goal, listed above and summarize the basics for any possible homogenization procedures. Details of the multipole model applied

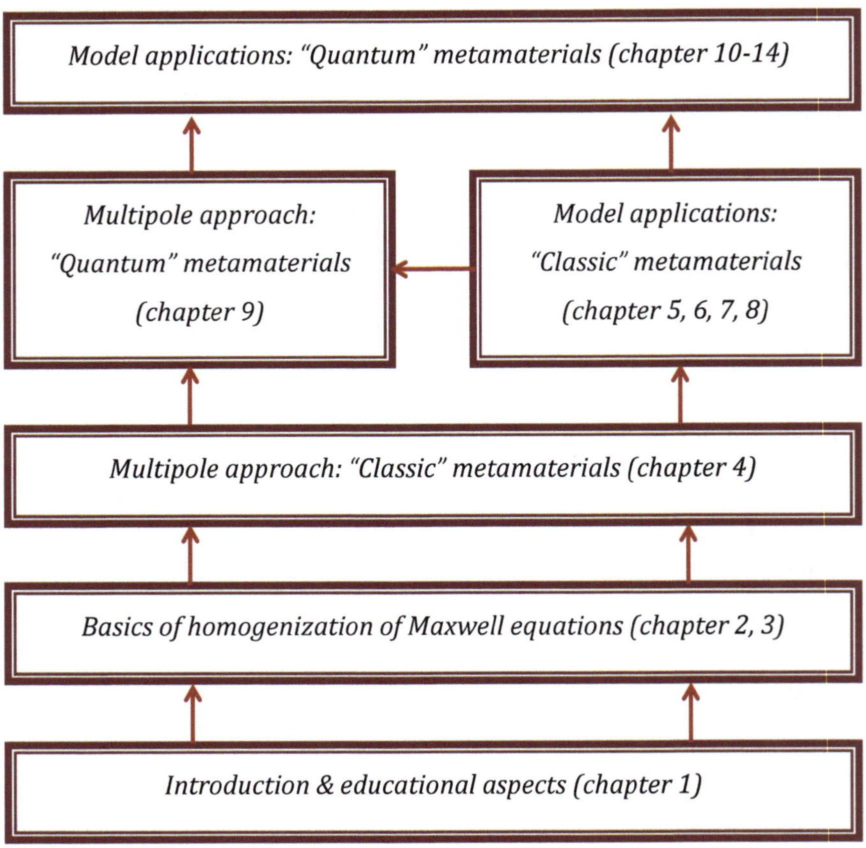

Fig. 1.7 Structure and logic of the presented work "Analytical modeling of optical MM: scientific and educational aspects"

to the optical MM are given in Chap. 4 (response to the second goal of the work), which is followed by the demonstration of several applications for description of the linear (Chaps. 5, 6 and 7) and nonlinear (Chap. 8) properties of the MM (response to the third goal). Extension of the model on the quantum MM is presented in Chap. 9 (the fourth goal), and the nonlinear properties of the quantum MM are demonstrated in Chap. 10 (the fifth goal). Chapter 11 presents the results of the dynamics of the single MAs consisting of coupled plasmonic nanoresonators and quantum system (the fifth goal), which serves as a prerequisite for the consideration of the propagation of the plane wave in the MM with gain in Chap. 12 (the fifth goal). Chapter 13 summarizes the obtained results from the point of view of the problem facing modern educational courses in area of the optical MM. Results of the work are summarized in Chap. 14 "Conclusions".

References

1. V. Veselago, Sov. Phys. Usp. **10**, 509 (1968)
2. H. Lamb, Negative phase velocity and its consequence in hydrodynamics: "On group velocity". Proc. Lond. Math. Soc. **1**, 473–479 (1904)
3. A. Schuster, *Negative Phase Velocity and Its Consequence in Optics: An Introduction to the Theory of Optics* (Edward Arnold, 1904)
4. L. Mandelshtam, Optical properties of the left-handed media: the 4th lecture of L. I. Mandel'shtam given at Moscow State University (05/05/1944). Nauka **5**, 461 (1994)
5. N.A. Khizhnyak, Anomalously large effective dielectric and magnetic constants for the resonant regimes of elementary scatterers: artificial anisotropic dielectrics formed from two-dimensional lattices of infinite bars and rods. Sov. Phys. Technol. Phys. **29**, 604–614 (1959)
6. R. Shelby, D. Smith, S. Schultz, Experimental verification of a negative index of refraction. Science **292**, 77–79 (2001)
7. A.N. Grigorenko, A.K. Geim, H.F. Gleeson, Y. Zhang, A.A. Firsov, I.Y. Khrushchev, J. Petrovic, Nanofabricated media with negative permeability at visible frequencies. Nature **438**, 335–338 (2005)
8. V. Shalaev, W. Cai, U. Chettiar, H.-K. Yuan, A. Sarychev, V. Drachev, A. Kildishev, Negative index of refraction in optical metamaterials. Opt. Lett. **30**, 3356 (2005)
9. H.-K. Yuan, W. Cai, S. Xiao, V. Drachev, V. Shalaev, Opt. Lett. **32**, 1671 (2007)
10. G. Dolling, M. Wegener, C. Soukoulis, Opt. Lett. **32**, 53 (2007)
11. J. Valentine, S. Zhang, T. Zentgraf, G. Ulin-Avila, D. Genov, X. Zhang, Nature **455**, 376 (2008)
12. J. Valentine, J. Li, T. Zentgraf, G. Bartal, X. Zhang, Nat. Mater. **8**, 568 (2009)
13. A. Alu, N. Engheta, Phys. Rev. Lett. **102**, 1 (2009)
14. Y. Lai, J. Ng, H. Chen, D. Han, J. Xiao, Z.-Q. Zhang, C.T. Chan, PRL **102**, 1 (2009)
15. M. Farhat, S. Guenneau, S. Enoch, PRL **103**, 1 (2009)
16. B. Justice, S. Cummer, J. Pendry, A. Starr, Science **314**, 977 (2006)
17. U. Leonhardt, Science **312**, 1777 (2006)
18. E.E. Narimanov, A.V. Kildishev, APL **95**, 041106 (2009)
19. S. Vukovic, I. Shadrivov, Y. Kivshar, Appl. Phys. Lett. **95**, 041902 (2009)
20. D.Ö. Göuney, D.A. Meyer, Phys. Rev. A **79**, 1 (2009)
21. N. Papasimakis, V. Fedotov, N. Zheludev, PRL **101**, 253903 (2008)
22. N. Liu, L. Langguth, J.K.T. Weiss, M. Fleischhauer, T. Pfau, H. Giessen, Nat. Mater. **8**, 758 (2009)

23. C. Helgert, C. Menzel, C. Rockstuhl, E. Pshenay-Severin, E.B. Kley, A. Chipouline, A. Tunnermann, F. Lederer, T. Pertsch, Opt. Lett. **34**, 704 (2009)
24. C. Garcia-Meca, R. Ortuno, F.J. Rodriguez-Fortuno, J. Marti, A. Martinez, Opt. Lett. **34**, 1603 (2009)
25. M. Thiel, G. von Freymann, S. Linden, M. Wegener, Opt. Lett. **34**, 19 (2009)
26. B. Bai, Y. Svirko, J. Turunen, T. Vallius, Phys. Rev. A **76**, 023811 (2007)
27. L. Arnaut, J. Electromagn. Waves Appl. **11**, 1459 (1997)
28. J. Reyes, A. Lakhtakia, Opt. Commun. **266**, 565 (2006)
29. S. Prosvirnin, N. Zheludev, J. Opt. A: Pure Appl. Opt. **11**, 074002 (2009)
30. S. Tretyakov, I. Nefedov, A. Shivola, S. Maslovski, C. Simovski, J. Electromagn. Waves Appl. **17**, 695 (2003)
31. V. Fedotov, P. Mladyonov, S. Prosvirnin, A.V. Rogacheva, Y. Chen, N. Zheludev, PRL **97**, 167401 (2006)
32. S. Zhukovsky, A. Novitsky, V. Galynsky, Opt. Lett. **34**, 1988 (2009)
33. J. Pendry, Science **306**, 1353 (2004)
34. S. Tretyakov, A. Sihvola, L. Jylhä, Photonics Nanostruct. Fundam. Appl. **3**, 107 (2005)
35. J. Zhou, J. Dong, B. Wang, T. Koschny, M. Kafesaki, C. Soukoulis, Phys. Rev. B **79**, 1 (2009)
36. K. Yee, IEEE Trans. Antennas Propag. **14**, 302 (1966)
37. A. Taflove, S.C. Hagness, *Computational Electrodynamics*, 3rd edn. (Artech House, Boston, 2005)
38. L. Li, J. Opt. Soc. Am. A **14**, 2758 (1997)
39. C. Rockstuhl, M.G. Salt, H.P. Herzig, JOSA A **20**, 1969 (2003)
40. B.T. Draine, P.J. Flatau, JOSA A **11**, 1491 (1994)
41. C. Hafner, *The Generalized Multipole Technique for Computational, Electromagnetics* (Artech House, Boston, 1990)
42. V. Podolskiy, A. Sarychev, E. Narimanov, V. Shalaev, J. Opt. A: Pure Appl. Opt. **7**, 32 (2005)
43. A. Podolskiy, A. Sarychev, V. Shalaev, Opt. Express **11**, 735 (2003)
44. A. Sarychev, G. Shvets, V. Shalaev, Phys. Rev. E **73**, 036609 (2006)
45. A.N. Lagarkov, A.K. Sarychev, Phys. Rev. B **53**, 6318 (1996)
46. L. Panina, A. Grigorenko, D. Makhnovskiy, Phys. Rev. B **66**, 155411 (2002)
47. T.P. Meyrath, T. Zentgraf, H. Giessen, Phys. Rev. B **75**, 205102 (2007)
48. N. Zheludev, The road ahead for metamaterials. Science **328**, 582 (2010)
49. N. Zheludev, A roadmap for metamaterials. OPN Opt. Photonics News **31** (2011)
50. M. Stockman, Nanoplasmonics: past, present, and glimpse into future. Opt. Express **19**, 22029 (2011)
51. C. Soukoulis, M. Wegener, Past achievements and future challenges in the development of three-dimensional photonic metamaterials. Nat. Photonics **5**, 523–530 (2011)
52. N. Lindquist, P. Nagpal, K. McPeak, D. Norris, S.-H. Oh, Engineering metallic nanostructures for plasmonics and nanophotonics. Rep. Prog. Phys. **75**, 036501 (2012)
53. M. Stockman, Spaser explained. Nat. Photonics **2**, 327 (2008)
54. C. Kurter, P. Tassin, L. Zhang, T. Koschny, A. Zhuravel, A. Ustinov, S. Anlage, C. Soukoulis, Classical analogue of electromagnetically induced transparency with a metal-superconductor hybrid metamaterial. PRL **107**, 043901 (2011)
55. T. Kaelberer, V.A. Fedotov, N. Papasimakis, D.P. Tsai, N.I. Zheludev, Science **330**, 1510 (2010)
56. G. Afanasiev, Vector solutions of the Laplace equation and the influence of helicity on Aharonov-Bohm scattering. J. Phys. A: Math. Gen. **27**, 2143 (1994)
57. K. Marinov, A.D. Boardman, V.A. Fedotov, N. Zheludev, Toroidal metamaterial. New J. Phys. **9**, 324 (2007)
58. V.A. Fedotov, A. Rogacheva, V. Savinov, D. Tsai, N.I. Zheludev, Resonant transparency and non-trivial non-radiating excitations in toroidal metamaterials. Sci. Rep. **3**, 2967 (2013)

59. B. Ögüt, N. Talebi, R. Vogelgesang, W. Sigle, P.A. van Aken, Toroidal plasmonic eigenmodes in oligomer nanocavities for the visible. Nano Lett. **12**, 5239 (2012)
60. I.B. Zeldovich, Electromagnetic interaction with parity violation. JETP **33**, 1531 (1957)
61. G. Afanasiev, Simplest source of electromagnetic fields as a tool for testing the reciprocity-like theorems. J. Phys. D: Appl. Phys. **34**, 539 (2001)
62. A. Miroshnichenko, A. Evlyukhin, Y.F. Yu, R. Bakker, A. Chipouline, A. Kuznetsov, B. Luk'yanchuk, B. Chichkov, Y. Kivshar, Observation of an anapole with dielectric nanoparticles. Nat. Commun. **6**, 8069 (2015)
63. V. Agranovich, V. Ginzburg, *Kristallooptika s Uchetom Prostranstvennoi Dispersii i Teoriya Eksitonov (Crystal Optics with Spatial Dispersion, and Excitons)* (Nauka, Moscow, 1965) [Translated into English (Springer, Berlin, 1984)]
64. V. Agranovich, Yu. Gartstein, Electrodynamics of metamaterials and the Landau-Lifshitz approach to the magnetic permeability. Metamaterials **3**, 1 (2009)
65. V. Agranovich, Yu. Gartstein, Spatial dispersion and negative refraction of light. Phys. Usp. **49**(10), 1029 (2006)
66. A. Andryieuski, S. Ha, A. Sukhorukov, Y. Kivshar, A. Lavrinenko, Bloch-mode analysis for retrieving effective parameters of metamaterials. Phys. Rev. B **86**, 035127 (2012)
67. C. Simovski, On electromagnetic characterization and homogenization of nanostructured metamaterials. J. Opt. **13**, 013001 (2011)
68. S. Tretyakov, *Analytical Modeling in Applied Electromagnetics* (Artech House, Boston, 2003)
69. A. Vinogradov, *Electrodynamics of Compound Media* (Scientific and Educational Literature Publisher, Russian Federation, 2001). ISBN 5-8360-0283-5 (in Russian)
70. P. Mazur, B. Nijboer, On the statistical mechanics of matter in an electromagnetic field. I. Physica **XIX**, 971 (1953)
71. N. Papasimakis, V. Fedotov, K. Marinov, N. Zheludev, Gyrotropy of a metamolecule: wire on a torus. PRL **103**, 093901 (2009)
72. A. Vinogradov, A. Merzlikin, Comment on "Basics of averaging of the Maxwell equations for bulk materials". Metamaterials **6**, 121 (2012)
73. C. Simovski, Material parameters of metamaterials (a review). Opt. Spectrosc. **107**, 726 (2009)
74. J.D. Jackson, *Classical Electrodynamics*, 3rd edn. (Wiley, New York, 1999)
75. C. Simovski, *Weak Spatial Dispersion in Composite Media* (Polytechnika, St. Petersburg, 2003) (in Russian)
76. E.M. Purcell, Phys. Rev. **69**, 681 (1946)

Chapter 2
Homogenization of Maxwell Equations—Macroscopic and Microscopic Approaches

2.1 Microscopic Maxwell Equations and Averaging Procedure

We consider as a starting point a system of microscopic MEs in the following form:

$$\begin{cases} \operatorname{rot}\vec{e} = \dfrac{i\omega}{c}\vec{h} \\ \operatorname{div}\vec{h} = 0 \\ \operatorname{div}\vec{e} = 4\pi\rho \\ \operatorname{rot}\vec{h} = -\dfrac{i\omega}{c}\vec{e} + \dfrac{4\pi}{c}\vec{j} \end{cases} \qquad \begin{cases} \rho = \sum\limits_i q_i \delta(\vec{r} - \vec{r}_i) \\ \vec{j} = \sum\limits_i \vec{v}_i q_i \delta(\vec{r} - \vec{r}_i) \\ \dfrac{\mathrm{d}\vec{p}_i}{\mathrm{d}t} = q_i\vec{e} + \dfrac{q_i}{c}\left[\vec{v}_i * \vec{h}\right] \end{cases} \tag{2.1}$$

Here $\vec{e}$ and $\vec{h}$ are the microscopic electric and magnetic fields, respectively, ρ is the charge density, $\vec{q}_i$, $\vec{p}_i$, $\vec{r}_i$ and $\vec{v}_i$ are the charges, pulses, coordinates and velocities of charges, $\vec{j}$ is the microscopic current density, ω and c are the frequency and the velocity of light in vacuum. It is assumed that system (2.1) is strictly valid without any approximations. Actually, system (2.1) can be elaborated in the framework of the minimum action approach [1]; nevertheless, one should remember that the minimum action principle does not give an unambiguous form of the MEs (2.1), but instead gives a set of different forms which satisfy the requirement of relativistic invariance. The "right" form can be chosen based on the evident requirement of correspondence of the results of the final system of equations to the observed physical effects. One should also mention that in the framework of the minimum action approach the final equations are written for "potentials + particles", not for "fields + particles"; the respective equations are:

A. Chipouline and F. Küppers, *Optical Metamaterials: Qualitative Models*, Springer Series in Optical Sciences 211, https://doi.org/10.1007/978-3-319-77520-3_2

$$\begin{cases} \frac{\mathrm{d}\vec{p}_i}{\mathrm{d}t} = -\frac{q_i}{c}\frac{\partial \vec{A}}{\partial t} - q_i \nabla \varphi + \frac{q_i}{c}\left[\vec{v}_i * \mathrm{rot}\vec{A}\right] \\ \frac{\partial F^{ik}}{\partial x^k} = -\frac{4\pi}{c} j^i, \quad F_{ik} = \frac{\partial A_k}{\partial x_i} - \frac{\partial A_i}{\partial x_k} \end{cases} \tag{2.2}$$

Here $\vec{A}$ and φ are the components of the 4-vector potential, and the relations between the microscopic fields and the potentials are given by:

$$\begin{cases} \vec{e} = -\frac{1}{c}\frac{\partial \vec{A}}{\partial t} - \nabla \varphi \\ \vec{h} = \mathrm{rot}\vec{A} \end{cases} \tag{2.3}$$

Form (2.2) will not be used in the following discussions and is presented here just for methodological reasons.

It has to be mentioned that the basic formulation of electrodynamics is still under discussion [2]. Here it is assumed that system (2.1) fully describes the electromagnetic phenomena and further discussion about validity of (2.1) is left out.

One more remark has to be given. Maxwell equations (2.1) assume that the charge dynamics is described in the frame of relativistic, but classical mechanics. Rigorously speaking, in case of quantum mechanics Maxwell equations (2.1) is no more valid and have to be elaborated differently. Further discussion of this fundamental question is out of the scope of this book.

We consider propagation of an electromagnetic plane wave interacting with the medium in case when the classical dynamics is supposed to be valid and the bulk material fills the whole space; the system (2.1) in this case can be formally averaged over a physically small volume (or through statistic averaging), which results in:

$$\begin{cases} \mathrm{rot}\langle\vec{e}\rangle = \frac{i\omega}{c}\langle\vec{h}\rangle \\ \mathrm{div}\langle\vec{h}\rangle = 0 \\ \mathrm{div}\langle\vec{e}\rangle = 4\pi\langle\rho\rangle \\ \mathrm{rot}\langle\vec{h}\rangle = -\frac{i\omega}{c}\langle\vec{e}\rangle + \frac{4\pi}{c}\langle\vec{j}\rangle \\ \rho = \sum_i q_i \delta(\vec{r} - \vec{r}_i) \\ \langle\vec{j}\rangle = \sum_i q_i \langle\vec{v}_i \delta(\vec{r} - \vec{r}_i)\rangle \\ \frac{\mathrm{d}\vec{v}_i}{\mathrm{d}t} = \frac{q_i}{m_i}\vec{e} + \frac{q_i}{m_i c}\left[\vec{v}_i * \vec{h}\right] \end{cases} \xrightarrow[\langle\vec{e}\rangle = \vec{E}]{\langle\vec{h}\rangle = \vec{B}} \begin{cases} \mathrm{rot}\vec{E} = \frac{i\omega}{c}\vec{B} \\ \mathrm{div}\vec{B} = 0 \\ \mathrm{div}\vec{E} = 4\pi\langle\rho\rangle \\ \mathrm{rot}\vec{B} = -\frac{i\omega}{c}\vec{E} + \frac{4\pi}{c}\langle\vec{j}\rangle \\ \langle\rho\rangle = \sum_i q_i \delta(\vec{r} - \vec{r}_i) = \langle\rho\rangle(\vec{E}, \vec{B}) \\ \langle\vec{j}\rangle = \sum_i q_i \langle\vec{v}_i \delta(\vec{r} - \vec{r}_i)\rangle = \langle\vec{j}\rangle(\vec{E}, \vec{B}) \end{cases} \tag{2.4}$$

The averaging is usually performed in case of a large number of atoms/molecules in the volume of averaging; from the other side the volume is supposed to be small in comparison with the wavelength of the electromagnetic wave, propagating in the medium.

The main problem here is to find the averaged current and charge distribution as functions of the averaged electric and magnetic fields:

$$\begin{aligned}\langle\vec{j}\rangle &= \langle\vec{j}\rangle(\vec{E},\vec{B})\\ \langle\rho\rangle &= \langle\rho\rangle(\vec{E},\vec{B})\end{aligned} \tag{2.5}$$

Averaged (or macroscopic) fields result from the averaging of the actual, microscopic fields which are produced by some external sources and charged particles. The averaging procedure itself is rarely considered in phenomenological models of macroscopic Maxwell equations. In order to set the applicability limitations for the averaging procedure, it is necessary to determine the procedure itself. For example, in the frame of the Lorenz-Lorentz concept, the relation between the wavelength and averaging volume is chosen so that the particle size and the distances between them are 2–3 orders of magnitude smaller, than the wavelength (in Fig. 2.1 this means $L_{\text{inter}}, L_{\text{intra}} \leq 10^{-2\,\text{to}\,-3}\lambda$). This requirement fails in case of composites even for ultrahigh frequencies, but nevertheless the Lorenz-Lorentz approach still gives

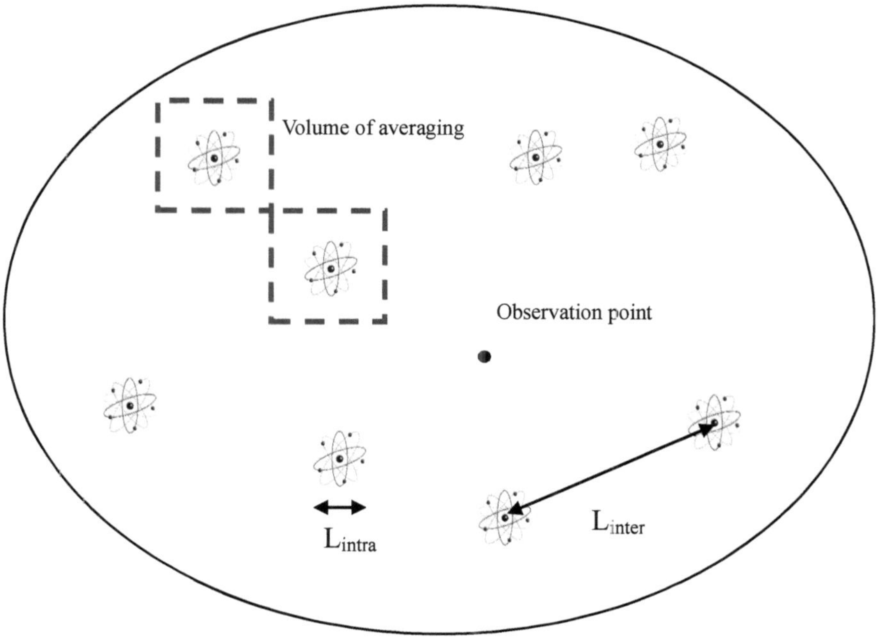

Fig. 2.1 Schematic representation of the media which is subject of the averaging with typical lengths, involved in consideration

very good results. It turns out that the requirement $L_{\text{inter}}, L_{\text{intra}} \leq 10^{-2\,\text{to}\,-3}\lambda$ can be significantly relaxed, and the main question in this case becomes: up to which relations between particle size and the wavelength the averaging procedure will still make sense.

First, the averaging volume (typical size of this volume) has to be in any case less than the wavelength, otherwise the field (which is supposed to be harmonic in space) disappears. Roughly speaking, the typical size of the averaging volume has to be in any case 5–10 times smaller than the wavelength. Taking into account that the particle size and the distance between particles have to be only order of magnitude less, than the wavelength, it results in no more than one particle per averaged volume. Hence, the averaging concept for this case is required to be qualitatively different in compare with the classical (for example, Lorenz-Lorentz) one.

The key notion for the introduced averaging procedure is the volume of averaging—the unit cell of the procedure (see Fig. 2.1). All volume is subdivided into cells (the "volume of averaging" in Fig. 2.1), and each of them contains at least one particle. Moreover, it is supposed that the composite is homogeneous, i.e. there are no empty cells and there are no cells containing more than one particle. It means that the considered media is regular, but the particle in the cell is not necessary positioned at the center of the cell; each cell contributes to the fields at the point of consideration, from which all averaged characteristics are supposed to be depended on. The average volume is taken equal to the volume of the elementary cell (has been first proposed in [3]).

There are two main approaches to the averaging, namely statistical one [4–6], where averaging is taking over the ensemble of realization; and spatial, where the main questions are volume of the averaging and the averaging function, which set the minimal macroscopic scale at which change of the macroscopic (averaged) functions is still significant [7].

A significant advantage of the statistical approach is the absence of characteristic scales, like the volume of the unit cell, because the averaging is performed over all possible realisations rather than over physical volume. A drawback of the statistical approach is in relatively complex math required for the elaboration of the model. The concept of the statistical averaging, being well developed for the electrostatics and magnetostatics, appears not to be completed for the microwave frequencies and optical spectra [8].

One more approach, called scaling algorithm has to be mentioned [9]. This is a generalized method of spatial averaging, where at the first step the averaging is performed over small-scale cells, then over bigger cells, which include many small cells. The procedure is repeated before the averaged functions converge to their asymptotical values. The method is based on specially introduced multipoles (different from the usually defined ones), which are assigned to the cells at each step of the averaging. The method (similarly to [5]) is a universal one for diluted and dense composites, because it does not require introduction of the local field, but is relatively complicated and not constructive, e.g. the method itself does not suggest constructive averaging procedure.

More comprehensive general review of different approaches to the problem of averaging can be found in [10].

2.2 System Under Consideration

In spite of the fact, that in this work general properties of ME (irrespective to a particular medium) are of interest, it seems to be methodologically appropriate to determine from the beginning the type of MMs which will be considered and keep this type in mind throughout the text.

In this work a medium consisting of artificial MAs embedded in a dielectric matrix will be considered—see Fig. 2.2, where only one layer of the considered material is presented.

The MAs are assumed to be complex plasmonic structures, possessing so called symmetric and anti-symmetric eigenmodes—see Fig. 2.3, where one possible structure (coupled nanowires, in general of different sizes) is shown. The structure consists of two nanowires with typical lengths (for optical domain) of tens to hundreds nanometers, placed one under another with the distance of several tens of nanometers, ensuring strong near field interaction between both nanowires [11].

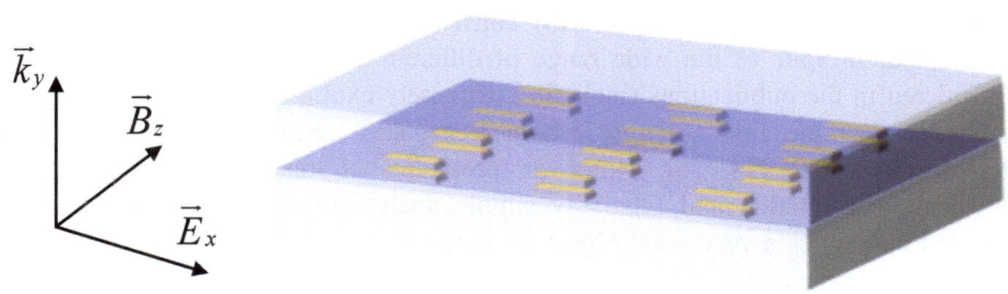

Fig. 2.2 Artificial MAs (plasmonic nanoresonators) embedded in a dielectric matrix form a MM (only one layer is presented). Polarization of the electric and magnetic fields, and direction of the wave vector are shown

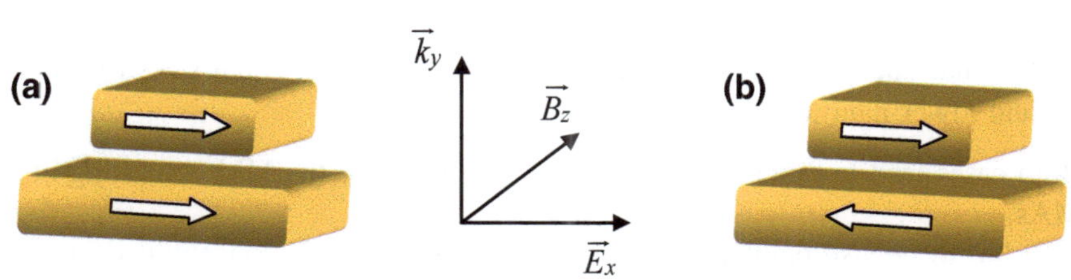

Fig. 2.3 One of the possible shape of MAs, possessing **a** symmetric and **b** anti symmetric modes. Electric field $\vec{E}_y$ of the incoming wave, propagating along y axis excites eigenmodes of the plasmonic MA

Considering both nanoresonators as two coupled harmonic oscillators, it becomes clear that the structure possess two fundamental eigenmodes, namely symmetric (Fig. 2.3a) which produces effective dielectric response, and anti-symmetric (Fig. 2.3b) which is responsible for the magnetic response of the media.

Here the incoming electromagnetic wave interacts with the electrons of the plasmonic structure and effectively excites symmetric and asymmetric oscillations, provided that the frequency of the incoming wave is close to the respective eigenfrequency of the eigenmodes. In case of symmetric mode electrons in both nanoresonators oscillate in phase, while in case of anti-symmetric mode—out of phase (Fig. 2.3a, b, respectively).

It is also intuitively clear, that in case of symmetric oscillation (Fig. 2.3a) the MA formed by two wires of equal length does not produce any magnetic effects, but rather exhibits extra dipole moments leading finally to the change of the permittivity of the MM. The main interest to the MM is stipulated by the possibility to excite the antisymmetric modes (Fig. 2.3b). In this case the structure presents to the first approximation a circle current, which provides (as it is well known from the school course of physics) a magnetic response. The fantastic peculiarity of the MMs is in the fact, that the MMs provide magnetic response at optical frequencies, where no natural media has similar properties. This fundamentally distinguishes MMs from any natural materials, and makes such MMs (among others) extremely interesting objects for both fundamental research and various applications [12].

In most parts of this work, a MM consisting of coupled nanowires will be considered. In spite of the wide range of different shapes of the nanoresonators considered in the publications (including extremely exotic ones [13]), the coupled nanowire was the first structure whom a negative refractive index has been demonstrated with [14, 15]. Moreover, properties of this structure allow clear physical interpretation and relatively simple analytical treatment, which makes coupled nanowires a very good object for discussion of different physical models. One of the simple and at the same time rather good examples is in consideration of necessary conditions for the existence of anti-symmetric modes. Remembering the mentioned above interpretation of the dynamics of the coupled nanowires in terms of coupled harmonic oscillators, it is clear that the anti-symmetric mode can be excited due to the following reasons:

- First, asymmetric excitation (for example, retardation at the wave propagation between lower and upper nanowires), and,
- Second, due to asymmetric shape of the structure (not equal sizes of the upper and lower nanowires).

Both cases lead to rather different optical properties of the MM consisting of such MAs.

It has to be also emphasized that the presented book is concentrated mainly on homogenization of bulk material—multilayer (in Fig. 2.2 only one layer is presented) MM.

Comparing Fig. 2.1 with Fig. 2.2 and identifying the MAs as artificial atoms one can immediately conclude, that the basic requirements for the homogenization procedure are satisfied rather poorly. For example, typical sizes of the MAs, having the resonance wavelength of about 1 μm, are of the order of several hundreds of nanometers; the distances between the MAs are of the same order. It is clear, that classical approach for homogenization which results in known expressions like ones for small concentrations and small permittivity variation [1, 16, 17], Lorenz-Lorentz and Clausius-Mossotti equation [18–21], and Bruggeman equation [22] are no more valid. All these mentioned above theories have been elaborated in the framework of the static approximation, where electric and magnetic effects are separated. In the case of MM the main effect is in appearance of a magnetic response under the interaction of the MAs with the electric field, which is not included in the static approximation. An attempt to include dynamic corrections into the Lorenz-Lorentz expression, obtained in [10, 23] loses, as it was admitted by author, its physical meaning for the MM frequencies ranges [17].

2.3 Frequency Range of Homogenization

It is methodologically important to distinguish between MM and other forms of compound structures like, for example, photonic crystals. In addition to the typical sizes (see Fig. 2.1) it is useful to introduce two characteristic wavelengths, namely:

1. Wavelength of the plane wave in the effective medium $\lambda_{\text{eff}} = 2\pi/\text{Re}(k)$.
2. The resonance wavelength λ_{res} of the internal resonances of the inclusions.

We are by definition interested in the wavelength region where the internal resonances of the inclusions exist $\lambda_{\text{res}} \sim \lambda_{\text{eff}}$. The relation between the effective wavelength and the typical distance between inclusions $\lambda_{\text{eff}} > L_{\text{inter}}/2$ guaranties that the effective wavelength is safely longer than the wavelength of the Bragg resonances, otherwise the effect of mutual interference of the scattered waves dominates and the media is no more optically dense. Nevertheless, it is possible to create media where one of the typical sizes of the inclusions exceeds the effective wavelength significantly but the media can be homogenized anyway. For example, the media consisting of long and short wires can be homogenized and thus far can be considered as MM [24–27].

The question about possible frequencies where the homogenization is possible has been considered in details in [17, 28] (see also references therein). First, the effect of anisotropy which leads to different wave vectors for different propagation directions and which is typical for crystals, has to be taken into account. It was shown that the effect of anisotropy leads to two possible families of isofrequency contours, namely ellipsoids and hyperboloids, depending on the sign of the ratios $\varepsilon_{xx}/\varepsilon_{zz}$ and $\varepsilon_{yy}/\varepsilon_{zz}$ (here the consideration is carried out in the main axis of the intrinsic coordinate system of the crystal) [29–31]. From the other side, the effective parameters themselves do not depend on the wave vector (spatial dispersion is negligible). In media with spatial dispersion the effective parameters depend on the wave vector considerably, therefore making the shape of the isofrequency contours arbitrary. Comparison of isofrequencies of MM obtained numerically with ellipsoids and hyperboloids makes it possible to identify frequency intervals, where the effective parameters do not depend on the wave vector. The conclusion in these papers is that the homogenization is possible only for the frequencies possessing isofrequency curves of ellipsoid or hyperboloid types in isofrequency contours—see, for example, Fig. 2.1 in [17]. In other words, it was concluded that homogenization is possible only in case of absence of spatial dispersion. This conclusion is questionable. Keeping in mind, that the goal of the bulk homogenization is in finding a dispersion relation it has to be admitted, that this goal is fully achieved by all (not only by ellipsoid/hyperboloid) curves in the isofrequency contour. In other words, the fact that the dispersion relation (and the respective refractive index) depends on the direction does not contradict to the possibility of homogenization.

In case of periodic media there is a frequency range that corresponds to the regime of a single propagating mode (understood as the absence of higher-order propagating Bloch modes). According to [7, 17, 32–35], this is the requirement of possibility of homogenization. This assumption appeared to be questionable as well: in general, the presence of several propagating modes at the same frequency does not mean that the media could not be substituted by a homogeneous one with the same dispersion characteristics. The presence of several possible modes just mean that the propagating field will be presented as a sum of these modes with appropriate weighting coefficients; the question about relation between these coefficients is related to the problem of excitation of these modes, which in turn concerns the boundary condition problem, which is out of the scope of this paper.

Nevertheless, for the sake of simplicity and in order to fix the notations, it is assumed here that all characteristic sizes of the media (the sizes of the MAs and the distances between them) are smaller than one half of the effective wavelength in the media, which separates the problem of homogenization of MMs from the same problem for photonic crystal structures.

2.4 Different Representations of Material Equation

The last equation in (2.4) is not averaged and describes the microscopic dynamics which is supposed to be substituted in $\langle\vec{j}\rangle$, $\langle\rho\rangle$ and averaged in order to get (2.5). Relations (2.5) in turn use information about microscopic dynamics as a function of microscopic fields which get averaged after substitution into equations for $\langle\vec{j}\rangle$, $\langle\rho\rangle$. In fact, there is only one model (a multipole model [4]) where the averaging procedure for the $\langle\vec{j}\rangle$, $\langle\rho\rangle$ as functions of microscopic dynamics is performed rigorously, all other models do not even try to make this step; the last equation in (2.4) is usually left out completely, and the necessary equivalent information about charge dynamics is brought to the model phenomenologically. It has to be realized that this gap in the theory [inability to average rigorously the dynamic equation for particles in (2.4)] causes all the problems in further consideration and leads to all appearing contradictions and ambiguities.

In this paper we do not suggest any new ways of inclusion of this dynamic equation [the mentioned last equation in (2.4)] into consideration; instead, after this fact (ignorance of the dynamic equation from the averaging procedure) has been recognized, we try to create an approach to homogenization rigorously in terms of a logical chain with clearly recognized steps, assumptions, and approximations.

The system of (2.4) and (2.5) is rather useless in practice until we find analytical expressions for (2.5). Nevertheless, even without finding of an analytical form for (2.5), the averaged MEs can be analysed and important conclusions can be made.

It is worth noticing that if we assume some analytical form for (2.5) (see, for example, [36] and references herein) then the averaging problem is basically fixed (or, better to say, bypassed). System (2.4) becomes self-consistent and can be solved for the electric and magnetic fields $\vec{E}$, $\vec{B}$. Any further considerations (including introduction of $\vec{D}$, $\vec{H}$ in different representations, as well as the permittivity and permeability) in this case are no more required. Thus, in what follows we assume that there are no explicit forms of (2.5) and it is necessary to elaborate (2.5) further in order to find some reasonable analytical expressions for the averaged charge and current densities. It has to be emphasized as well that both these characteristics—the averaged charge and current densities—are measurable quantities and they do not change in math transformations; we will see below, that these characteristics keep their form in different ME representations.

First, following [1] we consider a volume with charges and fields (see again Fig. 2.1). The averaged charge in (2.5) can be represented through another function taking into account that the total charge of the considered volume is zero:

$$\int \langle\rho\rangle \, \mathrm{d}V = 0 \tag{2.6}$$

It means that the averaged density of charges can be presented as a divergence of another unknown function $\vec{P}_{\text{full}}$ (see more details in [1, 16]):

$$\langle\rho\rangle = -\mathrm{div}\vec{P}_{\text{full}} \tag{2.7}$$

The function $\vec{P}_{\text{full}}$ is supposed to be zero outside the volume of integration in (2.6) [1]. In addition, this function is introduced with the accuracy of "rot" from any other arbitrary differentiable function $\vec{F}_1$:

$$\begin{cases} \langle\rho\rangle = -\mathrm{div}\vec{P}_{\text{full}} = -\mathrm{div}\left(\vec{P} + \mathrm{rot}\,\vec{F}_1\right) \\ \vec{P}_{\text{full}} = \vec{P} + \mathrm{rot}\,\vec{F}_1 \end{cases} \tag{2.8}$$

The averaged current is connected with the averaged charge density through the continuity relation [37], which remains valid for the macroscopic representation:

$$\begin{cases} i\omega\langle\rho\rangle = \operatorname{div}\langle\vec{j}\rangle \\ \langle\rho\rangle = -\operatorname{div}\vec{P}_{\text{full}} \end{cases} \tag{2.9}$$

which gives:

$$\operatorname{div}\left(\langle\vec{j}\rangle + i\omega\vec{P}_{\text{full}}\right) = 0 \tag{2.10}$$

This means, that the averaged current can be introduced with the accuracy of "rot" of one more arbitrary function $\vec{F}_2$:

$$\langle\vec{j}\rangle = -i\omega\vec{P}_{\text{full}} = -i\omega\vec{P}_{\text{full}} + \operatorname{rot}\vec{F}_2 \tag{2.11}$$

or, taking into account (2.8):

$$\langle\vec{j}\rangle = -i\omega\vec{P}_{\text{full}} + \operatorname{rot}\vec{F}_2 = -i\omega\vec{P} + \operatorname{rot}\left(-i\omega\vec{F}_1 + \vec{F}_2\right) \tag{2.12}$$

It turns out that the material equations (2.3) can be written through one new function $\vec{P}$ with the accuracy of two more arbitrary functions $\vec{F}_1$ and $\vec{F}_2$:

$$\begin{cases} \langle\rho\rangle = -\operatorname{div}\left(\vec{P} + \operatorname{rot}\vec{F}_1\right) \\ \langle\vec{j}\rangle = -i\omega\vec{P} + \operatorname{rot}\left(-i\omega\vec{F}_1 + \vec{F}_2\right) \end{cases} \tag{2.13}$$

This approach assumes that the introduced function $\vec{P}_{\text{full}}$ is zero outside the volume of integration in (2.6), and, moreover, the function $\vec{P}_{\text{full}}$ does not depend on the chosen integration volume. Another approach has been proposed in [9] and is called the scaling algorithm. The developed approach is based on a lemma proving that any field can be represented as a sum of three terms which are called "electric dipole", "magnetic dipole", and "electric quadrupole" moments (in the frequency domain) [9]:

$$J_i = -i\omega p_i + ce_{ijk}\frac{\partial m_k}{\partial x_j} + i\omega c\frac{\partial}{\partial x_k}q_{ik} = J_i^{(p)} + J_i^{(m)} + J_i^{(q)}$$

$$\begin{cases} m_i\left(x_j,\, J_k\right) = \dfrac{1}{2c}e_{ijk}x_jJ_k \\[2ex] -\, i\omega\, q_{ij}\left(x_j,\, J_k\right) = \dfrac{1}{2c}\left(x_jJ_i + x_iJ_j\right) \\[2ex] -\, i\omega\, p_i(x_i,\, J_k) = -\left(x_j\dfrac{\partial J_k}{\partial x_k}\right) \end{cases} \tag{2.14}$$

This lemma leads to the possibility to represent, for example, the averaged current in the following form [9]:

$$\langle \vec{j} \rangle = -i\omega\left(\vec{P} - \mathrm{div}Q\right) + c\,\mathrm{rot}\,\vec{M} \tag{2.15}$$

which basically repeats the second equation in (2.13). The equation for the averaged current $\langle\rho\rangle$ then becomes:

$$\langle \rho \rangle = -\mathrm{div}\left(\vec{P} - \mathrm{div}Q\right) \tag{2.16}$$

which is similar to the first equation in (2.13) if we assume that $\vec{P} \to \vec{P} - \mathrm{div}Q$. It is also rather straightforward to extend the approach (2.14) and include an analogy with the arbitrary functions $\vec{F}_1$ and $\vec{F}_2$. First, the function $\vec{M}$ in (2.15) appears to be the same as the function $\vec{F}_2$ (basically, $\vec{F}_2 = c\vec{M}$), and the function $\vec{F}_1$ can be included in (2.14) as extra terms for the magnetic [first equation in (2.14)] and dipole [last equation in (2.14)] terms:

$$J_i = -i\omega p_i + c e_{ijk}\frac{\partial m_k}{\partial x_j} + i\omega c\frac{\partial}{\partial x_k}q_{ik} = J_i^{(p)} + J_i^{(m)} + J_i^{(q)}$$

$$\begin{cases} m_i\left(x_j, J_k\right) = \dfrac{1}{2c}e_{ijk}x_jJ_k + \dfrac{i\omega}{c}F_{1,i} \\ -i\omega\, q_{ij}\left(x_j, J_k\right) = \dfrac{1}{2c}\left(x_jJ_i + x_iJ_j\right) \\ -i\omega\, p_i(x_i, J_k) = -\left(x_j\dfrac{\partial J_k}{\partial x_k}\right) - e_{ijk}\dfrac{\partial F_{1,k}}{\partial x_j} \end{cases} \tag{2.17}$$

It is worth noting that (2.14)–(2.17) have been obtained without any additional assumptions about integration which have been used at the elaboration (2.7) from (2.6). The question about the physical meaning of the functions in (2.14)–(2.17) and (2.13) remains opened.

Now we come back to the consideration of (2.13). Both functions $\vec{F}_1$ and $\vec{F}_2$ are arbitrary and independent. This means that it is possible to consider different possibilities and to impose any additional requirements on them. There are different but countable number of choices for the possible representations of (2.13). The most general case is when both $\vec{F}_1$ and $\vec{F}_2$ are non-zero functions, namely:

$$\begin{cases} \vec{P}_{\mathrm{full}} = \vec{P}_{\mathrm{C}} = \vec{P} + \mathrm{rot}\,\vec{F}_1 \\ \vec{F}_2 = c * \vec{M}_{\mathrm{C}} \end{cases} \tag{2.18}$$

which leads to the so called Casimir (subscript "C" stands for Casimir) form of material equations:

$$\begin{cases} \langle \rho \rangle = -\mathrm{div}\vec{P}_{\mathrm{C}} \\ \langle \vec{j} \rangle = -i\omega\,\vec{P}_{\mathrm{C}} + c\,\mathrm{rot}\,\vec{M}_{\mathrm{C}} \end{cases} \qquad \begin{cases} \vec{D} = \vec{E} + 4\pi\vec{P}_{\mathrm{C}} \\ \vec{H} = \vec{B} - 4\pi\vec{M}_{\mathrm{C}} \end{cases} \tag{2.19}$$

In this case MEs include four functions $\vec{E}$, $\vec{B}$, $\vec{D}$, $\vec{H}$:

$$\begin{cases} \operatorname{rot}\vec{E} = \frac{i\omega}{c}\vec{B} \\ \operatorname{div}\vec{B} = 0 \\ \operatorname{div}\vec{D} = 0 \\ \operatorname{rot}\vec{H} = -\frac{i\omega}{c}\vec{D} \end{cases} \tag{2.20}$$

Note, that the case $\vec{F}_1 = 0$ and $\vec{F}_2 = c * \vec{M}_C$ leads to the same form (2.19), where the curl part of the full polarizability $\vec{P}_{\text{full}}$ [see (2.8)] is excluded (the physical meaning of this part $\vec{F}_1$—presence of anapoles [38]—will be considered below). Alternatively to (2.18), we can set:

$$\begin{cases} \vec{P}_{\text{full}} = \vec{P}_{\text{LL}} = \vec{P} + \operatorname{rot}\vec{F}_1 \\ \vec{F}_2 = 0 \end{cases} \tag{2.21}$$

which leads according to (2.13) to the so called Landau&Lifshitz (subscript "L&L" stands for Landau&Lifshitz) [39] form of material equations:

$$\begin{cases} \langle\rho\rangle = -\operatorname{div}\vec{P}_{\text{LL}} \\ \langle\vec{j}\rangle = -i\omega\vec{P}_{\text{LL}} \end{cases} \qquad \begin{cases} \vec{D} = \vec{E} + 4\pi\vec{P}_{\text{LL}} \\ \vec{B} = \vec{B} \end{cases} \tag{2.22}$$

In this case MEs contain three functions $\vec{E}$, $\vec{B}$, $\vec{D}$:

$$\begin{cases} \operatorname{rot}\vec{E} = \frac{i\omega}{c}\vec{B} \\ \operatorname{div}\vec{B} = 0 \\ \operatorname{div}\vec{D} = 0 \\ \operatorname{rot}\vec{B} = -\frac{i\omega}{c}\vec{D} \end{cases} \tag{2.23}$$

Note that the form (2.22) does not assume that the averaged current $\langle\vec{j}\rangle$ does not contain any curl part—this part is included in $\langle\vec{j}\rangle$ through $\vec{F}_1$ [see (2.21)]. The main difference between "C" and "L&L" representation is in the absence in the latter any stationary (not proportional to ω) part of the curl part of $\langle\vec{j}\rangle$, described by $\vec{M}_C$. In case of the absence of the stationary magnetization both representations have to be equivalent.

Finally, we assume that the full polarizability $\vec{P}_{\text{full}}$ contains only the curl part, namely:

$$\begin{cases} \vec{P}_{\text{full}} = \operatorname{rot}\vec{F}_1 \\ \vec{F}_2 = c * \vec{M}_C \end{cases} \tag{2.24}$$

which leads, according to (2.13), to the case which we call here Toroidal (subscript "T" stands for Toroid) form of material equations:

$$\begin{cases} \langle \rho \rangle = 0 \\ \langle \vec{j} \rangle = -i\omega \,\mathrm{rot}\, \vec{F}_1 + c \,\mathrm{rot}\, \vec{M}_\mathrm{C} = c \,\mathrm{rot}\, \vec{M}_\mathrm{T} \end{cases} \qquad \begin{cases} \vec{D} = \vec{E} + 4\pi \,\mathrm{rot}\, \vec{F}_1 \\ \vec{H} = \vec{B} - 4\pi \vec{M}_A \end{cases} \tag{2.25}$$

In this case the system of MEs contains three functions $\vec{E}$, $\vec{B}$, $\vec{H}$ and reads:

$$\begin{cases} \mathrm{rot}\, \vec{E} = \dfrac{i\omega}{c} \vec{B} \\ \mathrm{div} \vec{B} = 0 \\ \mathrm{div}\, \vec{E} = 0 \\ \mathrm{rot}\, \vec{H} = -\dfrac{i\omega}{c} \vec{E} \end{cases} \tag{2.26}$$

We stress that this set of equations can be used only in very special cases where the averaged charge density [not only total charge—compare with (2.6)] is zero. Thus, in general the "T" form cannot be used *instead* of the "C" or "L&L" forms.

The physical object corresponding to such representation is a toroid [16, 38], which is now of great interest in connection with potential possibility of design of such structures at nanoscales for optical wavelength region application [40]. It is seen, that the presence of toroid is responsible for the function $\vec{F}_1$, and fixes the functions $\langle \rho \rangle$, $\langle \vec{j} \rangle$ in general representation (2.13).

Physical interpretation of the three mentioned above representations can be done based on the types of atoms/molecules (or MAs/metamolecules) which the considered media consist of. In the most general case "C" form is preferable. In the case of absence of stationary magnetisation (but presence of all others) the "C" and

Table 2.1 Possible forms of representation of material equations and the respective forms of the macroscopic MEs

Casimir form ("C" form) $\vec{E}, \vec{B}, \vec{D}, \vec{H}$	Landau & Lifshitz form ("LL" form) $\vec{E}, \vec{B}, \vec{D}, \cancel{\vec{H}}$	Toroidal form ("T" form) $\vec{E}, \vec{B}, \vec{D}, \cancel{\vec{H}}$
$\begin{cases} \langle \rho \rangle = -\mathrm{div} \vec{P}_\mathrm{C} \\ \langle \vec{j} \rangle = -i\omega \vec{P}_\mathrm{C} + c \,\mathrm{rot}\, \vec{M}_\mathrm{C} \end{cases}$ $\begin{cases} \vec{D} = \vec{E} + 4\pi \vec{P}_\mathrm{C} \\ \vec{H} = \vec{B} - 4\pi \vec{M}_\mathrm{C} \end{cases}$ $\begin{cases} \mathrm{rot}\, \vec{E} = \frac{i\omega}{c} \vec{B} \\ \mathrm{div} \vec{B} = 0 \\ \mathrm{div}\, \vec{D} = 0 \\ \mathrm{rot}\, \vec{H} = -\frac{i\omega}{c} \vec{D} \end{cases}$	$\begin{cases} \langle \rho \rangle = -\mathrm{div} \vec{P}_\mathrm{LL} \\ \langle \vec{j} \rangle = -i\omega \vec{P}_\mathrm{LL} \end{cases}$ $\begin{cases} \vec{D} = \vec{E} + 4\pi \vec{P}_\mathrm{LL} \\ \vec{H} = \vec{B} \end{cases}$ $\begin{cases} \mathrm{rot}\, \vec{E} = \frac{i\omega}{c} \vec{B} \\ \mathrm{div} \vec{B} = 0 \\ \mathrm{div}\, \vec{D} = 0 \\ \mathrm{rot}\, \vec{B} = -\frac{i\omega}{c} \vec{D} \end{cases}$	$\begin{cases} \langle \rho \rangle = 0 \\ \langle \vec{j} \rangle = c \,\mathrm{rot}\, \vec{M}_\mathrm{T} \end{cases}$ $\begin{cases} \vec{D} = \vec{E} \\ \vec{H} = \vec{B} - 4\pi \vec{M}_\mathrm{T} \end{cases}$ $\begin{cases} \mathrm{rot}\, \vec{E} = \frac{i\omega}{c} \vec{B} \\ \mathrm{div} \vec{B} = 0 \\ \mathrm{div}\, \vec{E} = 0 \\ \mathrm{rot}\, \vec{H} = -\frac{i\omega}{c} \vec{E} \end{cases}$

"L&L" representations have to be equivalent. In case of absence of the non curl part of magnetization (presence of only toroids and maybe stationary magnetization) the "T" form is appropriate.

It is important to realize that there are no other choices for the material equations. Any homogenization model has to start from the statement in which representation it will be developed; arbitrary mixing between several representations is not acceptable, as it will be seen below.

The possible representations are summarized in Table 2.1.

2.5 Serdyukov-Fedorov Transformations

The Serdyukov-Fedorov transformations (SFT) are relations between two sets of four vectors $\vec{E}$, $\vec{B}$, $\vec{D}$, $\vec{H}$ and $\vec{E}'$, $\vec{B}'$, $\vec{D}'$, $\vec{H}'$, where both sets satisfy Maxwell's equations. The SFT are usually written in the following form (for "C" representation):

$$\begin{cases} \operatorname{rot}\vec{E} = \dfrac{i\omega}{c}\vec{B} \\ \operatorname{div}\vec{B} = 0 \\ \operatorname{div}\vec{D} = 0 \\ \operatorname{rot}\vec{H} = -\dfrac{i\omega}{c}\vec{D} \end{cases} \qquad \begin{cases} \vec{B} = \vec{B}' + \operatorname{rot}\vec{T}_1 \\ \vec{E} = \vec{E}' + \dfrac{i\omega}{c}\vec{T}_1 \\ \vec{D} = \vec{D}' + \operatorname{rot}\vec{T}_2 \\ \vec{H} = \vec{H}' - \dfrac{i\omega}{c}\vec{T}_2 \end{cases} \tag{2.27}$$

The SFT are composed from two parts—"field" transformations for $\vec{E}$, $\vec{B}$ and "material" transformations for $\vec{D}$, $\vec{H}$). Below SFT are used to find relations between the three forms of material relations.

First of all, let us consider the system of averaged MEs in the following form:

$$\begin{cases} \operatorname{rot}\vec{E} = \dfrac{i\omega}{c}\vec{B} \\ \operatorname{div}\vec{B} = 0 \\ \operatorname{div}\vec{E} = 4\pi\langle\rho\rangle \\ \operatorname{rot}\vec{B} = -\dfrac{i\omega}{c}\vec{E} + \dfrac{4\pi}{c}\langle\vec{j}\rangle \end{cases} \tag{2.28}$$

It is easy to see that the first two equations are invariant under the following transformation:

$$\begin{cases} \vec{B} = \vec{B}' + \operatorname{rot}\vec{T}_1 \\ \vec{E} = \vec{E}' + \dfrac{i\omega}{c}\vec{T}_1 \end{cases} \tag{2.29}$$

The new primed fields satisfy the same form of the first two MEs (2.28). It corresponds exactly to the SFT in parts of the field transformations in (2.27).

Substituting (2.29) into the last two equations of (2.28), it is easy to find relations between new and old averaged charge density and current:

$$\begin{cases} \langle \rho \rangle = \langle \rho \rangle' + \dfrac{i\omega}{4\pi c} \mathrm{div} \vec{T}_1 \\ \langle \vec{j} \rangle = \langle \vec{j} \rangle' + \dfrac{c}{4\pi} \mathrm{rot}\, \mathrm{rot}\, \vec{T}_1 - \dfrac{\omega^2}{4\pi c} \vec{T}_1 \end{cases} \tag{2.30}$$

Transformations (2.29) and (2.30) give relations between fields and charge and current densities so that the functional form of the MEs remains the same. It should be emphasized that at this stage absolutely no assumptions about possible distributions of the charge and current densities have been made.

Physical interpretation of the found transformations is pretty obvious: they just link two possible solutions of MEs. For example, from a known (non-primed) solution of MEs one can find new (primed) one if we know the connection between the non-primed and primed averaged charge and current densities (2.30). A more important consequence of (2.30) is that in case of the field transformation (2.29) it is necessary to transform charge and current densities (2.30), if we require that the new primed values satisfy the MEs.

Previously it has been found that the material equations (2.5) can be written through two new functions $\vec{P}$ and $\vec{M}$, which for the most general "C" representation takes the form [see (2.19)]:

$$\begin{cases} \langle \rho \rangle = -\mathrm{div} \vec{P} \\ \langle \vec{j} \rangle = -i\omega \vec{P} + c\, \mathrm{rot}\, \vec{M} \end{cases} \tag{2.31}$$

From (2.31) one can immediately obtain that in case of transformation (2.30) functions $\vec{P}$ and $\vec{M}$ have to be transformed as well:

$$\begin{cases} \vec{P} = \vec{P}' - \dfrac{i\omega}{4\pi c} \vec{T}_1 \\ \vec{M} = \vec{M}' + \dfrac{1}{4\pi} \mathrm{rot}\, \vec{T}_1 \end{cases} \tag{2.32}$$

Substituting (2.31) into MEs we obtain:

$$\begin{cases} \mathrm{rot}\, \vec{E} = \dfrac{i\omega}{c} \vec{B} \\ \mathrm{div} \vec{B} = 0 \\ \mathrm{div} \left(\vec{E} + 4\pi \vec{P} \right) = 0 \\ \mathrm{rot}\, \left(\vec{B} - 4\pi \vec{M} \right) = -\dfrac{i\omega}{c} \left(\vec{E} + 4\pi \vec{P} \right) \end{cases} \tag{2.33}$$

It is clear that the last two equations allow similar transformation with the use of one more arbitrary function $\vec{T}_2$, as in the second set of (2.27). Namely:

$$\begin{cases} \vec{P} = \vec{P}' - \dfrac{i\omega}{4\pi c}\vec{T}_1 + \operatorname{rot}\vec{T}_2 \\ \vec{M} = \vec{M}' + \dfrac{1}{4\pi}\operatorname{rot}\vec{T}_1 + \dfrac{i\omega}{c}\vec{T}_2 \end{cases} \tag{2.34}$$

which also does not change the form of the MEs.

It should be emphasized that function $\vec{T}_2$ does not appear in the expressions for transformation of averaged charge and current densities; in other words, measurable values are not changed under transformation (2.34).

As discussed above, the physical meaning of the transformations (2.29), (2.30) is in just an algebraic link between two possible solutions of MEs: if the averaged charge and current densities are transformed according to (2.30), then the new fields can be obtained without the necessity to solve MEs, but with the use of (2.29). The physical interpretation of (2.34) with $\vec{T}_1 = 0$ is different. This transformation changes neither fields nor averaged densities of charges and currents, but rather redistributes the representations of them between $\vec{P}$ and $\vec{M}$. In other words, this transformation describes the same physical situation by different representations, e.g. by different pairs of $\vec{P}$, $\vec{M}$. It is clear (and it will be used below) that this corresponds to transformations between different representations of MEs ("C", "L&L", and "T" forms).

Alternatively, in (2.32) one can introduce transformation not for $\vec{P}$, $\vec{M}$, but for combinations $\vec{E} + 4\pi\vec{P}$, $\vec{B} - 4\pi\vec{M}$, namely:

$$\begin{cases} \vec{E} + 4\pi\vec{P} = \vec{E}' + 4\pi\vec{P}' + \operatorname{rot}\vec{T}_2 \\ \vec{B} - 4\pi\vec{M} = \vec{B}' - 4\pi\vec{M}' - \dfrac{i\omega}{c}\vec{T}_2 \end{cases} \tag{2.35}$$

which is equivalent to the last two equations in (2.27):

$$\begin{cases} \vec{D} = \vec{D}' + \operatorname{rot}\vec{T}_2 \\ \vec{H} = \vec{H}' - \dfrac{i\omega}{c}\vec{T}_2 \end{cases} \tag{2.36}$$

Combining all elaborated expressions, one can finally obtain, that the transformations

$$\begin{cases} \operatorname{rot}\vec{E} = \dfrac{i\omega}{c}\vec{B} \\ \operatorname{div}\vec{B} = 0 \\ \operatorname{div}(\vec{E} + 4\pi\vec{P}) = 0 \\ \operatorname{rot}(\vec{B} - 4\pi\vec{M}) = -\dfrac{i\omega}{c}(\vec{E} + 4\pi\vec{P}) \\ \langle\rho\rangle = -\operatorname{div}\vec{P} \\ \langle\vec{j}\rangle = -i\omega\vec{P} + c\operatorname{rot}\vec{M} \end{cases} \qquad \begin{cases} \vec{B} = \vec{B}' + \operatorname{rot}\vec{T}_1 \\ \vec{E} = \vec{E}' + \dfrac{i\omega}{c}\vec{T}_1 \\ \vec{P} = \vec{P}' - \dfrac{i\omega}{4\pi c}\vec{T}_1 + \operatorname{rot}\vec{T}_2 \\ \vec{M} = \vec{M}' + \dfrac{1}{4\pi}\operatorname{rot}\vec{T}_1 + \dfrac{i\omega}{c}\vec{T}_2 \\ \langle\rho\rangle = \langle\rho\rangle' + \dfrac{i\omega}{4\pi c}\operatorname{div}\vec{T}_1 \\ \langle\vec{j}\rangle = \langle\vec{j}\rangle' + \dfrac{c}{4\pi}\operatorname{rot}\operatorname{rot}\vec{T}_1 - \dfrac{\omega^2}{4\pi c}\vec{T}_1 \end{cases} \tag{2.37}$$

are equivalent to the SFT (2.27). The difference between SFT in form (2.27) and (2.37) is in that in the developed here approach functions $\vec{P}$, $\vec{M}$ are used instead of $\vec{D}$, $\vec{H}$. The latter variant appears to be convenient due to the fact that in this case transformation for the fields $\vec{E}$, $\vec{B}$ is independent from the transformation for $\vec{D}$, $\vec{H}$. Nevertheless, it is important to remind that in order to come to the representation using vectors $\vec{D}$, $\vec{H}$, it is necessary first to introduce $\vec{P}$, $\vec{M}$, after that regroup the respective terms in MEs, and then introduce vectors $\vec{D}$, $\vec{H}$. It is also possible to postulate the macroscopic MEs directly in form (2.27), but in this case the logical transition between microscopic and macroscopic MEs is lost.

Concluding this part, we can state that the SFT provide transformations between two different realizable physical situations ($\vec{T}_1 \neq 0$, $\vec{T}_2 = 0$) or between two representations of the same physical situation ($\vec{T}_1 = 0$, $\vec{T}_2 \neq 0$).

2.6 Transformations Between Different Representations

Let us consider the relation between different representations of MEs, and start from the "C" form (2.19). It is known that the "C" form is invariant with respect to the SFT (2.27), (2.37). Here the SFT will be applied (following [16]) and possible conclusions which can be made based on the application of the SFT will be considered.

"C" to "L&L" Transformation

Let us start from "C" to "L&L" transformation. For the material equations in "C" (not primed) and "L&L" (primed) forms one can respectively write:

$$\begin{cases} \vec{B}_{\mathrm{C}} = \vec{B}'_{\mathrm{LL}} + \mathrm{rot}\,\vec{T}_1 \\ \vec{E}_{\mathrm{C}} = \vec{E}'_{\mathrm{LL}} + \dfrac{i\omega}{c}\vec{T}_1 \end{cases} \qquad \begin{cases} \vec{P}_{\mathrm{C}} = \vec{P}'_{\mathrm{LL}} - \dfrac{i\omega}{4\pi c}\vec{T}_1 + \mathrm{rot}\,\vec{T}_2 \\ \vec{M}_{\mathrm{C}} = \vec{M}'_{\mathrm{LL}} + \dfrac{i\omega}{c}\vec{T}_2 + \dfrac{\mathrm{rot}\,\vec{T}_1}{4\pi} \\ \langle\rho\rangle_{\mathrm{C}} = \langle\rho\rangle'_{\mathrm{LL}} + \dfrac{i\omega}{4\pi c}\mathrm{div}\vec{T}_1 \\ \langle\vec{j}\rangle_{\mathrm{C}} = \langle\vec{j}\rangle'_{\mathrm{LL}} + \dfrac{c}{4\pi}\,\mathrm{rot}\,\mathrm{rot}\,\vec{T}_1 - \dfrac{\omega^2}{4\pi c}\vec{T}_1 \end{cases} \tag{2.38}$$

In order to get the MEs for the new fields in "L&L" form in the primed system we have to require that:

$$\vec{M}'_{\mathrm{LL}} = 0 \tag{2.39}$$

which leads to:

$$\vec{T}_2 = \frac{ic}{\omega}\left(\mathrm{rot}\,\vec{T}_1 - \vec{M}_{\mathrm{C}}\right) \tag{2.40}$$

Substituting the last equation into (2.38), we finally have:

$$\begin{cases} \vec{B}_{\mathrm{C}} = \vec{B'}_{\mathrm{LL}} + \mathrm{rot}\,\vec{T}_1 \\ \vec{E}_{\mathrm{C}} = \vec{E'}_{\mathrm{LL}} + \dfrac{i\omega}{c}\vec{T}_1 \end{cases} \qquad \begin{cases} \vec{P'}_{\mathrm{LL}} = \vec{P}_{\mathrm{C}} + \dfrac{i\omega}{4\pi c}\vec{T}_1 + \dfrac{ic}{\omega}\,\mathrm{rot}\left(\mathrm{rot}\,\vec{T}_1 - \vec{M}_{\mathrm{C}}\right) \\ \vec{M'}_{\mathrm{LL}} = 0 \\ \langle\rho\rangle'_{\mathrm{LL}} = \langle\rho\rangle_{\mathrm{C}} - \dfrac{i\omega}{4\pi c}\mathrm{div}\vec{T}_1 \\ \langle\vec{j}\rangle'_{\mathrm{LL}} = \langle\vec{j}\rangle_{\mathrm{C}} - \dfrac{c}{4\pi}\,\mathrm{rot}\,\mathrm{rot}\,\vec{T}_1 + \dfrac{\omega^2}{4\pi c}\vec{T}_1 \end{cases} \tag{2.41}$$

which gives us the MEs in form of "L&L". If, in addition, we require that the electric and magnetic fields remain the same for both representations (which is reasonable, because both fields are assumed to be physically measurable), we obtain by setting $\vec{T}_1$ to zero:

$$\begin{cases} \vec{B}_{\mathrm{C}} = \vec{B'}_{\mathrm{LL}} \\ \vec{E}_{\mathrm{C}} = \vec{E'}_{\mathrm{LL}} \end{cases} \qquad \begin{cases} \vec{P'}_{\mathrm{LL}} = \vec{P}_{\mathrm{C}} - \dfrac{ic}{\omega}\,\mathrm{rot}\,\vec{M}_{\mathrm{C}} \\ \vec{M'}_{\mathrm{LL}} = 0 \\ \langle\rho\rangle_{\mathrm{C}} = \langle\rho\rangle'_{\mathrm{LL}} \\ \langle\vec{j}\rangle_{\mathrm{C}} = \langle\vec{j}\rangle'_{\mathrm{LL}} \end{cases} \tag{2.42}$$

We see that starting from "C" representation, we can unambiguously reduce the MEs to the "L&L" form. It is important to emphasize, that in general ($\vec{T}_1 \neq 0$) both electric and magnetic fields are transformed and lose their initial physical meanings. The requirement of keeping the electric and magnetic fields the same in both representations is an additional one with respect to the SFT.

"L&L" to "C" Transformation

Let us consider the inverse transformation ("L&L" to "C" representation), namely we start from system (2.22) and write the SFT in this case:

$$\begin{cases} \vec{B}_{\mathrm{LL}} = \vec{B'}_{\mathrm{C}} + \mathrm{rot}\,\vec{T}_1 \\ \vec{E}_{\mathrm{LL}} = \vec{E'}_{\mathrm{C}} + \dfrac{i\omega}{c}\vec{T}_1 \end{cases} \qquad \begin{cases} \vec{P}_{\mathrm{LL}} = \vec{P'}_{\mathrm{C}} - \dfrac{i\omega}{4\pi c}\vec{T}_1 + \mathrm{rot}\,\vec{T}_2 \\ 0 = \vec{M'}_{\mathrm{C}} + \dfrac{i\omega}{c}\vec{T}_2 + \dfrac{\mathrm{rot}\,\vec{T}_1}{4\pi} \\ \langle\rho\rangle_{\mathrm{LL}} = \langle\rho\rangle'_{\mathrm{C}} + \dfrac{i\omega}{4\pi c}\mathrm{div}\vec{T}_1 \\ \langle\vec{j}\rangle_{\mathrm{LL}} = \langle\vec{j}\rangle'_{\mathrm{C}} + \dfrac{c}{4\pi}\,\mathrm{rot}\,\mathrm{rot}\,\vec{T}_1 - \dfrac{\omega^2}{4\pi c}\vec{T}_1 \end{cases} \tag{2.43}$$

It is seen, that in general the functions $\vec{T}_1$, $\vec{T}_2$ cannot be unambiguously found.

As like as for direct ("C" to "L&L") transformation, we require that the field in both representations remain the same $\vec{T}_1 = 0$:

$$\begin{cases} \vec{B}_{\mathrm{LL}} = \vec{B}'_{\mathrm{C}} \\ \vec{E}_{\mathrm{LL}} = \vec{E}'_{\mathrm{C}} \end{cases} \qquad \begin{cases} \vec{P}'_{\mathrm{C}} = \vec{P}_{\mathrm{LL}} - \operatorname{rot} \vec{T}_2 \\ \vec{M}'_{\mathrm{C}} = -\dfrac{i\omega}{c} \vec{T}_2 \\ \langle \rho \rangle'_{\mathrm{C}} = \langle \rho \rangle_{\mathrm{LL}} \\ \langle \vec{j} \rangle'_{\mathrm{C}} = \langle \vec{j} \rangle_{\mathrm{LL}} \end{cases} \tag{2.44}$$

It is seen that the function $\vec{T}_2$ cannot be unambiguously determined, and the reverse "C" to "L&L" transformation is in general undetermined as well. This is because splitting of the total polarization current into two parts cannot be made uniquely and some additional physical requirements are needed to define, as is usually done, the parts corresponding to electric polarization density and magnetization current.

Let us emphasize again, that starting from the "C" form it is possible to arrive to the "L&L" form, but starting from the "L&L" form it is impossible to reduce MEs to the "C" form using the SFT unambiguously. There are unlimited number of "C" forms which correspond to the same "L&L" form.

"C" to "T" Transformation

Let us now consider the "C" to "T" transformation. In order to get the MEs for the new fields in "T" form (2.25) in the primed system we have to require that:

$$\langle \rho \rangle_{\mathrm{T}} = 0 \tag{2.45}$$

Writing the SFT for this case, we have:

$$\begin{cases} \vec{B}_{\mathrm{C}} = \vec{B}'_{\mathrm{T}} + \operatorname{rot} \vec{T}_1 \\ \vec{E}_{\mathrm{C}} = \vec{E}'_{\mathrm{T}} + \dfrac{i\omega}{c} \vec{T}_1 \end{cases} \qquad \begin{cases} \vec{P}_{\mathrm{C}} = \vec{P}'_{\mathrm{T}} - \dfrac{i\omega}{4\pi c} \vec{T}_1 + \operatorname{rot} \vec{T}_2 \\ \vec{M}_{\mathrm{C}} = \vec{M}'_{\mathrm{T}} + \dfrac{i\omega}{c} \vec{T}_2 + \dfrac{\operatorname{rot} \vec{T}_1}{4\pi} \\ \langle \rho \rangle_{\mathrm{C}} = \dfrac{i\omega}{4\pi c} \operatorname{div} \vec{T}_1 \\ \langle \vec{j} \rangle_{\mathrm{C}} = \langle \vec{j} \rangle'_{\mathrm{T}} + \dfrac{c}{4\pi} \operatorname{rot} \operatorname{rot} \vec{T}_1 - \dfrac{\omega^2}{4\pi c} \vec{T}_1 \end{cases} \tag{2.46}$$

In addition, if we require the same values for the electric and magnetic fields in both representations ($\vec{T}_1 = 0$), then:

$$\begin{cases} \vec{B}_{\mathrm{C}} = \vec{B}'_{\mathrm{T}} \\ \vec{E}_{\mathrm{C}} = \vec{E}'_{\mathrm{T}} \end{cases} \qquad \begin{cases} \vec{P}'_{\mathrm{T}} = \vec{P}_{\mathrm{C}} - \operatorname{rot} \vec{T}_2 \\ \vec{M}'_{\mathrm{T}} = \vec{M}_{\mathrm{C}} - \dfrac{i\omega}{c} \vec{T}_2 \\ \langle \rho \rangle'_{\mathrm{T}} = \langle \rho \rangle_{\mathrm{C}} = 0 \\ \langle \vec{j} \rangle'_{\mathrm{T}} = \langle \vec{j} \rangle_{\mathrm{C}} \end{cases} \tag{2.47}$$

One can conclude that the "C" to "T" transformation in general cannot be performed. Let us stress that this reflects the fact that the toroidal form can be used only if the average charge density is zero, which is of course generally not the case. If $\langle\rho\rangle_C = 0$ is initially satisfied, then (2.46) gives "redistribution" between dielectric and magnetic responses in full analogy with "L&L" to "C" transformation (2.44).

"T" to "C" Transformation
In this case the SFT has the following form:

$$\begin{cases} \vec{B}_T = \vec{B}'_C + \mathrm{rot}\,\vec{T}_1 \\ \vec{E}_T = \vec{E}'_C + \dfrac{i\omega}{c}\vec{T}_1 \end{cases} \qquad \begin{cases} \vec{P}'_C = \vec{P}_T + \dfrac{i\omega}{4\pi c}\vec{T}_1 - \mathrm{rot}\,\vec{T}_2 \\ \vec{M}'_C = \vec{M}_T - \dfrac{i\omega}{c}\vec{T}_2 - \dfrac{\mathrm{rot}\,\vec{T}_1}{4\pi} \\ \langle\rho\rangle'_C = -\dfrac{i\omega}{4\pi c}\mathrm{div}\vec{T}_1 \\ \langle\vec{j}\rangle_T = \langle\vec{j}\rangle'_C + \dfrac{c}{4\pi}\,\mathrm{rot}\,\mathrm{rot}\,\vec{T}_1 - \dfrac{\omega^2}{4\pi c}\vec{T}_1 \end{cases} \tag{2.48}$$

In case of the same fields $\vec{T}_1 = 0$ and consequently:

$$\begin{cases} \vec{B}_T = \vec{B}'_C \\ \vec{E}_T = \vec{E}'_C \end{cases} \qquad \begin{cases} \vec{P}'_C = \vec{P}_T - \mathrm{rot}\,\vec{T}_2 \\ \vec{M}'_C = \vec{M}_T - \dfrac{i\omega}{c}\vec{T}_2 \\ \langle\rho\rangle'_C = 0 \\ \langle\vec{j}\rangle'_C = \langle\vec{j}\rangle_T \end{cases} \tag{2.49}$$

which is again no more than the "redistribution" between dielectric and magnetic responses in the frame of "C" representation.

"L&L" to "T" Transformation
In this case the SFT reads:

$$\begin{cases} \vec{B}_{LL} = \vec{B}'_T + \mathrm{rot}\,\vec{T}_1 \\ \vec{E}_{LL} = \vec{E}'_T + \dfrac{i\omega}{c}\vec{T}_1 \end{cases} \qquad \begin{cases} \vec{P}'_T = \vec{P}_{LL} + \dfrac{i\omega}{4\pi c}\vec{T}_1 - \mathrm{rot}\,\vec{T}_2 \\ \vec{M}'_T = -\dfrac{i\omega}{c}\vec{T}_2 - \dfrac{\mathrm{rot}\,\vec{T}_1}{4\pi} \\ \langle\rho\rangle_{LL} = \dfrac{i\omega}{4\pi c}\mathrm{div}\vec{T}_1 \\ \langle\vec{j}\rangle'_T = \langle\vec{j}\rangle_{LL} - \dfrac{c}{4\pi}\,\mathrm{rot}\,\mathrm{rot}\,\vec{T}_1 + \dfrac{\omega^2}{4\pi c}\vec{T}_1 \end{cases} \tag{2.50}$$

and for the case of the same fields $\vec{T}_1 = 0$:

$$\begin{cases} \vec{B}_{LL} = \vec{B}'_T \\ \vec{E}_{LL} = \vec{E}'_T \end{cases} \qquad \begin{cases} \vec{P'}_T = \vec{P}_{LL} - \operatorname{rot}\vec{T}_2 \\ \vec{M'}_T = -\dfrac{i\omega}{c}\vec{T}_2 \\ \langle\rho\rangle_{LL} = 0 \\ \langle\vec{j}\rangle'_T = \langle\vec{j}\rangle_{LL} \end{cases} \tag{2.51}$$

The transformation, similarly to the case "C" to "T", in general is not determined due to the fact that in general $\langle\rho\rangle_{LL} \neq 0$.

"T" to "L&L" Transformation

The reverse transformation follows the equations:

$$\begin{cases} \vec{B}_T = \vec{B}'_{LL} + \operatorname{rot}\vec{T}_1 \\ \vec{E}_T = \vec{E}'_{LL} + \dfrac{i\omega}{c}\vec{T}_1 \end{cases} \qquad \begin{cases} \vec{P'}_{LL} = \vec{P}_T + \dfrac{i\omega}{4\pi c}\vec{T}_1 - \operatorname{rot}\vec{T}_2 \\ \vec{M}_T = \dfrac{i\omega}{c}\vec{T}_2 + \dfrac{1}{4\pi}\operatorname{rot}\vec{T}_1 \\ \langle\rho\rangle'_{LL} = -\dfrac{i\omega}{4\pi c}\operatorname{div}\vec{T}_1 \\ \langle\vec{j}\rangle'_{LL} = \langle\vec{j}\rangle_T - \dfrac{c}{4\pi}\operatorname{rot}\operatorname{rot}\vec{T}_1 + \dfrac{\omega^2}{4\pi c}\vec{T}_1 \end{cases} \tag{2.52}$$

and for the same fields $\vec{T}_1 = 0$:

$$\begin{cases} \vec{B}_T = \vec{B}'_{LL} \\ \vec{E}_T = \vec{E}'_{LL} \end{cases} \qquad \begin{cases} \vec{P'}_{LL} = \vec{P}_T - \dfrac{c}{i\omega}\operatorname{rot}\vec{M}_T \\ \dfrac{c}{i\omega}\vec{M}_T = \vec{T}_2 \\ \langle\rho\rangle'_{LL} = 0 \\ \langle\vec{j}\rangle'_{LL} = \langle\vec{j}\rangle_T \end{cases} \tag{2.53}$$

In this case the transformation is determined unambiguously.

The mutual transformations between different representations are presented in Fig. 2.4.

From Fig. 2.4 one can see that the "L&L" form occupies a special place in the elaborated hierarchy—the other two forms can be reduced to the "L&L" form, while the "L&L" form itself cannot be transformed to the other unambiguously. Another specific position is occupied by the "T" representation: this representation is valid only if the averaged charge density is identically zero and it cannot be achieved in general from two other ones.

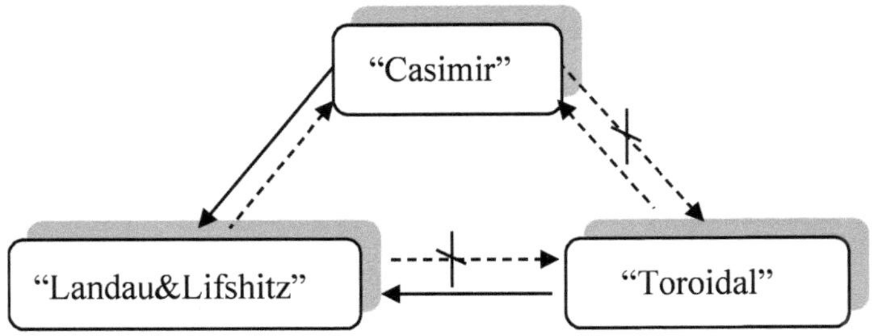

Fig. 2.4 Possibility of mutual transformations between different representations. Crossed dashed lines between "Casimir" and "Toroidal" and "Landau&Lifshitz" and "Toroidal" show impossible transformations, dashed lines between "L&L" and "C" and "Toroidal" and "Casimir" show not unique transformations

2.7 Conclusion

All the considerations above did not answer the question "How to get the unknown functions for the polarizability ("L&L" form) or polarization and magnetization ("C" form) starting from the microscopic picture?"

The main problem is to develop a model, which would give us a recipe for finding analytical expressions for $\vec{P}$ and $\vec{M}$ as functions of the averaged fields—it has to be also pointed out, that the expressions have to be presented as functions of the averaged (macroscopic), and not the microscopic fields; only in this case we can formulate the MEs as a self-consistent system. Nevertheless, it is important to realize, that whatever model is developed, it can be presented only in "C", "L&L", or "T" form with the respective consequences, described above.

The conclusions for the presented above chapter are:

1. In this chapter the main goal of this work has been formulated in form of defining the functional dependence (2.5) or, equivalently, (2.19), (2.22), (2.25). The goal has been formulated based on the microscopic Maxwell equations in form of fields (2.4) as a starting point for all considerations.
2. The frequency range where the homogenization procedure can be performed has been determined.
3. The macroscopic approach to the homogenization has been outlined resulting in three possible representations of ME—"C", "L&L", and "T" forms.
4. The Serdyukov-Fedorov transformations (SFT) have been reformulated and used to establish relationships between the three ME representations.

References

1. L.D. Landau, E.L. Lifshitz, *Electrodynamics of Continuous Media*, 2nd edn. (Pergamon Press, New York, 1960) (Chapter IX)
2. V. Dubovik, S. Shabanov, in *Essays on the Formal Aspects of Electromagnetic Theory*, ed. by A. Lakhakia. The Gauge Invariance, Toroid Order Parameters and Radiation in Electromagnetic Theory, vol. 399 (World Scientific, Singapore, New Jersey, London, Hong-Kong, 1993)
3. J. Schwinger, L.L. De-Raad, K. Milton, W. Tsai, *Classical Electrodynamics* (Perseus Books, Reading, MA, 1998)
4. P. Mazur, B. Nijboer, On the statistical mechanics of matter in an electromagnetic field. I. Physica **XIX**, 971 (1953)
5. G. Rusakoff, A derivation of the macroscopic Maxwell equations. Am. J. Phys. **38**(10), 1188 (1970)
6. A. Maradudin, D.L. Mills, Phys. Rev. B **7**, 2787 (1973)
7. J.D. Jackson, *Classical Electrodynamics*, 3rd edn. (Wiley, New York, 1999)
8. S. Maslovsky, Electrodynamics of composite materials with pronounced spatial dispersion. Ph.D. thesis (St. Petersburg Polytechnic University, Russia, 2004)
9. A. Vinogradov, A. Aivazyan, Scaling theory of homogenization of the Maxwell equations. Phys. Rev. E **60**, 987 (1999)
10. C. Simovski, Weak spatial dispersion in composite media. Polytechnika (St. Petersburg, 2003) (in Russian)
11. Y. Svirko, N. Zheludev, M. Osipov, Layered chiral metallic microstructures with inductive coupling. APL **78**, 498 (2001)
12. N. Zheludev, The road ahead for metamaterials. Science **328**, 582 (2010)
13. N. Papasimakis, V. Fedotov, K. Marinov, N. Zheludev, Gyrotropy of a metamolecule: wire on a torus. PRL **103**, 093901 (2009)
14. A.N. Grigorenko, A.K. Geim, H.F. Gleeson, Y. Zhang, A.A. Firsov, I.Y. Khrushchev, J. Petrovic, Nanofabricated media with negative permeability at visible frequencies. Nature **438**, 335–338 (2005)
15. V. Shalaev, W. Cai, U. Chettiar, H.-K. Yuan, A. Sarychev, V. Drachev, A. Kildishev, Negative index of refraction in optical metamaterials. Opt. Lett. **30**, 3356 (2005)
16. A. Vinogradov, *Electrodynamics of Compound Media* (Scientific and Educational Literature Publisher, Russian Federation, 2001). ISBN 5-8360-0283-5 (in Russian)
17. C. Simovski, Material parameters of metamaterials (a review). Opt. Spectrosc. **107**, 726 (2009)
18. R. Clausius, *Mechanische Warmetheorie*, vol. 2, 2nd edn. (Braunschweig, 1878), p. 62
19. O. Mossotti, Mem. Soc. Sci. Modena **14**, 49 (1850)
20. H. Lorentz, Proc. Acad. Sci. Amsterdam **13**, 92 (1910)
21. P. Ewald, Ann. Phys. **64**, 2943 (1921)
22. D. Bruggeman, Ann. Phys. Lpz. **24**, 636 (1935)
23. C. Simovski, Radiotekh. Elecktron. **52**, 1031 (2007)
24. G. Shvets, Photonic approach to make a surface wave accelerator. AIP Conf. Proc. **647**, 371 (2002)
25. P. Belov, R. Marques, M. Silveirinha, I. Nefedov, C. Simovski, S. Trtyakov, Strong spatial dispersion in wire media in the very long wavelength limit. Phys. Rev. B **70**, 113103 (2003)
26. M. Silveirinha, Nonlocal homogenization model for a periodic array of epsilon-negative rods. Phys. Rev. E **73**, 046612 (2006)
27. E. Pshenay-Severin, A. Chipouline, J. Petschulat, U. Huebner, A. Tuennermann, T. Pertsch, Optical properties of metamaterials based on asymmetric double-wire structures. Opt. Express **19**, 6269 (2011)

28. C. Simovski, On electromagnetic characterization and homogenization of nanostructured metamaterials. J. Opt. **13**, 013001 (2011)
29. A. Sarychev, V. Shalaev, *Electrodynamics of Metamaterials* (World Scientific, Singapore, 2007)
30. D. Smith, D. Shurig, PRL **90**, 077405 (2003)
31. P. Belov, C. Simovski, Phys. Rev. E **72**, 026615 (2005)
32. J.D. Joannopoulos, R.D. Meade, J.N. Winn, *Photonic Crystals: Molding the Flow of Light* (1995). ISBN-13: 978-0-691-03744-8
33. C. Rockstuhl, T. Paul, F. Lederer, T. Pertsch, T. Zentgraf, T. Meyrath, H. Giessen, Phys. Rev. B **77**, 035126 (2008)
34. M. Born, K. Huang, *Dynamic Theory of Crystal Lattices* (Oxford University Press, Oxford, 1954)
35. M. Silveirinha, Time domain homogenization of metamaterials. Phys. Rev. B **77**, 035126 (2011)
36. C. Menzel, T. Paul, C. Rockstuhl, T. Pertsch, S. Tretyakov, F. Lederer, Validity of effective material parameters for optical fishnet metamaterials. Phys. Rev. B **81**, 035320 (2010)
37. M. Bredov, V. Rumyantcev, I. Toptygin, *Classical Electrodynamics* (Nauka, 1985) (in Russian)
38. I.B. Zeldovich, Electromagnetic interaction with parity violation. JETP **33**, 1531 (1957)
39. L. Mandelshtam, Full collection of publications. Publ. Acad. Sci. USSR **1**, 162 (1957) (in Russian)
40. T. Kaelberer, V.A. Fedotov, N. Papasimakis, D.P. Tsai, N.I. Zheludev, Science **330**, 1510 (2010)

Chapter 3
Phenomenological Versus Multipole Models

In order to further develop the homogenization procedure, we should find an analytical form for the functions $\vec{P}$ and $\vec{M}$ in case of “C” representation, $\vec{P}$ in case of “L&L” representation, or $\vec{M}$ in case of “T” representation. There are two principal ways to do it: introduce a functional form following the phenomenological approach, or try to create a microscopic model which would result finally in the required forms for $\vec{P}$ and $\vec{M}$. Below, both ways will be considered.

The most results of this chapter have been originally obtained in [1, 2] with new, partially previously unpublished interpretations.

3.1 Phenomenological Model (“L&L” and “C” Representations)

3.1.1 “L&L” Representation

Case of Strong Spatial Dispersion

There is a commonly accepted integral form of $\vec{P}_{\mathrm{LL}}$, which can be written according to the causality principle (which imposes limitations on the frequency dispersion form) and assuming that the physical processes at some point depend on the fields at other points (which gives rise to spatial dispersion) (e.g., [3]):

$$\langle\vec{j}\rangle(\vec{r},\omega) = -i\omega\int\limits_V R_{\mathrm{LL}}(\vec{r},\vec{r}',\omega)\vec{E}(\vec{r}',\omega)\mathrm{d}r' \qquad (3.1)$$

This equation in case of translational invariance can be written as:

$$\langle\vec{j}\rangle(\vec{r},\omega) = -i\omega\int\limits_V R_{\mathrm{LL}}(\vec{r}-\vec{r}',\omega)\vec{E}(\vec{r}',\omega)\mathrm{d}r' \qquad (3.2)$$

A. Chipouline and F. Küppers, *Optical Metamaterials: Qualitative Models*, Springer Series in Optical Sciences 211, https://doi.org/10.1007/978-3-319-77520-3_3

It has to be realized that (3.2) assumes translational symmetry of the considered object, which in turn means that (3.2) and any consequences cannot be directly applied to the consideration of the boundary condition problems. Transforming (3.2) to the spatial Fourier domain, we get (summation over repeated indexes is assumed as usual):

$$\langle j\rangle_\alpha\left(\vec{k},\,\omega\right) = -i\omega R_{\mathrm{LL},\,\alpha\beta}\left(\vec{k},\,\omega\right)E_\beta\left(\vec{k},\,\omega\right) \tag{3.3}$$

Form (3.3) is an expansion of the averaged current and fields over plane waves. In principle, the dependence of the averaged current on the magnetic field can be explicitly included in (3.3) as well. Nevertheless taking into account that in a plane wave the magnetic field can always be expressed through the electric field $B_\alpha = \frac{c}{\omega}e_{\alpha\beta l}k_\beta E_l$ ($e_{\alpha\beta l}$ is the Levi-Chivita tensor), one can leave out the dependence on the magnetic field without loss of generality.

From (2.22) we obtain:

$$\begin{cases} P_{\mathrm{LL},\,\alpha}\left(\vec{k},\,\omega\right) = R_{\mathrm{LL},\,\alpha\beta}\left(\vec{k},\,\omega\right)E_\beta\left(\vec{k},\,\omega\right) \\ D_{\mathrm{LL},\,\alpha}\left(\vec{k},\,\omega\right) = E_\alpha\left(\vec{k},\,\omega\right) + 4\pi P_{\mathrm{LL},\,\alpha}\left(\vec{k},\,\omega\right) = \left(\delta_{\alpha\beta} + 4\pi R_{\mathrm{LL},\,\alpha\beta}\left(\vec{k},\,\omega\right)\right)E_\beta\left(\vec{k},\,\omega\right) \\ \qquad = \varepsilon_{\mathrm{LL},\,\alpha\beta}\left(\vec{k},\,\omega\right)E_\beta\left(\vec{k},\,\omega\right) \end{cases} \tag{3.4}$$

Here the effective permittivity has been introduced:

$$\varepsilon_{\mathrm{LL},\,\alpha\beta}\left(\vec{k},\,\omega\right) = \delta_{\alpha\beta} + 4\pi R_{\mathrm{LL},\,\alpha\beta}\left(\vec{k},\omega\right) \tag{3.5}$$

The respective dispersion relation for the plane waves is:

$$\det\left[e_{\alpha\beta l}k_\beta e_{l\beta\gamma}k_\beta + \frac{\omega^2}{c^2}\varepsilon_{\mathrm{LL},\,\alpha\beta}(\vec{k},\,\omega)\right] = 0 \tag{3.6}$$

The introduced above function $\varepsilon_{\mathrm{LL},\,\alpha\beta}(\vec{k},\omega)$—effective permittivity—is a tensor with components depending on both the frequency and the wave vector. It is worth noticing that the effective permittivity introduced this way depends not only on the MA properties but on the excitation conditions as well, und thus far it cannot be called "material parameter"—i.e., a parameter which depends on the material properties only (this issue will be discussed in details later).

Nevertheless, it should be stated that the knowledge of the effective permittivity fully solves the problem of propagation of plane waves in bulk media. Equation (3.6) can have several solutions for the same propagation direction and the same polarization state [4].

When we use this representation, it has to be clearly realized that:

1. We are working with the “L&L” representation where there is no magnetization [the magnetic response is included through spatial dispersion of the electric polarization (3.4)].
2. Form (3.2) assumes translational invariance of the media, which means that the form is acceptable for homogeneous material far from the boundaries (bulk materials).
3. In the framework of (3.1) it is impossible to introduce any permeability, because of in “L&L” representation there is no magnetization. All magnetic effects are included in the effective permittivity. In order to introduce permeability it is necessary to transit the “L&L” to the “C” form, which cannot be done unambiguously (see Sect. 2.6).

Case of Weak Spatial Dispersion

The phenomenological model (3.1) is widely used in different branches of physics like plasma physics or physics of crystals. In the vast majority of the considered problems, the function $R(\vec{k},\,\omega)$ (or equivalently $\varepsilon_{\mathrm{LL}}(\vec{k},\omega)$) is expanded into the Taylor series up to the second order, namely:

$$\begin{aligned} R_{\mathrm{LL},\alpha\beta}\left(\vec{k},\omega\right) \approx R_{\mathrm{LL},\alpha\beta}\left(\vec{k}_0,\omega\right) + \left.\frac{\partial R_{\mathrm{LL},\alpha\beta}\left(\vec{k},\omega\right)}{\partial k_\gamma}\right|_{\vec{k}=\vec{k}_0} \left(k_\gamma - k_{0,\gamma}\right) \\ + \frac{1}{2}\left.\frac{\partial^2 R_{\mathrm{LL},\alpha\beta}\left(\vec{k},\,\omega\right)}{\partial k_\gamma \partial k_\delta}\right|_{\vec{k}=\vec{k}_0} \left(k_\gamma - k_{0,\gamma}\right)\left(k_\delta - k_{0,\delta}\right) \end{aligned} \tag{3.7}$$

The functions $R_{\mathrm{LL},\alpha\beta}(\vec{k}_0,\omega)$, $\left.\frac{\partial R_{\mathrm{LL},\alpha\beta}(\vec{k},\omega)}{\partial k_\gamma}\right|_{\vec{k}=\vec{k}_0}$, $\frac{1}{2}\left.\frac{\partial^2 R_{\mathrm{LL},\alpha\beta}(\vec{k},\omega)}{\partial k_\gamma \partial k_\delta}\right|_{\vec{k}=\vec{k}_0}$ are supposed to be found from experiments or rigorous microscopic calculations.

It should be pointed out that the Taylor expansion itself can be performed around any $\vec{k}_0$, not necessarily $\vec{k}_0 = 0$, provided that expansion over angles and wavelengths is properly done; in other words, the expansion formally can be written for small spatial dispersion and for strong spatial dispersion, but in the last case only for waves with the wave numbers close to $\vec{k}_0$. The math in this case does not impose any limitations.

After having all this said, the final form of the $R_{\mathrm{LL},\alpha\beta}$ in the “L&L” representation in case of weak spatial dispersion can be written as:

$$R_{\mathrm{LL},\alpha\beta}\left(\vec{k},\,\omega\right) \approx R_{\mathrm{LL},\alpha\beta}(0,\omega) + \left.\frac{\partial R_{\mathrm{LL},\alpha\beta}\left(\vec{k},\omega\right)}{\partial k_\gamma}\right|_{\vec{k}=0} k_\gamma + \frac{1}{2}\left.\frac{\partial^2 R_{\mathrm{LL},\alpha\beta}\left(\vec{k},\,\omega\right)}{\partial k_\gamma \partial k_\delta}\right|_{\vec{k}=0} k_\gamma k_\delta \tag{3.8}$$

and the respective relations are:

$$\begin{cases} P_{\mathrm{LL},\alpha}\left(\vec{k},\,\omega\right) = \left(R_{\mathrm{LL},\alpha\beta}(0,\omega) + \left.\dfrac{\partial R_{\mathrm{LL},\alpha\beta}\left(\vec{k},\omega\right)}{\partial k_\gamma}\right|_{\vec{k}=0} k_\gamma \right. \\ \qquad \left. + \dfrac{1}{2}\left.\dfrac{\partial^2 R_{\mathrm{LL},\alpha\beta}\left(\vec{k},\omega\right)}{\partial k_\gamma \partial k_\delta}\right|_{\vec{k}=0} k_\gamma k_\delta \right) E_\beta\left(\vec{k},\omega\right) \\ D_{\mathrm{LL},\alpha}\left(\vec{k},\omega\right) = E_\alpha\left(\vec{k},\omega\right) + 4\pi P_{\mathrm{LL},\alpha}\left(\vec{k},\omega\right) \\ \qquad = \left(\varepsilon^{(0)}_{\mathrm{LL},\alpha\beta}(\omega) + \varepsilon^{(1)}_{\mathrm{LL},\alpha\beta\gamma}(\omega)k_\gamma + \varepsilon^{(2)}_{\mathrm{LL},\alpha\beta\gamma\delta}(\omega)k_\gamma k_\delta\right) E_\beta\left(\vec{k},\omega\right) \\ \varepsilon^{(0)}_{\mathrm{LL},\alpha\beta}(\omega) = \left(\delta_{\alpha\beta} + 4\pi R_{\mathrm{LL},\alpha\beta}(0,\omega)\right) \\ \varepsilon^{(1)}_{\mathrm{LL},\alpha\beta\gamma}(\omega) = 4\pi \left.\dfrac{\partial R_{\mathrm{LL},\alpha\beta}\left(\vec{k},\omega\right)}{\partial k_\gamma}\right|_{\vec{k}=0} \\ \varepsilon^{(2)}_{\mathrm{LL},\alpha\beta\gamma\delta}(\omega) = 2\pi \left.\dfrac{\partial^2 R_{\mathrm{LL},\alpha\beta}\left(\vec{k},\omega\right)}{\partial k_\gamma \partial k_\delta}\right|_{\vec{k}=0} \\ \langle j\rangle_\alpha\left(\vec{k},\omega\right) = -\dfrac{i\omega}{4\pi}\left(\varepsilon^{(0)}_{\mathrm{LL},\alpha\beta}(\omega) - \delta_{\alpha\beta} + \varepsilon^{(1)}_{\mathrm{LL},\alpha\beta\gamma}(\omega)k_\gamma \right. \\ \qquad \left. + \varepsilon^{(2)}_{\mathrm{LL},\alpha\beta\gamma\delta}(\omega)k_\gamma k_\delta\right) E_\beta\left(\vec{k},\,\omega\right). \end{cases} \tag{3.9}$$

We see that the expansion of function $R_{\mathrm{LL},\alpha\beta}(\vec{k},\omega)$ up to the second order corresponds to the expansion of the averaged current $\langle j\rangle_\alpha(\vec{k},\omega)$ up to the second order as well; the expression for the averaged current $\langle j\rangle_\alpha(\vec{k},\,\omega)$ is useful, because the averaged current is invariant in all representations (see (2.37) in case of untransformed fields, e.g. $\vec{T}_1 = 0$) and can be used to compare the weak dispersion expansions in all representations.

Now it is worth considering relation of expansion (3.6) to the other representations for the spatial dispersion, used in various publications.

Expansion (3.9) is rather well known [5], tensors $\varepsilon^{(0)}_{\mathrm{LL},\alpha\beta}(\omega)$, $\varepsilon^{(1)}_{\mathrm{LL},\alpha\beta\gamma}(\omega)$, and $\varepsilon^{(2)}_{\mathrm{LL},\alpha\beta\gamma\delta}(\omega)$ depend on the symmetry properties of the considered system and satisfy symmetry principles of Onsager coefficients.

It should be noted that in the framework of the "L&L" representation for the case of weak dispersion (up to the second order) three tensor functions—$\varepsilon^{(0)}_{\mathrm{LL},\alpha\beta}(\omega)$, $\varepsilon^{(1)}_{\mathrm{LL},\alpha\beta\gamma}(\omega)$, and $\varepsilon^{(2)}_{\mathrm{LL},\alpha\beta\gamma\delta}(\omega)$—have been obtained, but the functions are not independent because they are originated from the same function $R_{\mathrm{LL},\alpha\beta}(\vec{k},\omega)$.

3.1.2 "C" Representation

Case of Strong Spatial Dispersion

Expressions (3.1), (3.4) and (3.5) in the L&L formalism are valid for the case of arbitrary strong spatial dispersion, as the only assumption is the linearity and causality of the medium. In the "C" representation, we introduce two functions for polarizability $P_{C,\alpha}$ and magnetization $M_{C,\alpha}$:

$$\begin{cases} P_{C,\alpha}\left(\vec{k},\omega\right) = R_{C,\alpha\beta}\left(\vec{k},\omega\right)E_\beta\left(\vec{k},\omega\right) + R'_{C,\alpha\beta}\left(\vec{k},\omega\right)B_\beta\left(\vec{k},\omega\right) \\ M_{C,\alpha}\left(\vec{k},\omega\right) = F'_{C,\alpha\beta}\left(\vec{k},\omega\right)E_\beta\left(\vec{k},\omega\right) + F_{C,\alpha\beta}\left(\vec{k},\omega\right)B_\beta\left(\vec{k},\omega\right) \end{cases} \tag{3.10}$$

which gives the most general form of expressions for $D_{C,\alpha}$:

$$\begin{aligned} D_{C,\alpha}\left(\vec{k},\omega\right) = {} & \left(\delta_{\alpha\beta} + 4\pi R_{C,\alpha\beta}\left(\vec{k},\omega\right)\right)E_\beta\left(\vec{k},\omega\right) \\ & + 4\pi R'_{C,\alpha\beta}\left(\vec{k},\omega\right)B_\beta\left(\vec{k},\omega\right) \end{aligned} \tag{3.11}$$

The fact that in this representation there are more "material parameters" than in the L&L formalism does not mean that this form is somehow more general and is able to capture more physical effects. As it was shown above, the forms are equivalent. Basically, in this formalism different parts of the induced polarization are modeled by separate polarizability coefficients. Care should be taken in using this form, so that the same physical effect is not included in the model twice. For example, chirality of the medium microstructure can be modeled either by a second-order anti-symmetric part of the permittivity in the L&L formalism or by including a contribution to electric polarization induced by magnetic induction in the "C" representation. Including in (3.10) both second-order curl terms into $R_{C,\alpha\beta}$ and a k-independent term in $R'_{C,\alpha\beta}$ appears to be redundant. As for the magnetic material relations, the commonly assumed step would be an introduction of $H_\alpha = B_\alpha - 4\pi M_{C,\alpha}$:

$$\begin{aligned} H_{C,\alpha}\left(\vec{k},\omega\right) = {} & -4\pi F'_{C,\alpha\beta}\left(\vec{k},\omega\right)E_\beta\left(\vec{k},\omega\right) \\ & + \left(\delta_{\alpha\beta} + 4\pi F_{C,\alpha\beta}\left(\vec{k},\omega\right)\right)B_\beta\left(\vec{k},\omega\right) \end{aligned} \tag{3.12}$$

At this point it is worth noting that formally, using the respective transformations between the electric and magnetic fields $B_\alpha = \frac{c}{\omega}e_{\alpha\beta l}k_\beta E_l$, the expressions for $D_{C,\alpha}$ and $H_{C,\alpha}$ can be written as:

$$\begin{cases} D_{C,\alpha}\left(\vec{k},\,\omega\right) = \varepsilon_{C,\alpha\beta}\left(\vec{k},\,\omega\right)E_\beta\left(\vec{k},\,\omega\right) \\ P_{C,\alpha}\left(\vec{k},\,\omega\right) = \left(R_{C,\alpha\beta}\left(\vec{k},\,\omega\right) + \frac{c}{\omega}R'_{C,\alpha l}\left(\vec{k},\,\omega\right)e_{l\gamma\beta}k_\gamma\right)E_\beta\left(\vec{k},\,\omega\right) \\ \varepsilon_{C,\alpha\beta}\left(\vec{k},\,\omega\right) = \delta_{\alpha\beta} + 4\pi R_{C,\alpha\beta}\left(\vec{k},\,\omega\right) + \frac{4\pi c}{\omega}R'_{C,\alpha l}\left(\vec{k},\,\omega\right)e_{l\gamma\beta}k_\gamma \\ M_{C,\alpha}\left(\vec{k},\,\omega\right) = \left(F'_{C,\alpha\beta}\left(\vec{k},\,\omega\right) + \frac{c}{\omega}F_{C,\alpha l}\left(\vec{k},\,\omega\right)e_{\alpha l\beta}k_l\right)E_\beta \\ H_\alpha\left(\vec{k},\omega\right) = \xi_{\alpha\beta}\left(\vec{k},\omega\right)E_\beta\left(\vec{k},\omega\right) \\ \xi_{\alpha\beta}\left(\vec{k},\omega\right) = -\left(4\pi F'_{C,\alpha\beta}\left(\vec{k},\omega\right) + \frac{c}{\omega}\left(\delta_{\alpha l} + 4\pi F_{C,\alpha l}\left(\vec{k},\omega\right)\right)\left(e_{l\gamma\beta}k_\gamma\right)\right) \end{cases} \tag{3.13a}$$

Using the representation (3.13a) it is rather straightforward to write down a "propagation equation"—analogy of the well-established Helmholz equation for the plane electromagnetic wave propagation [6]. The last equation in (2.23) after substitution $D_{C,\alpha}(\vec{k},\omega)$ and $H_{C,\alpha}(\vec{k},\omega)$ from (3.13a) becomes a "propagation equation":

$$e_{\alpha\beta\gamma}k_\beta H_\gamma = D_\alpha \Rightarrow e_{\alpha\beta\gamma}k_\beta\xi_{\gamma p}E_p = \varepsilon_{\alpha\beta}E_\beta \Rightarrow \det\left|e_{\alpha p\gamma}k_p\xi_{\gamma\beta} - \varepsilon_{\alpha\beta}\right| = 0 \tag{3.13b}$$

Propagation equation and respective dispersion relation (3.13b) appear to be much more simple and natural in compare with the usually used Helmholz equation; moreover form (3.13a) does not allow us to introduce a kind of permeability μ straightforwardly. In order to introduce μ, instead of substitution the magnetic field $\vec{B}$ in terms of the electric field $\vec{E}$, we should perform the opposite operation and express $\vec{E}$ through $\vec{B}$. In order to do it, we have to solve the equation:

$$\left[\vec{k}\times\vec{E}\right] = -\frac{\omega}{c}\vec{B} \tag{3.14}$$

assuming that the wave vector $\vec{k}$ and the magnetic field $\vec{B}$ are known and considering the electric field $\vec{E}$ as a variable. From the vector analysis solution of this problem is known, namely if there are three vectors $\vec{x}$, $\vec{a}$, and $\vec{b}$ so that $[\vec{x}\times\vec{a}\times\vec{b}] \neq 0$ and $[\vec{x}\times\vec{a}] = \vec{b}$ then the solution for $\vec{x}$ is:

$$\begin{cases} \vec{x} = \vec{a}\frac{\gamma}{|a|^2} + \left[\vec{a}\times\vec{b}\right]\frac{1}{|a|^2} \\ \gamma = (\vec{x}\times\vec{a}) \end{cases} \tag{3.15}$$

or, in other words, this requires the knowledge of one more constant γ.

In our case the last requirement is given by the Maxwell equation $(\vec{k}\times\vec{E}) = -4\pi(\vec{k}\times\vec{P})$ and the final solution of (3.14) is:

$$\vec{E} = -\vec{k}\frac{4\pi\left(\vec{k}\times\vec{P}\right)}{|k|^2} + \frac{\omega}{c}\left[\vec{k}\times\vec{B}\right]\frac{1}{|k|^2} \tag{3.16}$$

or, in terms of vector components:

$$E_\alpha = -4\pi\, k_\alpha\left(\frac{k_i P_{\mathrm{C},i}}{k_j k_j^*}\right) + \frac{\omega}{c} e_{\alpha\beta\gamma} k_\beta B_\gamma\left(\frac{1}{k_m k_m^*}\right) \tag{3.17}$$

Substituting (3.17) into (3.12) we have:

$$\begin{aligned} H_{\mathrm{C},\alpha}\left(\vec{k},\omega\right) &= -(4\pi)^2 F'_{\mathrm{C},\alpha\beta}\left(\vec{k},\omega\right)k_\beta\left(\frac{k_i P_{\mathrm{C},i}}{k_j k_j}\right) \\ &+ \left[\left(\delta_{\alpha\beta}+4\pi\, F_{\mathrm{C},\alpha\beta}\left(\vec{k},\omega\right)\right) + \frac{4\pi\omega}{c} F'_{\mathrm{C},\alpha\varsigma}\left(\vec{k},\omega\right) e_{\varsigma\gamma\beta} k_\gamma\left(\frac{1}{k_m k_m^*}\right)\right] B_\beta\left(\vec{k},\omega\right) \end{aligned} \tag{3.18}$$

It is clear that in this case any attempt to introduce proportionality between $H_{\mathrm{C},\alpha}$ and B_α in form $B_\alpha = \mu_{\alpha\beta}(\vec{k},\,\omega)H_{\mathrm{C},\alpha}$ fails if polarization $P_{\mathrm{C},\alpha}$ and the wave vector k_α are not perpendicular to each other $k_\alpha P_{\mathrm{C},\alpha} \neq 0$. In general, the polarizability $P_{\mathrm{C},\alpha}$ is not perpendicular to k_α and permeability cannot be introduced at all. It is worth noting that the problem arises from the fact that the magnetic response is stipulated by an interaction with the electric field, not with the magnetic one. Obviously, introduction of magnetic constant in form of a proportionality coefficient between $H_{\mathrm{C},\alpha}$ and B_α is neither logical nor necessary—the form (3.13a) $H_\alpha(\vec{k},\omega) = \xi_{\alpha\beta}(\vec{k},\omega)E_\beta(\vec{k},\omega)$ is much more physically justified than the form $B_\alpha = \mu_{\alpha\beta}(\vec{k},\,\omega)H_{\mathrm{C},\alpha}$, which is unconditionally suitable only for the case of interaction of a system with the magnetic field. Nevertheless, in case $k_\alpha P_{\mathrm{C},\alpha} = 0$ the electric field can be unambiguously and straightforwardly presented as a function of the magnetic field $E_\alpha = \frac{\omega}{c} e_{\alpha\beta\gamma} k_\beta B_\gamma (\frac{1}{k_m k_m^*})$ and (3.18) can be rewritten as:

$$\begin{cases} H_{\mathrm{C},\alpha}\left(\vec{k},\omega\right) = \left[\mu_{\alpha\beta}\left(\vec{k},\omega\right)\right]^{-1} B_\beta\left(\vec{k},\omega\right) \\ \left[\mu_{\alpha\beta}\left(\vec{k},\omega\right)\right]^{-1} = \left(\delta_{\alpha\beta}+4\pi\, F_{\mathrm{C},\alpha\beta}\left(\vec{k},\omega\right)\right) + \dfrac{4\pi\omega}{c} F'_{\mathrm{C},\alpha\varsigma}\left(\vec{k},\omega\right) e_{\varsigma\gamma\beta} k_\gamma\left(\dfrac{1}{k_m k_m^*}\right) \end{cases} \tag{3.19}$$

It has to be noted, that even in this case (when the proportionality between $H_{\mathrm{C},\alpha}$ and B_α can be established), introduction of $[\mu_{\alpha\beta}(\vec{k},\omega)]|^{-1}$ (not $\mu_{\alpha\beta}(\vec{k},\omega)$!) appears to be the logical step in the elaboration of the homogenization model. Here it is seen also, that if magnetization is caused by the electric field ($F'_{\mathrm{C},\alpha\varsigma}(\vec{k},\omega) \neq 0$), then the introduced this way $\mu_{\alpha\beta}(\vec{k},\omega)$ is spatially dispersive even in case of non-spatially dispersive response $F'_{\mathrm{C},\alpha\beta}(\vec{k},\omega) = F'_{\mathrm{C},\alpha\beta}(\omega)$. In fact, as it will be clear later from the consideration of the multipole model $F'_{\mathrm{C}}(\vec{k},\omega) \sim k$, which results in absence of spatial dispersion for $\mu_{\alpha\beta}(\vec{k},\omega) = \mu_{\alpha\beta}(\omega)$ under weak (up to the second order) spatial dispersion approximation.

Considering a plane wave propagating into z direction in a media with tensor character of polarizability, one can easily see that in order to satisfy the condition $k_\alpha P_{\mathrm{C},\alpha} = 0$ (which guarantees a possibility to introduce the permeability) we have to have tensor $\varepsilon_{\mathrm{C},\alpha\beta}(\vec{k},\omega)$ in the following form:

$$\begin{pmatrix} \varepsilon_{\mathrm{C},xx}\left(\vec{k},\omega\right) & \varepsilon_{\mathrm{C},xy}\left(\vec{k},\omega\right) & 0 \\ \varepsilon_{\mathrm{C},yx}\left(\vec{k},\omega\right) & \varepsilon_{\mathrm{C},yy}\left(\vec{k},\omega\right) & 0 \\ 0 & 0 & 0 \end{pmatrix} \tag{3.20}$$

It is interesting to demonstrate a design which, according to (3.20) does not allow introducing of the permeability due to the appearance of the polarization $P_{\mathrm{C},\alpha}$, parallel to the wave vector k_α, so that $k_\alpha P_{\mathrm{C},\alpha} \neq 0$. This could be a SRR structure placed with its top part parallel to the wave vector k_α, as it is shown in Fig. 3.1.

Now it is methodologically interesting to find relations between "C" and "L&L" representation in the frame of the strong spatial dispersion. It can be easily performed by equating the relations for the averaged current in both representations (let us remind, that according to SFT the averaged current is not changed provided fields remain unchanged as well), which gives after some algebra:

$$\begin{aligned} R_{\mathrm{LL},\alpha\beta}\left(\vec{k},\omega\right) = R_{\mathrm{C},\alpha\beta}\left(\vec{k},\omega\right) + \frac{c}{\omega}\left(R'_{\mathrm{C},\alpha\beta}\left(\vec{k},\omega\right)e_{l\gamma\beta} + F'_{\mathrm{C},l\beta}\left(\vec{k},\omega\right)e_{\alpha\gamma l}\right)k_\gamma \\ + \left(\frac{c}{\omega}\right)^2 F_{\mathrm{C},\gamma l}\left(\vec{k},\omega\right)e_{\alpha m\gamma}e_{lp\beta}k_m k_\beta \end{aligned} \tag{3.21}$$

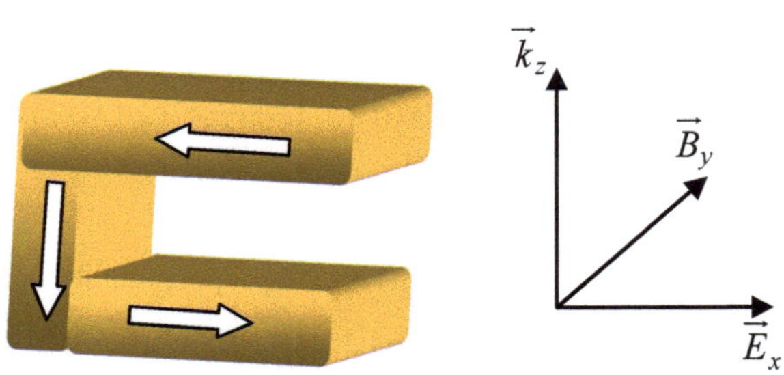

Fig. 3.1 Design and positioning of a MA which does not allow introducing of permeability. Appearance of polarization parallel to the wave vector is caused by the short cut between parts of the MA

It is easy to see that even in the case of dispersion-free response in “C” representation ($R_{C,\alpha\beta}(\vec{k},\omega) = R_{C,\alpha\beta}(\omega)$, $R'_{C,\alpha\beta}(\vec{k},\omega) = R'_{C,\alpha\beta}(\omega)$, $F'_{C,l\beta}(\vec{k},\omega) = F'_{C,l\beta}(\omega)$, $F_{C,\gamma l}(\vec{k},\omega) = F_{C,\gamma l}(\omega)$), spatial dispersion unavoidably appears in “L&L” representation. Moreover, if “non-eigen” responses are not zero ($R'_{C,\alpha\beta}(\vec{k},\omega) \neq 0$, $F'_{C,l\beta}(\vec{k},\omega) \neq 0$), then $R_{LL,\alpha\beta}(\vec{k},\omega)$ is proportional at least to the first order of the wave vector (linear spatial dispersion); if the magnetization is caused by an interaction with the magnetic field ($F_{C,l\beta}(\vec{k},\omega) \neq 0$) then $R_{LL,\alpha\beta}(\vec{k},\omega)$ is proportional at least to the second order of the wave vector (quadratic spatial dispersion). Thus, one can conclude that the nature of interaction (dependence of the polarizability and magnetization on the electric and/or magnetic fields) is the basic question, which determines the whole theoretical construction of the homogenization of the Maxwell equations. The nature of interaction of the electromagnetic wave with the charges in metallic nanoresonator (in case of MMs based on plasmonic structures, for example in the optical domain) is the interaction of electrons with the electric field only, because typical velocities of electrons are far from the velocity of light:

$$\frac{d\vec{p}_i}{dt} = q_i\vec{e} + \frac{q_i}{c}\left[\vec{v}_i * \vec{h}\right] \approx q_i\vec{e} \tag{3.22}$$

Here $\vec{e}$ and $\vec{h}$ are the microscopic electric and magnetic fields, respectively, $\vec{q}_i$, $\vec{p}_i$ and $\vec{v}_i$ are the charges, pulses, and velocities of charges, c is the speed of light. Hence there is no reason to assume that the magnetic field should appear in the phenomenological expression for the polarizability and magnetization, i.e. $R'_{C,\alpha\beta}(\vec{k},\omega) = 0$, $F_{C,\gamma l}(\vec{k},\omega) = 0$. In this case the elaborated above expressions can be summarized one more time:

$$\begin{cases} P_{C,\alpha}\left(\vec{k},\omega\right) = R_{C,\alpha\beta}\left(\vec{k},\omega\right)E_\beta\left(\vec{k},\omega\right) \\ M_{C,\alpha}\left(\vec{k},\omega\right) = F'_{C,\alpha\beta}\left(\vec{k},\omega\right)E_\beta\left(\vec{k},\omega\right) \end{cases} \tag{3.23}$$

$$\begin{cases} D_{C,\alpha}\left(\vec{k},\omega\right) = \varepsilon_{C,\alpha\beta}\left(\vec{k},\omega\right)E_\beta\left(\vec{k},\omega\right) \\ \varepsilon_{C,\alpha\beta}\left(\vec{k},\omega\right) = \delta_{\alpha\beta} + 4\pi R_{C,\alpha\beta}\left(\vec{k},\omega\right) \\ H_\alpha\left(\vec{k},\omega\right) = \xi_{\alpha\beta}\left(\vec{k},\omega\right)E_\beta\left(\vec{k},\omega\right) \\ \xi_{\alpha\beta}\left(\vec{k},\omega\right) = -\left(4\pi F'_{C,\alpha\beta}\left(\vec{k},\omega\right) + \frac{c}{\omega}e_{\alpha\gamma\beta}k_\gamma\right) \end{cases} \tag{3.24}$$

$$\begin{cases} D_{\mathrm{C},\alpha}\left(\vec{k},\omega\right) = \varepsilon_{\mathrm{C},\alpha\beta}\left(\vec{k},\omega\right)E_\beta\left(\vec{k},\omega\right) \\ \varepsilon_{\mathrm{C},\alpha\beta}\left(\vec{k},\omega\right) = \delta_{\alpha\beta} + 4\pi R_{\mathrm{C},\alpha\beta}\left(\vec{k},\omega\right) \\ H_{\mathrm{C},\alpha}\left(\vec{k},\omega\right) == \left[\mu_{\alpha\beta}\left(\vec{k},\omega\right)\right]^{-1} B_\beta\left(\vec{k},\omega\right) \\ \left[\mu_{\alpha\beta}\left(\vec{k},\omega\right)\right]^{-1} = \left[\delta_{\alpha\beta} + \frac{4\pi\omega}{c}F'_{\mathrm{C},\alpha\varsigma}\left(\vec{k},\omega\right)e_{\varsigma\gamma\beta}k_\gamma\left(\frac{1}{k_m k_m^*}\right)\right] \end{cases} \tag{3.25}$$

In the scientific literature there is a widely known and used representation of permittivity in the following form [7, 8]:

$$\varepsilon_{\mathrm{C},\alpha\beta}(\vec{k},\,\omega) = \varepsilon_{\mathrm{C}}^{\mathrm{tr}}(\vec{k},\,\omega)\left(\delta_{\alpha\beta} - \frac{k_\alpha k_\beta}{k^2}\right) + \varepsilon_{\mathrm{C}}^{l}(\vec{k},\,\omega)\frac{k_\alpha k_\beta}{k^2} \tag{3.26}$$

which is basically a special case of (3.24) or (3.25).

There is a hypothesis [9] that the permeability obeys the same type of expression:

$$\mu_{\mathrm{C},\alpha\beta}(\vec{k},\,\omega) = \mu_{\mathrm{C}}^{\mathrm{tr}}(\vec{k},\,\omega)\left(\delta_{\alpha\beta} - \frac{k_\alpha k_\beta}{k^2}\right) + \mu_{\mathrm{C}}^{l}(\vec{k},\,\omega)\frac{k_\alpha k_\beta}{k^2} \tag{3.27}$$

Exhausting consideration of relations between the coefficients $\varepsilon_{\mathrm{C}}^{\mathrm{tr}}(\omega)$, $\varepsilon_{\mathrm{C}}^{l}(\omega)$, $\mu_{\mathrm{C}}^{\mathrm{tr}}(\omega)$, $\mu_{\mathrm{C}}^{l}(\omega)$ from one side and $\varepsilon_{\mathrm{LL}}^{\mathrm{tr}}(\omega)$, and $\varepsilon_{\mathrm{LL}}^{\mathrm{tr}}(\omega)$ from another side can be found in [8].

It is necessary to mention, that some authors try to subdivide function $P_{\mathrm{C},\alpha}(\vec{k},\omega)$ on a dipole and a quadrupole parts within the frame of the phenomenological approach. This option is supported by a proven in [10] possibility to express the averaged current in the "C" representation through the dipole, quadrupole, and magnetic dipole parts, namely:

$$\langle j\rangle_\alpha = -i\omega\left(P_{\mathrm{C},\alpha} - k_\beta Q_{\alpha\beta}\right) + ice_{\alpha\beta l}k_\beta M_{\mathrm{C},l} \tag{3.28}$$

Here $P_{\mathrm{C},\alpha}$, $Q_{\alpha\beta}$, and $M_{\mathrm{C},l}$ are the dipole, qudrupole, and magnetic dipole contributions [see (2.14), (2.15)]. After that the three mentioned parts have to be reformulated in analogy with (3.10) through electric and (possibly) magnetic fields, which leads us finally to the same kind of expressions for $D_{\mathrm{C},\alpha}$ and H_α as in (3.24) and (3.25). It is seen, that in the framework of the phenomenological approach it does not make sense to subdivide function $P_{\mathrm{C},\alpha}(\vec{k},\omega)$ into dipole and quadrupole parts. In order to benefit from the representation (3.28) it is necessary to have a functional form for all three parts $P_{\mathrm{C},\alpha}$, $Q_{\alpha\beta}$, and $M_{\mathrm{C},l}$ [11–13], which is not possible within the frameworks of the phenomenological model; an attempt to get these functions will be performed in Sect. 3.2.

Case of Weak Spatial Dispersion

In the case of weak spatial dispersion functions $R_{C,\alpha\beta}$, $R'_{C,\alpha\beta}$, $F_{C,\alpha\beta}$, $F'_{C,\alpha\beta}$ can be expanded into the Taylor series (we limit our consideration by the second order of expansion):

$$\begin{cases} P_{C,\alpha}\left(\vec{k},\omega\right) = R_{C,\alpha\beta}\left(\vec{k},\omega\right)E_\beta\left(\vec{k},\omega\right) + R'_{C,\alpha\beta}\left(\vec{k},\omega\right)B_\beta\left(\vec{k},\omega\right) \\ M_{C,\alpha}\left(\vec{k},\omega\right) = F'_{C,\alpha\beta}\left(\vec{k},\omega\right)E_\beta\left(\vec{k},\omega\right) + F_{C,\alpha\beta}\left(\vec{k},\omega\right)B_\beta\left(\vec{k},\omega\right) \\ R_{C,\alpha\beta}\left(\vec{k},\omega\right) \approx R_{C,\alpha\beta}(0,\omega) + \left.\frac{\partial R_{C,\alpha\beta}\left(\vec{k},\omega\right)}{\partial k_\gamma}\right|_{\vec{k}=0} k_\gamma + \frac{1}{2}\left.\frac{\partial^2 R_{C,\alpha\beta}\left(\vec{k},\omega\right)}{\partial k_\gamma \partial k_\delta}\right|_{\vec{k}=0} k_\gamma k_\delta \\ R'_{C,\alpha\beta}\left(\vec{k},\omega\right) \approx R'_{C,\alpha\beta}(0,\omega) + \left.\frac{\partial R'_{C,\alpha\beta}\left(\vec{k},\omega\right)}{\partial k_\gamma}\right|_{\vec{k}=0} k_\gamma + \frac{1}{2}\left.\frac{\partial^2 R'_{C,\alpha\beta}\left(\vec{k},\omega\right)}{\partial k_\gamma \partial k_\delta}\right|_{\vec{k}=0} k_\gamma k_\delta \\ F_{C,\alpha\beta}\left(\vec{k},\omega\right) \approx F_{C,\alpha\beta}(0,\omega) + \left.\frac{\partial F_{C,\alpha\beta}\left(\vec{k},\omega\right)}{\partial k_\gamma}\right|_{\vec{k}=0} k_\gamma + \frac{1}{2}\left.\frac{\partial^2 F_{C,\alpha\beta}\left(\vec{k},\omega\right)}{\partial k_\gamma \partial k_\delta}\right|_{\vec{k}=0} k_\gamma k_\delta \\ F'_{C,\alpha\beta}\left(\vec{k},\omega\right) \approx F'_{C,\alpha\beta}(0,\omega) + \left.\frac{\partial F'_{C,\alpha\beta}\left(\vec{k},\omega\right)}{\partial k_\gamma}\right|_{\vec{k}=0} k_\gamma + \frac{1}{2}\left.\frac{\partial^2 F'_{C,\alpha\beta}\left(\vec{k},\omega\right)}{\partial k_\gamma \partial k_\delta}\right|_{\vec{k}=0} k_\gamma k_\delta \end{cases} \tag{3.29}$$

which results for the respective expressions for $D_{C,\alpha}$ and H_α. It is also appropriate at this point to mention that it is enough to restrict the consideration by the first order in the expansion for $F_{C,\alpha\beta}$ and $F'_{C,\alpha\beta}$, because after substitution to MEs the first order of spatial dispersion corresponds to the second order of the other part due to the fact, that the expansion is placed under curl operator, that in turn brings one more order of spatial dispersion:

$$\begin{cases} D_{C,\alpha}\left(\vec{k},\omega\right) = \left(\varepsilon^{(0)}_{C,\alpha\beta}(\omega) + \varepsilon^{(1)}_{C,\alpha\beta\gamma}(\omega)k_\gamma + \varepsilon^{(2)}_{C,\alpha\beta\gamma\delta}(\omega)k_\gamma k_\delta\right)E_\beta\left(\vec{k},\omega\right) \\ \qquad + \left(\psi^{(0)}_{C,\alpha\beta}(\omega) + \psi^{(1)}_{C,\alpha\beta\gamma}(\omega)k_\gamma + \psi^{(2)}_{C,\alpha\beta\gamma\delta}(\omega)k_\gamma k_\delta\right)B_\beta\left(\vec{k},\omega\right) \\ \varepsilon^{(0)}_{C,\alpha\beta}(\omega) = \delta_{\alpha\beta} + 4\pi R_{C,\alpha\beta}(0,\omega); \quad \psi^{(0)}_{C,\alpha\beta}(\omega) = \delta_{\alpha\beta} + 4\pi R'_{C,\alpha\beta}(0,\omega) \\ \varepsilon^{(1)}_{C,\alpha\beta\gamma}(\omega) = 4\pi \left.\frac{\partial R_{C,\alpha\beta}\left(\vec{k},\omega\right)}{\partial k_\gamma}\right|_{\vec{k}=0}; \quad \psi^{(1)}_{C,\alpha\beta\gamma}(\omega) = 4\pi \left.\frac{\partial R'_{C,\alpha\beta}\left(\vec{k},\omega\right)}{\partial k_\gamma}\right|_{\vec{k}=0} \\ \varepsilon^{(2)}_{C,\alpha\beta\gamma\delta}(\omega) = 2\pi \left.\frac{\partial^2 R_{C,\alpha\beta}\left(\vec{k},\omega\right)}{\partial k_\gamma \partial k_\delta}\right|_{\vec{k}=0}; \quad \psi^{(2)}_{C,\alpha\beta\gamma\delta}(\omega) = 2\pi \left.\frac{\partial^2 R_{C,\alpha\beta}\left(\vec{k},\omega\right)}{\partial k_\gamma \partial k_\delta}\right|_{\vec{k}=0} \end{cases} . \tag{3.30}$$

$$\begin{cases} H_\alpha\left(\vec{k},\omega\right) = \left(\phi^{(0)}_{C,\alpha\beta}(\omega) + \phi^{(1)}_{C,\alpha\beta\gamma}(\omega)k_{\gamma\,\delta}\right)E_\beta\left(\vec{k},\omega\right) + \left(\mu^{(0)}_{C,\alpha\beta}(\omega) + \mu^{(1)}_{C,\alpha\beta\gamma}(\omega)k_\gamma\right)B_\beta\left(\vec{k},\omega\right) \\ \phi^{(0)}_{C,\alpha\beta}(\omega) = -4\pi F'_{C,\alpha\beta}(0,\omega); \qquad \mu^{(0)}_{C,\alpha\beta}(\omega) = \delta_{\alpha\beta} + 4\pi F_{C,\alpha\beta}(0,\omega) \\ \phi^{(1)}_{C,\alpha\beta\gamma}(\omega) = -4\pi \left.\frac{\partial F'_{C,\alpha\beta}\left(\vec{k},\omega\right)}{\partial k_\gamma}\right|_{\vec{k}=0}; \qquad \mu^{(1)}_{C,\alpha\beta\gamma}(\omega) = 4\pi \left.\frac{\partial F_{C,\alpha\beta}\left(\vec{k},\omega\right)}{\partial k_\gamma}\right|_{\vec{k}=0} \end{cases} \tag{3.31}$$

Depending on a particular situation one or more terms in this expansion can be set to zero, and relations between some terms can be established based on the symmetry conditions. Thus far, in order to complete the homogenization model in case of weak spatial dispersion it is necessary to fill in the following 3 × 4 matrix:

$$\begin{pmatrix} R_{C,\alpha\beta}(0,\omega) & \left.\frac{\partial R_{C,\alpha\beta}\left(\vec{k},\omega\right)}{\partial k_\gamma}\right|_{\vec{k}=0} & \left.\frac{\partial^2 R_{C,\alpha\beta}\left(\vec{k},\omega\right)}{\partial k_\gamma \partial k_\delta}\right|_{\vec{k}=0} \\ R'_{C,\alpha\beta}(0,\omega) & \left.\frac{\partial R'_{C,\alpha\beta}\left(\vec{k},\omega\right)}{\partial k_\gamma}\right|_{\vec{k}=0} & \left.\frac{\partial^2 R'_{C,\alpha\beta}\left(\vec{k},\omega\right)}{\partial k_\gamma \partial k_\delta}\right|_{\vec{k}=0} \\ F_{C,\alpha\beta}(0,\omega) & \left.\frac{\partial F_{C,\alpha\beta}\left(\vec{k},\omega\right)}{\partial k_\gamma}\right|_{\vec{k}=0} & 0 \\ F'_{C,\alpha\beta}(0,\omega) & \left.\frac{\partial F'_{C,\alpha\beta}\left(\vec{k},\omega\right)}{\partial k_\gamma}\right|_{\vec{k}=0} & 0 \end{pmatrix} \tag{3.32}$$

or, in terms of other notations:

$$\begin{pmatrix} \varepsilon^{(0)}_{C,\alpha\beta}(\omega) & \varepsilon^{(1)}_{C,\alpha\beta\gamma}(\omega) & \varepsilon^{(2)}_{C,\alpha\beta\gamma\delta}(\omega) \\ \psi^{(0)}_{C,\alpha\beta}(\omega) & \psi^{(1)}_{C,\alpha\beta\gamma}(\omega) & \psi^{(2)}_{C,\alpha\beta\gamma\delta}(\omega) \\ \mu^{(0)}_{C,\alpha\beta}(\omega) & \mu^{(1)}_{C,\alpha\beta\gamma}(\omega) & 0 \\ \phi^{(0)}_{C,\alpha\beta}(\omega) & \phi^{(1)}_{C,\alpha\beta\gamma}(\omega) & 0 \end{pmatrix} \tag{3.33}$$

For example, in [14] the authors arrive to the following representation based on qualitative consideration of physical processes appearing at the interaction of the electromagnetic field with MAs:

$$\begin{pmatrix} \varepsilon^{(0)}_{C,\alpha\beta}(\omega) & 0 & 0 \\ \psi^{(0)}_{C,\alpha\beta}(\omega) & 0 & 0 \\ \mu^{(0)}_{C,\alpha\beta}(\omega) & 0 & 0 \\ \phi^{(0)}_{C,\alpha\beta}(\omega) & \phi^{(1)}_{C,\alpha\beta\gamma}(\omega) & 0 \end{pmatrix} \tag{3.34a}$$

The coupling effect, described by tensors $\psi^{(0)}_{\alpha\beta} = -\phi^{(0)}_{\alpha\beta}$ is known in electromagnetism and is called bi-anisotropy [3]; the tensor is called magnetoelectric coupling parameter [15]. Two special cases of bianisotropic media are known: chirial media when tensor $\psi^{(0)}_{\alpha\beta}$ is symmetric and omega media when tensor $\psi^{(0)}_{\alpha\beta}$ is anti-symmetric. More details about classification of different media based on introduced above model can be found in [16]. The form of material equations accepted in [14] is similar to the one presented in [17], where matrix (3.33) is written in the following form:

$$\begin{pmatrix} \varepsilon^{(0)}_{\mathrm{C},\alpha\beta}(\omega) & 0 & 0 \\ \psi^{(0)}_{\mathrm{C},\alpha\beta}(\omega) & 0 & 0 \\ \mu^{(0)}_{\mathrm{C},\alpha\beta}(\omega) & 0 & 0 \\ \phi^{(0)}_{\mathrm{C},\alpha\beta}(\omega) & 0 & 0 \end{pmatrix} \tag{3.34b}$$

and differs from (3.34a) by the absence of linear spatial dispersion in the equation for magnetization, which is equivalent to neglecting of anisotropy.

In both papers [14, 17] and many others the functional forms of the expressions for P_α and M_α are introduced based not on the developed here phenomenological approach, but using the multipole approach, considered in the next chapter. The usual way is to calculate the dipole and magnetic dipole moments using expressions known from electrostatics. It is believed, that in applications in the optical domain this approach has at least three drawbacks, namely:

1. It is clear from the physical point of view, that magnetic field does not affect charge dynamics and should not to be included in the basic considerations; it has been shown above, that the introduction of response to magnetic field through the electric one using one of the Maxwell equations is not straightforward.
2. Authors often do not distinguish the local and averaged fields when the charge dynamics in MAs is considered, that hides in some cases the role of spatial dispersion.
3. Practically all authors, taking into account magnetic moment, do not include into consideration quadrupole moment, which is in most cases mandatory due to the fact that magnetic moment and quadrupole moment are of the same order of the multipole expansion. Negligence of the quadrupole moment leads in turn to incorrect material equation representations, for example artificial exclusion of the first-order spatial dispersion term. This in turn excludes from the consideration some effects of anisotropy. The quadrupole moment effects can be neglected as compared with the magnetic moment influence only for specific geometries of inclusions. For example, the fundamental mode of double split ring resonators widely used in microwave MMs is characterized by a strong magnetic moment but negligible electric quadrupole moment, because the total current along two rings is nearly uniform around the whole ring structure.

Below a more consistent way of introduction of material equations in the frame of the phenomenological approach will be presented. As it was mentioned above, magnetization for the considered MM is evidently proportional only to the electric field and its spatial derivatives—see (3.23), (3.24), and (3.25). In this case of weak spatial dispersion the summarized expressions are:

$$\begin{cases} D_{C,\alpha}\left(\vec{k},\omega\right) = \left(\varepsilon^{(0)}_{C,\alpha\beta}(\omega) + \varepsilon^{(1)}_{C,\alpha\beta\gamma}(\omega)k_\gamma + \varepsilon^{(2)}_{C,\alpha\beta\gamma\delta}(\omega)k_\gamma k_\delta\right)E_\beta\left(\vec{k},\omega\right) \\ \varepsilon^{(0)}_{C,\alpha\beta}(\omega) = \delta_{\alpha\beta} + 4\pi R_{C,\alpha\beta}(0,\omega); \\ \varepsilon^{(1)}_{C,\alpha\beta\gamma}(\omega) = 4\pi \left.\dfrac{\partial R_{C,\alpha\beta}\left(\vec{k},\omega\right)}{\partial k_\gamma}\right|_{\vec{k}=0}; \\ \varepsilon^{(2)}_{C,\alpha\beta\gamma\delta}(\omega) = 2\pi \left.\dfrac{\partial^2 R_{C,\alpha\beta}\left(\vec{k},\omega\right)}{\partial k_\gamma \partial k_\delta}\right|_{\vec{k}=0}; \end{cases} \tag{3.35}$$

$$\begin{cases} H_\alpha\left(\vec{k},\omega\right) = \left(\phi^{(0)}_{C,\alpha\beta}(\omega) + \phi^{(1)}_{C,\alpha\beta\gamma}(\omega)k_\gamma\right)E_\beta\left(\vec{k},\omega\right) \\ \phi^{(0)}_{C,\alpha\beta}(\omega) = \delta_{\alpha\beta} + 4\pi F'_{C,\alpha\beta}(0,\omega) \\ \phi^{(1)}_{C,\alpha\beta\gamma}(\omega) = 4\pi \left.\dfrac{\partial F'_{C,\alpha\beta}\left(\vec{k},\omega\right)}{\partial k_\gamma}\right|_{\vec{k}=0} \end{cases} \tag{3.36}$$

Hence, in order to complete the homogenization model it is necessary to fill in the following 3 × 2 matrix:

$$\begin{pmatrix} R_{C,\alpha\beta}(0,\omega) & \left.\frac{\partial R_{C,\alpha\beta}\left(\vec{k},\omega\right)}{\partial k_\gamma}\right|_{\vec{k}=0} & \left.\frac{\partial^2 R_{C,\alpha\beta}\left(\vec{k},\omega\right)}{\partial k_\gamma \partial k_\delta}\right|_{\vec{k}=0} \\ F'_{C,\alpha\beta}(0,\omega) & \left.\frac{\partial F'_{C,\alpha\beta}\left(\vec{k},\omega\right)}{\partial k_\gamma}\right|_{\vec{k}=0} & 0 \end{pmatrix} \tag{3.37}$$

or, in terms of other notations:

$$\begin{pmatrix} \varepsilon^{(0)}_{C,\alpha\beta}(\omega) & \varepsilon^{(1)}_{C,\alpha\beta\gamma}(\omega) & \varepsilon^{(2)}_{C,\alpha\beta\gamma\delta}(\omega) \\ \phi^{(0)}_{C,\alpha\beta}(\omega) & \phi^{(1)}_{C,\alpha\beta\gamma}(\omega) & 0 \end{pmatrix} \tag{3.38}$$

Referring again to the paper [14] one can conclude that the suggested there representation is equivalent to:

$$\begin{pmatrix} \varepsilon^{(0)}_{C,\alpha\beta}(\omega) & 0 & \varepsilon^{(2)}_{C,\alpha\beta\gamma\delta}(\omega) \\ \phi^{(0)}_{C,\alpha\beta}(\omega) & \phi^{(1)}_{C,\alpha\beta\gamma}(\omega) & 0 \end{pmatrix} \tag{3.39}$$

and the representation accepted in [17] is equivalent to:

$$\begin{pmatrix} \varepsilon^{(0)}_{\mathrm{C},\alpha\beta}(\omega) & 0 & \varepsilon^{(2)}_{\mathrm{C},\alpha\beta\gamma\delta}(\omega) \\ \phi^{(0)}_{\mathrm{C},\alpha\beta}(\omega) & 0 & 0 \end{pmatrix} \tag{3.40}$$

3.1.3 Transformation Between "C" and "L&L" Representations in Case of Strong Spatial Dispersion

The relation between "L&L" and "C" representations can be obtained from (3.21) taking into account the fact that the direct interaction with the magnetic field is absent:

$$R_{\mathrm{LL},\alpha\beta}\left(\vec{k},\omega\right) = R_{\mathrm{C},\alpha\beta}\left(\vec{k},\omega\right) + \frac{c}{\omega} F'_{\mathrm{C},l\beta}\left(\vec{k},\omega\right) e_{\alpha\gamma l} k_\gamma \tag{3.41}$$

From the other side, it would be interesting to find connections between the commonly used permittivity and permeability in both representations. From (3.5) we have:

$$R_{\mathrm{LL},\alpha\beta}\left(\vec{k},\omega\right) = \frac{\varepsilon_{\mathrm{LL},\alpha\beta}\left(\vec{k},\omega\right) - \delta_{\alpha\beta}}{4\pi} \tag{3.42}$$

From the other side, "C" representation possesses two forms, namely (3.24) and (3.25). Starting from the more widely used form (3.25), we get:

$$\begin{cases} R_{\mathrm{C},\alpha\beta}\left(\vec{k},\omega\right) = \dfrac{\varepsilon_{\mathrm{C},\alpha\beta}\left(\vec{k},\omega\right) - \delta_{\alpha\beta}}{4\pi} \\ \left[\mu_{\alpha\beta}\left(\vec{k},\omega\right)\right]^{-1} = \left[\delta_{\alpha\beta} + \dfrac{4\pi\omega}{c} F'_{\mathrm{C},\alpha\varsigma}\left(\vec{k},\omega\right) e_{\varsigma\gamma\beta} k_\gamma \left(\dfrac{1}{k_m k_m^*}\right)\right] \end{cases} \tag{3.43}$$

From the second equation of (3.43) one can express $F'_{\mathrm{C},\alpha\beta}(\vec{k},\omega)$:

$$F'_{\mathrm{C},\alpha\beta}\left(\vec{k},\omega\right) = \frac{c e_{\beta l\gamma} k_l}{4\pi\omega}\left(\left[\mu_{\alpha\gamma}\left(\vec{k},\omega\right)\right]^{-1} - \delta_{\alpha\gamma}\right) \tag{3.44}$$

and finally get for the permittivity:

$$\varepsilon_{\mathrm{LL},\alpha\beta}\left(\vec{k},\omega\right) = \varepsilon_{\mathrm{C},\alpha\beta}\left(\vec{k},\omega\right) + \left(\frac{c}{\omega}\right)^2 e_{\alpha\gamma l} e_{\beta pm} k_\gamma k_p \left(\left[\mu_{lm}\left(\vec{k},\omega\right)\right]^{-1} - \delta_{lm}\right) \tag{3.45}$$

Following the same arguments, for the representation (3.24) we have:

$$\varepsilon_{\mathrm{LL},\alpha\beta}\left(\vec{k},\omega\right) = \varepsilon_{\mathrm{C},\alpha\beta}\left(\vec{k},\omega\right) + \frac{c}{\omega}e_{\alpha\gamma l}k_{\gamma}\left(\frac{c}{\omega}e_{lp\beta}k_{\mathrm{p}} - \xi_{l\beta}\left(\vec{k},\omega\right)\right) \tag{3.46}$$

The permittivity in the "L&L" representation is proportional at least to the second order of the wave vector (quadratic spatial dispersion); $\xi_{l\beta}(\vec{k},\omega) \sim k$ in order to model magnetic response.

It has to be noted again, that in "L&L" representation there is only one function (actually, one family of functions) which fully determines the averaged optical response of media. From the other side, in "C" representation there are two functions [in both cases (3.24) and (3.25)]. It is clear that from the known functions in "C" representation it is possible to construct one function in "L&L" representation, while the opposite transformation could not be done unambiguously.

The form (3.24) remains valid for any structures, while (3.25) can be used only in case when the permeability can be introduced—see (3.18), (3.19). Actually, the representation (3.24) is not only more general, but also is more convenient, because it fully reflects the physical nature of the processes, namely interaction of the MAs with the electric (not magnetic!) field.

It is believed, that the form (3.24) generally should be used in case when the basic processes causing magnetization are stipulated by the electric field; in other words, the form (3.24) has to be used in cases when there are no natural magnetic moments (like magnetic moments of natural atoms or molecules in ferromagnetic, for example), which can directly interact with magnetic fields at low frequencies.

After consideration of the phenomenological models of homogenization one can conclude that:

1. In both "L&L" and "C" representations it is possible to develop a phenomenological approach and reduce the homogenization procedure to several effective parameters with, in general, unknown functions/coefficients.
2. The effective parameters in general depend not only on the properties of media, but on the wave vector as well.
3. Due to the phenomenological nature of the presented here approach, it is in general impossible to separate in the effective parameters the parts which depend on the properties of media only from the parts, which contain dependence on the wave vector. Nevertheless, in case of a weak spatial dispersion (expansion of the respective functions up to the second order over the wave vector) it becomes possible to introduce effective material parameters in both cases of "L&L" and "C" representations.

3.1.4 Reduction to Material Equations for Bianisotropic Media in Case of Weak Spatial Dispersion

The presented here phenomenological approach in the case of weak spatial dispersion results in a system of material equations in form (3.35), (3.36), namely:

$$\begin{cases} D_{C,\alpha}\left(\vec{k},\omega\right) = \left(\varepsilon^{(0)}_{C,\alpha\beta}(\omega) + \varepsilon^{(1)}_{C,\alpha\beta\gamma}(\omega)k_\gamma + \varepsilon^{(2)}_{C,\alpha\beta\gamma\delta}(\omega)k_\gamma k_\delta\right)E_\beta\left(\vec{k},\omega\right) \\ H_\alpha\left(\vec{k},\omega\right) = \left(\phi^{(0)}_{C,\alpha\beta}(\omega) + \phi^{(1)}_{C,\alpha\beta\gamma}(\omega)k_\gamma\right)E_\beta\left(\vec{k},\omega\right) \end{cases} \tag{3.47}$$

One can show that system (3.47) can be reduced to the form similar to the well-known and widely used in the literature material equations for bianisotropic media [18]. In order to perform the necessary transformation, we use the known from the tensor algebra theorem which proves that any third-rank tensor can be presented as a sum of its symmetric and anti-symmetric parts. The latter in turn can be presented as a tensor of the second rank multiplied by the Levi-Chivita tensor. Using this theorem, we can write:

$$\begin{cases} \varepsilon^{(1)}_{C,\alpha\beta\gamma}(\omega) = \varepsilon^{(1,\mathrm{sym})}_{C,\alpha\beta\gamma}(\omega) + \varepsilon^{(1,\mathrm{asym})}_{C,\alpha\beta\gamma}(\omega) = \varepsilon^{(1,\mathrm{sym})}_{C,\alpha\beta\gamma}(\omega) + G^{(\varepsilon)}_{C,\alpha p}(\omega)e_{p\beta\gamma} \\ \phi^{(1)}_{C,\alpha\beta\gamma}(\omega) = \phi^{(1,\mathrm{sym})}_{C,\alpha\beta\gamma}(\omega) + \phi^{(1,\mathrm{asym})}_{C,\alpha\beta\gamma}(\omega) = \phi^{(1,\mathrm{sym})}_{C,\alpha\beta\gamma}(\omega) + G^{(\phi)}_{C,\alpha p}(\omega)e_{p\beta\gamma} \end{cases} \tag{3.48}$$

Substituting (3.48) into (3.47) and taking into account that $e_{\alpha\beta l}k_\beta E_l = \frac{\omega}{c}B_\alpha$, system (3.47) becomes:

$$\begin{cases} D_{C,\alpha}\left(\vec{k},\omega\right) = \left(\varepsilon^{(0)}_{C,\alpha\beta}(\omega) + \varepsilon^{(1,\mathrm{sym})}_{C,\alpha\beta\gamma}(\omega)k_\gamma + \varepsilon^{(2)}_{C,\alpha\beta\gamma\delta}(\omega)k_\gamma k_\delta\right)E_\beta\left(\vec{k},\omega\right) \\ \qquad + G^{(\varepsilon)}_{C,\alpha\beta}(\omega)B_\beta\left(\vec{k},\omega\right) \\ H_\alpha\left(\vec{k},\omega\right) = G^{(\phi)}_{C,\alpha\beta}(\omega)B_\beta\left(\vec{k},\omega\right) + \left(\phi^{(0)}_{C,\alpha\beta}(\omega) + \phi^{(1,\mathrm{sym})}_{C,\alpha\beta\gamma}(\omega)k_\gamma\right)E_\beta\left(\vec{k},\omega\right) \end{cases} \tag{3.49}$$

The last form is rather close to the usually used Post [18] form in case when the consideration is restricted to the first-order spatial dispersion $\varepsilon^{(2)}_{C,\alpha\beta\gamma\delta}(\omega) = 0$, $\phi^{(1,\mathrm{sym})}_{C,\alpha\beta\gamma}(\omega) = 0$:

$$\begin{cases} D_{C,\alpha}\left(\vec{k},\omega\right) = \left(\varepsilon^{(0)}_{C,\alpha\beta}(\omega) + \varepsilon^{(1,\mathrm{sym})}_{C,\alpha\beta\gamma}(\omega)k_\gamma\right)E_\beta\left(\vec{k},\omega\right) \\ \qquad + G^{(\varepsilon)}_{C,\alpha\beta}(\omega)B_\beta\left(\vec{k},\omega\right) \\ H_\alpha\left(\vec{k},\omega\right) = G^{(\phi)}_{C,\alpha\beta}(\omega)B_\beta\left(\vec{k},\omega\right) + \phi^{(0)}_{C,\alpha\beta}(\omega)E_\beta\left(\vec{k},\omega\right) \end{cases} \tag{3.50}$$

It is important to emphasise that the final form (3.50) contains term $\varepsilon^{(1,\mathrm{sym})}_{C,\alpha\beta\gamma}(\omega)$ which manifests the fact that the first-order spatial dispersion has to be in general included in consideration as a separate term even in case of the Post material equations. In case of $\varepsilon^{(1,\mathrm{sym})}_{C,\alpha\beta\gamma}(\omega) = 0$ system (3.50) takes the form basically equivalent to the Post equations for bianisotropic media:

$$\begin{cases} D_{C,\alpha}\left(\vec{k},\omega\right) = \varepsilon^{(0)}_{C,\alpha\beta}(\omega)E_{\beta}\left(\vec{k},\omega\right) + G^{(\varepsilon)}_{C,\alpha\beta}(\omega)B_{\beta}\left(\vec{k},\omega\right) \\ H_{\alpha}\left(\vec{k},\omega\right) = G^{(\phi)}_{C,\alpha\beta}(\omega)B_{\beta}\left(\vec{k},\omega\right) + \phi^{(0)}_{C,\alpha\beta}(\omega)E_{\beta}\left(\vec{k},\omega\right) \end{cases} \tag{3.51}$$

The parameters in (3.51) can be further investigated for reciprocal and non-reciprocal media, for example, applying the reciprocity theorem [18], introducing the Tellegen parameter [19], etc.; these considerations are not done here. Note, that similar conclusion about necessity of introduction of additional terms in "traditional" material equations for bianisotropic media has been done in [14]—see (3.29), (3.30) there. It is worth noticing again, that the extra term $\varepsilon^{(1,\mathrm{sym})}_{C,\alpha\beta\gamma}(\omega)$ in (3.50) appears in the first-order spatial dispersion, while the extra terms in [14] correspond to the second order, and hence are responsible for different effects. It has to be also emphasised that the second-order spatial dispersion has to be in general taken into account in order to consider magnetic response (see [20]); in the frame of the phenomenological approach it corresponds to $\varepsilon^{(2)}_{C,\alpha\beta\gamma\delta}(\omega) \neq 0$, $\phi^{(1,\mathrm{sym})}_{C,\alpha\beta\gamma}(\omega) \neq 0$.

The first-order spatial dispersion term in (3.47), (3.48) could be compensated by the SFT—see (2.37). Nevertheless, it is easy to see, that SFT is able to compensate only the anti-symmetric part $\varepsilon^{(1,\mathrm{asym})}_{C,\alpha\beta\gamma}(\omega)$, but not the symmetric part $\varepsilon^{(1,\mathrm{sym})}_{C,\alpha\beta\gamma}(\omega)$, which has to be in general retained in the material equations (3.50).

3.2 Multipole Expansion ("C" Representation)

3.2.1 Multipole Approach

The multipole model was put forward in [21], and later developed in a similar form in [22]. The model is based on an averaging procedure using the Probability Distribution Function (PDF) for the positions and velocities of all charges, included in the consideration—statistical averaging, which is supposed to be equivalent to the originally assumed averaging over volume. Leaving alone the mathematical details of the model (which can be found in [21]), here it is worth to recall the main ideas of the elaboration of the model.

The essence of the developed in [21] averaging procedure is in summation of contributions from all atoms/molecules at the "Observation point" (see Fig. 2.1) using statistical math tools. In the framework of this approach each atom/molecule is considered as a cloud of positive and negative charges with some (a priori unknown) PDF over their coordinates and velocities. The condition $L_{\mathrm{intra}} \ll L_{\mathrm{inter}}$ allows us to use an expansion in the Taylor series of the potential, produced by each atom/molecule at the "Observation point". As a result, the total contribution can be expressed in terms of averaged moments, namely the total charge of the system (zero-order moment), the electric dipole (first-order moment), the quadrupole and magnetic dipole (second-order moments), etc.

It is worth noticing that the quadrupole and the magnetic dipole moments appear on the same level of Taylor expansion (second order) and therefore in general have the same order of magnitudes. In case of necessity to take into account a magnetic response of the atoms/molecules (as it takes place in case of MMs) it is usually necessary to use both moments together rather than voluntarily pick up just magnetic dipole moments and do not include into consideration the quardupole one. This requirement is stipulated by the fundamental principles of the multipole model and only in specific cases one can exclude the quadrupole moments of atoms/molecules from consideration.

The model results in constructive expressions for $\vec{P}_{\mathrm{C}}$ and $\vec{M}_{\mathrm{C}}$ presented through the averaged dynamics of the charges in "C" form [23]:

$$\begin{cases} \vec{P}(\vec{R},\omega) = \eta \left\langle \sum\limits_{s}^{\text{all charges}} q_s \vec{r}_s \right\rangle - \nabla \bullet Q(\vec{R},\omega) \\ Q_{ij}(\vec{R},\omega) = \dfrac{\eta}{2} \left\langle \sum\limits_{s}^{\text{all charges}} q_s r_{i,s} r_{j,s} \right\rangle \\ \vec{M}(\vec{R},t) = \dfrac{\eta}{2c} \left\langle \sum\limits_{s}^{\text{all charges}} q_s \left[\vec{r}_s, \dfrac{\partial \vec{r}_s}{\partial t} \right] \right\rangle \end{cases} \tag{3.52}$$

The definitions clearly distinguish between microscopic (**r**) and macroscopic (**R**) coordinates, q_k represents the charge, and η their density. The microscopic coordinates $\vec{r}$ designate the position vectors of the charges in a microscopic coordinate system, and $\vec{v}$ designate their velocities. The center of the microscopic coordinate system is chosen to be the center of symmetry of the charge distribution (consideration of the dependence on the origin of the coordinate system will be given later). The reason for the different coordinate systems derives from the averaging procedure for the averaged Maxwell equations [24]. The microscopic coordinates are functions of the electric field and do not appear explicitly in the final expressions. Only one coordinate system, namely the *macroscopic system of coordinates* $\vec{R}$ (i.e., the space coordinate) remains.

The functions $\vec{D}(\vec{R},\omega)$ and $\vec{H}(\vec{R},\omega)$ (2.19), (2.20) contain electric dipole, electric quadrupole, and magnetic dipole contributions:

$$\begin{cases} \vec{D}(\vec{R},\omega) = \vec{E}(\vec{R},\omega) + 4\pi\vec{P}(\vec{R},\omega) \\ \vec{H}(\vec{R},\omega) = \vec{B}(\vec{R},\omega) - 4\pi\vec{M}(\vec{R},\omega) \end{cases} \tag{3.53}$$

$\vec{P}(\vec{R},\omega)$, $Q_{ij}(\vec{R},\omega)$, and $\vec{M}(\vec{R},\omega)$ represent the electric polarization, the electric quadrupole tensor, and the magnetization, respectively. Capital letters $\vec{R}$ are used for macroscopic coordinates in the averaged Maxwell equations. The term:

$$\nabla \bullet Q(\vec{R}, \omega) = \frac{\partial Q_{ij}(\vec{R}, \omega)}{\partial R_j} \tag{3.54}$$

is the divergence of the quadrupole tensor. It is necessary to take into account both the electric quadrupole and themagnetic dipole terms, because they are of the same order in the multipole expansion series [6, 23].

It is important to realize that the formulas for the macroscopic polarization and magnetization are expressed in terms of the internal dynamics of the charges of the atoms/molecules, which are a priori functions of microscopic (not macroscopic!) fields. Even if we are able to write analytical forms for the dynamics, we will have to perform the averaging [see (3.52)] and express the functions $\vec{P}$ and $\vec{M}$ through the macroscopic fields. This finally closes the problem and makes from MEs a self-consistent system of equations, which can be (potentially) solved.

It should be emphasized that the multipole approach remains the only one, which allows us to create a logical connection from the microscopic to macroscopic forms of the MEs without any methodological gaps. The fact that finally this approach results in "C" form (one of the possible forms of MEs, obtained through the independent phenomenological consideration) serves as one more positive argument for the use of this model and its application to the problem of homogenization of MMs.

It has to be accepted that the basic conditions, under which system (3.52) has been elaborated are met for typical MMs in the optical domain rather poorly. Referring again to Fig. 2.1 and remembering typical experimental situations (for example, [25]), one can see that the distance between the MAs and the sizes of the MAs are of the same order, and the truncated Taylor expansion used in elaboration of (3.52) is not fully justified. Note that in contrast to the MM, in the case of natural materials system (3.52) works pretty well; the widely used dipole model for the permittivity is just the zero-order approximation of (3.52). Hence, the basic question about applicability of the multipole model to MMs remains open.

In spite of the fundamental doubts about its applicability, one can easily bring several arguments in favour of the multipole model:

1. The model offers a natural way to describe magnetization by introducing magnetic and quadrupole moments.
2. The model is physically clear and should be considered at least for the methodological reasons.
3. The model allows us to elaborate the functional forms for the introduced in phenomenological approach effective constants and fix the expressions for $\vec{P}$ and $\vec{M}$ as functions of the wave vector (in other words, find a functional form for spatial dispersion).
4. The model allows us to investigate the influence of the MA design on the optical properties of MMs.
5. The model allows us to investigate the influence of interactions between MAs on the optical properties of MMs.

6. The model allows us to investigate the influence of disorder (both spatial disorder in MA placements and disorder in eigen characteristics of the MAs) on the optical properties of MMs.
7. The model allows a natural extension beyond the purely plasmonic based MAs, for example to the case of combinations of plasmonic MAs and active quantum elements, or MAs consisting of purely quantum elements.

The multipole model created for MMs [26] contains parameters which can be tuned in order to compensate for the fundamentally stipulated discrepancies and finally fit the results of the model to the experimental and/or numerical data. It is believed, that the combination of the multipole approach with final tuning of these coefficients makes this model an extremely simple and versatile tool for investigation of optical properties of MMs [27]. In [20] analytical expressions for the effective permittivity and permeability have been elaborated for MMs based on double-wire structures. The charge dynamics has been treated using two coupled harmonic oscillator equations, possessing symmetric and anti-symmetric oscillation modes, excited by the electric field. Using the expressions for the symmetric and anti-symmetric modes, the dipole, quadrupole, and magnetic dipole terms (3.52) have been calculated as functions of the MAs and field parameters. Note, that in [20] only symmetric structures have been considered, extension to the case of asymmetric structures was performed in [28].

Below we will consider only 2D geometries (see Fig. 2.2), i.e., non-zero components of multipoles depend on the charge dynamics in the (x, y) plane and can be written as:

$$\begin{cases} \langle\vec{j}\rangle_x(y,t) = -i\omega P_x + c\dfrac{\partial M_z}{\partial y} \\ P_x(y,t) = \eta\left\langle \sum\limits_{\mathrm{s}}^{\text{all charges}} e_{\mathrm{s}} x_{\mathrm{s}} \right\rangle - \dfrac{\partial Q_{xy}}{\partial y} \\ Q_{xy}(y,t) = \dfrac{\eta}{2}\left\langle \sum\limits_{\mathrm{s}}^{\text{all charges}} q_{\mathrm{s}} x_{\mathrm{s}} y_{\mathrm{s}} \right\rangle \\ M_z(y,t) = \dfrac{\eta}{2c}\left\langle \sum\limits_{\mathrm{s}}^{\text{all charges}} q_{\mathrm{s}}\left(x_{\mathrm{s}}\dfrac{\partial y_{\mathrm{s}}}{\partial t} - y_{\mathrm{s}}\dfrac{\partial x_{\mathrm{s}}}{\partial t}\right)\right\rangle \end{cases} \tag{3.55}$$

Transition to the Fourier domain is not straightforward—both quadrupole and magnetic dipole moments depend nonlinearly on the coordinates. Nevertheless, it is assumed that to the first approximation any charge exhibits dynamics along just one direction: for example, in double wires it is the x direction, in case of split rings in two arms the charges move along the x direction and in the third arm—along the y direction, etc. In case of simultaneous (x, y) dynamics nonlinear response appears as a consequence of non-harmonic multipole dynamics [29].

A particular geometry of double wires, shown in Fig. 3.2, allows a purely linear description; the electric field is polarized along the x axis, magnetic field is polarized along the z axis, and the wave propagates along the y axis.

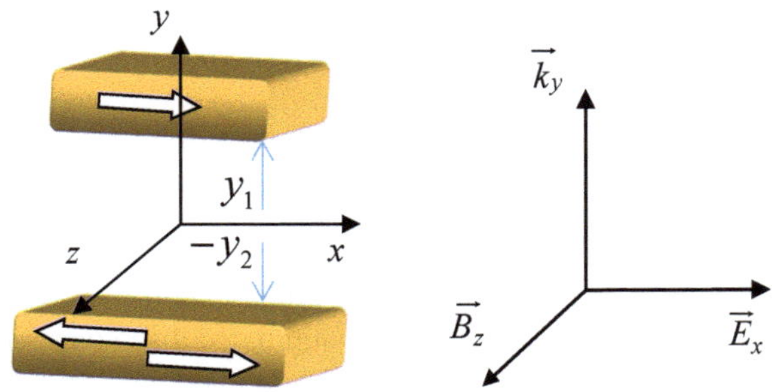

Fig. 3.2 Double wire (in general case asymmetric) structure possessing dipole, quadrupole, and magnetic moments with respective direction of plane wave propagation and field polarizations. The system of microscopic coordinates is chosen so that its origin is placed in general at different distances y_1 and y_2 from both wires

The wires are oriented along the x direction and are affected by the electric field which excites plasmonic modes in each wire; the coupled wires are considered as MAs, and multipole moments (3.52) are prescribed to these MAs. In this case (3.55) can be further simplified (no dynamics along the y axis) and straightforwardly transformed into the frequency domain:

$$\begin{cases} \langle \vec{j} \rangle_x (y,\omega) = -i\omega P_x + c\dfrac{\partial M_z}{\partial y} \\ P_x(y,\omega) = \eta \left\langle \sum\limits_s^{\text{all charges}} e_s x_s \right\rangle - \dfrac{\partial Q_{xy}}{\partial y} \\ Q_{xy}(y,\omega) = \dfrac{\eta}{2} \left\langle \sum\limits_s^{\text{all charges}} q_s x_s y_{s,0} \right\rangle \\ M_z(y,\omega) = \dfrac{i\eta\omega}{2c} \left\langle \sum\limits_s^{\text{all charges}} q_s x_s y_{s,0} \right\rangle \end{cases} \tag{3.56}$$

Note that $M_z(y,\omega) = \frac{i\omega}{c} Q_{xy}(y,\omega)$ and the magnetic dipole and the quadrupole moments have the same sign and are arranged together in the first equation in (3.56).

3.2.2 Dispersion Relation Elaboration

The derivation starts with the averaged (macroscopic) ME in "C" form in the frequency domain (2.19), (2.20). Here we consider MM consisting of double wire structures—one layer of such structure is presented in Fig. 2.2 on the top of a transparent substrate. The MM is formed by stacking of such layers in the y direction, and propagation of a plane wave along the y direction with the electric field polarized along the x direction is considered. This particular case illustrated in Fig. 2.2 is chosen in order to demonstrate main principles of the suggested model and avoid excessive math complications. The model is straightforwardly extendable to other MAs geometries, wave propagation directions, and field polarizations.

For a plane wave propagating along the positive y axis and with the electric field polarized along the x direction, the ME can be simplified and finally reduced to:

$$\begin{cases} P_x(y,\omega) = \eta\left\langle \sum_s^{\text{all charges}} e_s x_s \right\rangle - \dfrac{\partial Q_{xy}}{\partial y} \\ Q_{xy}(y,\omega) = \dfrac{\eta}{2}\left\langle \sum_s^{\text{all charges}} q_s x_s y_{s,0} \right\rangle \\ M_z(y,\omega) = \dfrac{i\eta\omega}{2c}\left\langle \sum_s^{\text{all charges}} q_s x_s y_{s,0} \right\rangle \end{cases} \quad \begin{cases} \dfrac{\partial E_x}{\partial y} = -\dfrac{i\omega}{c} B_z \\ \dfrac{\partial H_z}{\partial y} = -\dfrac{i\omega}{c} D_x \\ D_x = E_x(y,\omega) + 4\pi P_{C,x}(y,\omega) \\ H_z = B_z - 4\pi M_{C,z} \end{cases} \tag{3.57}$$

which provides a self-consistent equation for the x component of the electric field $E_x(y,\omega)$:

$$\frac{\partial^2 E_x(y,\omega)}{\partial y^2} + \frac{\omega^2}{c^2}\left(E_x(y,\omega) + 4\pi P_x(y,\omega)\right) + \frac{i4\pi\omega}{c}\frac{\partial M_z(y,\omega)}{\partial y} = 0 \tag{3.58}$$

It is interesting to note that the magnetic dipole and electric quadrupole contributions are identical. This is another proof that the electric quadrupole and magnetic dipole contributions have the same order of magnitude and both (not only the magnetic dipole term) have to be taken into account simultaneously [21].

Equation (3.58) after Fourier transformation over the spatial coordinate y becomes:

$$k_y^2 E_x(k_y,\omega) = \frac{\omega^2}{c^2}\left(E_x(k_y,\omega) + 4\pi P_x(k_y,\omega)\right) - \frac{4\pi k_y \omega}{c} M_z(k_y,\omega) \tag{3.59}$$

Assuming linear dependence of the multipole terms on the electric field (dependence on the magnetic field is negligible), one can write:

$$\begin{cases} P_x(k_y,\omega) = \left(p_x(k_y,\omega) - ik_y u_{xy}(k_y,\omega)\right) E_x(k_y,\omega) \\ Q_{xy}(k_y,\omega) = u_{xy}(k_y,\omega) E_x(k_y,\omega) \\ M_z(k_y,\omega) = m_x(k_y,\omega) E_x(k_y,\omega) \end{cases} \tag{3.60}$$

Substitution (3.60) into (3.59) results in the dispersion relation in the general form:

$$k_y^2 = \frac{\omega^2}{c^2}\left(1 + 4\pi p_x(k_y, \omega) - 4\pi i k_y u_{xy}(k_y, \omega)\right) - \frac{4\pi k_y \omega}{c} m_z(k_y, \omega) \tag{3.61a}$$

It is also useful to write down the material equations, corresponding to the form (3.61):

$$\begin{cases} D_x(k_y, \omega) = \left(1 + 4\pi p_x(k_y, \omega) - i4\pi k_y u_{xy}(k_y, \omega)\right) E_x(k_y, \omega) \\ H_z(k_y, \omega) = \left(-\frac{k_y c}{\omega} - 4\pi m_z(k_y, \omega)\right) E_x(k_y, \omega) \end{cases} \tag{3.61b}$$

$$\begin{cases} \varepsilon_x(k_y, \omega) = 1 + 4\pi p_x(k_y, \omega) - i4\pi k_y u_{xy}(k_y, \omega) \\ \mu_z(k_y, \omega) = \left(1 + \frac{4\pi\omega}{k_y c} m_z(k_y, \omega)\right)^{-1} \\ \xi_{zx}(k_y, \omega) = -\left(\frac{k_y c}{\omega} + 4\pi m_z(k_y, \omega)\right) \end{cases} \tag{3.61c}$$

Dispersion relation (3.61) is the main result of this part and basically solves the problem of propagation of plane waves in media with higher multipoles. It is worth noticing that (3.61) is pretty universal—the charge dynamics and multipoles in (3.61) can be expressed based on classical, quantum, or semi-classical approaches and hence can be applied to extremely wide range of various problems. Moreover, this unified approach is highly desirable for education courses in the area of nanophotonics and electrodynamics of MMs, because it provides logical, internally consistent, allowing in most cases analytical treatment approach giving deep understanding of physics of various problems using a single platform.

3.2.3 Physical Interpretation of Phenomenological Coefficients

From the performed above consideration a clear and unambiguous connection between different types of excited modes in MAs (symmetric or acourses in the area ofnti-symmetric) and the phenomenologically introduced in (3.35), (3.36) parameters can be revealed. It turned out [20], that the symmetric mode contributes to the dipole moment only, while the anti-symmetric mode is responsible for appearance of the quadrupole and magnetic dipole terms, and, consequently, for the magnetic response of media. The anti-symmetric mode can be excited in case of:

(a) asymmetric structure of MAs (for example, in case of double wires it means different lengths of the wires—see [28] for details),
(b) symmetric structure but inhomogeneous electric field (in [20] the retardation effect has been taken into account). In case of anti-symmetric structure and symmetric external electric field the resulted quadrupole and magnetic moments

do not contain wave vectors (are not spatially dispersive), while for the case of symmetric structure and inhomogeneous external electric field (for example, retarded field) the quadrupole and magnetic moments both depend on the wave vector, i.e. turn out to be spatially dispersive.

Comparison of the phenomenological (3.35), (3.36) and multipole (3.61) approaches allows us to prescribe clear physical interpretation of the coefficients in (3.35), (3.36); it requires more detailed MA structure consideration. For example, for symmetric double wires [20] and retarded at the scale of a MA field:

$$\begin{cases} p_x(k_y,\omega) = p_x(\omega) + p_x^{(2)}(\omega)k_y^2 \\ u_{xy}(k_y,\omega) = u_{xy}^{(1)}(\omega)k_y \\ m_z(k_y,\omega) = m_z^{(1)}(\omega)k_y \end{cases} \tag{3.62}$$

which gives for the coefficients of the phenomenological model (3.35), (3.36):

$$\begin{cases} \varepsilon^{(0)}_{\mathrm{C},\alpha\beta}(\omega) = 1 + 4\pi\, p_x(\omega) \\ \varepsilon^{(1)}_{\mathrm{C},\alpha\beta\gamma}(\omega) = 0 \\ \varepsilon^{(2)}_{\mathrm{C},\alpha\beta\gamma\delta}(\omega) = p_x^{(2)}(\omega) - i4\pi\, u_{xy}^{(1)}(\omega) \\ \phi^{(0)}_{\mathrm{C},\alpha\beta}(\omega) = 0 \\ \phi^{(1)}_{\mathrm{C},\alpha\beta\gamma}(\omega) = -\dfrac{c}{\omega} - 4\pi\, m_z^{(1)}(\omega) \end{cases} \tag{3.63}$$

At the same time, for asymmetric structure and not retarded at the scale of the MA field:

$$\begin{cases} p_x(k_y,\omega) = p_x(\omega) \\ u_{xy}(k_y,\omega) = u_{xy}^{(1)}(\omega) \\ m_z(k_y,\omega) = m_z^{(1)}(\omega) \end{cases} \tag{3.64}$$

$$\begin{cases} \varepsilon^{(0)}_{\mathrm{C},\alpha\beta}(\omega) = 1 + 4\pi p_x(\omega) \\ \varepsilon^{(1)}_{\mathrm{C},\alpha\beta\gamma}(\omega) = -i4\pi u_{xy}^{(1)}(\omega) \\ \varepsilon^{(2)}_{\mathrm{C},\alpha\beta\gamma\delta}(\omega) = 0 \\ \phi^{(0)}_{\mathrm{C},\alpha\beta}(\omega) = -4\pi m_z^{(1)}(\omega) \\ \phi^{(1)}_{\mathrm{C},\alpha\beta\gamma}(\omega) = -\dfrac{c}{\omega} \end{cases} \tag{3.65}$$

Based on the given above consideration, it is rather straightforward to prescribe clear physical mean to the coefficients introduced in (3.35), (3.36), namely:

$\varepsilon^{(0)}_{\mathrm{C},\alpha\beta}(\omega)$ is the standard dielectric permittivity due to induced dipoles;

$\varepsilon^{(1)}_{\mathrm{C},\alpha\beta\gamma}(\omega)$ is the term corresponding to the appearance of the quadrupole moment (anti-symmetric modes), where the quadrupole moment itself is not spatially dispersive. Physical situation is the MA with an anti-symmetric mode, which could be excited by a homogeneous field (in this case the MA itself has to be asymmetric); retardation on the size of MA is not necessary in order to excite an anti-symmetric mode of the MA. This term is responsible for bianisotropy [14, 17] (in notations, accepted in these and similar papers).

$\varepsilon^{(2)}_{\mathrm{C},\alpha\beta\gamma\delta}(\omega)$ is the term corresponding to the appearance of the quadrupole moment (anti-symmetric modes), where the quadrupole moment itself linearly depends on the wave vector (first-order spatial dispersion for the quadrupole moment) and dipole moment is proportional to the second order of the wave vector (second-order spatial dispersion for the dipole moment). Physically it corresponds to the situation when anti-symmetric modes are excited by an inhomogeneous (for example, retarded) electric field; in this case the MA itself can be symmetric.

$\phi^{(0)}_{\mathrm{C},\alpha\beta}(\omega)$ is the term that basically is also responsible for the bianisotropy (in notations of [14, 17]), and physically appears in case when the anti-symmetric mode of the MAs can be excited by a homogeneous field (in this case the MA itself has to be asymmetric), and thus corresponds to $\varepsilon^{(1)}_{\mathrm{C},\alpha\beta\gamma}(\omega)$.

$\phi^{(1)}_{\mathrm{C},\alpha\beta\gamma}(\omega)$ is the term where extra contribution (the term $-\frac{k_y c}{\omega}E_x(k_y,\omega)$ is just the magnetic field) appearing in the case when the anti-symmetric modes are excited by an inhomogeneous (for example, retarded) electric field.

It is worth noticing that the consideration of the problem here has been done based on the basic physical processes, appearing in the MAs, namely excitation of symmetric and anti-symmetric modes. It is useful to summarize the coefficients with respect to these properties—see Table 3.1.

It is interesting to note, that the developed approach establishes a way to determine the types of the modes exited in MM, provided the coefficients in

Table 3.1 Symmetry properties of MAs and exciting fields, and the respective coefficients responsible for particular modes

Symmetric properties of the MAs and fields	Type of excited modes	Respective coefficients
Symmetric MAs, homogeneous electric field	Symmetric	$\varepsilon^{(0)}_{\mathrm{C},\alpha\beta}(\omega)$
Asymmetric MAs, homogeneous electric field	Symmetric, anti-symmetric	$\varepsilon^{(1)}_{\mathrm{C},\alpha\beta\gamma}(\omega)$, $\phi^{(0)}_{\mathrm{C},\alpha\beta}(\omega)$
Symmetric MAs, inhomogeneous electric field	Symmetric, anti-symmetric	$\varepsilon^{(2)}_{\mathrm{C},\alpha\beta\gamma\delta}(\omega)$, $\phi^{(1)}_{\mathrm{C},\alpha\beta\gamma}(\omega)+\frac{c}{\omega}$

Table 3.1 could be experimentally determined with the help of a retrieval procedure. In case of $\varepsilon^{(1)}_{\mathrm{C},\alpha\beta\gamma}(\omega) \sim 0$, $\phi^{(0)}_{\mathrm{C},\alpha\beta}(\omega) \sim 0$ one can conclude that the structure itself appears to be symmetric, and the magnetic response (in case $\phi^{(1)}_{\mathrm{C},\alpha\beta\gamma}(\omega) + \frac{c}{\omega} \neq 0$) is caused by a gradient of the electric field. In opposite case $\varepsilon^{(2)}_{\mathrm{C},\alpha\beta\gamma\delta}(\omega) \sim 0$, $\phi^{(1)}_{\mathrm{C},\alpha\beta\gamma}(\omega) + \frac{c}{\omega} \sim 0$ and respectively $\phi^{(0)}_{\mathrm{C},\alpha\beta}(\omega) \neq 0$ (non-zero magnetic response) the plasmonic oscillation mode is excited due to the asymmetry of the structure itself. The possibility to make a conclusion about microscopic processes (type of the excited oscillation mode) based on the macroscopic measurements (assuming that the mentioned above retrieval procedure can be designed) looks rather attractive and undoubtedly deserves further investigations.

3.2.4 Origin Dependence of the Multipole Moments

It is known from the theory of multipoles (e.g., [23]) that the multipole moments, as they have been introduced in (3.56), depend on the origin of the microscopic system of coordinates (see the system of coordinates in Fig. 3.2). It has to be noted, that in case of zero of previous terms in multipole expansion, the next one does not depend on the origin. For example, in case of zero total charge, the dipole moment does not depend on the origin; in case of zero of total charge and dipole moment, the quadrupole/magnetic moments do not depend on the origin etc. We nevertheless assume here general case i.e. all terms (excepting total charge) are nonzero. In fact, if the origin is shifted by a vector $\vec{g}\,(g_x, g_y, g_z)$, the coordinates of all charges are shifted as well $\vec{r_s} = \vec{r}'_s + \vec{g}$, and multipoles (3.56) become:

$$\begin{cases} P_i(\vec{R},t) = P'_i(\vec{R},t) - \dfrac{1}{2}\dfrac{\partial\left(d_i g_j + d_j g_i\right)(\vec{R},t)}{\partial R_j} \\ Q_{ij}(\vec{R},t) = Q'_{ij}(\vec{R},t) + \left(d_i g_j + d_j g_i\right)(\vec{R},t) \\ M_i(\vec{R},t) = M'_i(\vec{R},t) + e_{ijk} g_j \langle \vec{j} \rangle_k (\vec{R},t) \\ d_i(\vec{R},t) = \eta \left\langle \displaystyle\sum_{s}^{\text{all charges}} q_s \vec{r}_{s,i} \right\rangle \\ \langle \vec{j} \rangle_i (\vec{R},t) = \dfrac{\eta}{2c} \left\langle \displaystyle\sum_{s}^{\text{all charges}} q_s \vec{v}_{s,i} \right\rangle \end{cases} \tag{3.66}$$

Here d_i are the components of the averaged dipole moment of the MA, $\langle \vec{j} \rangle_i$ are the components of the averaged current of the MA; it is also assumed that the MA is electrically neutral ($\sum_{s}^{\text{all charges}} q_s = 0$). For the sake of simplicity an only one particular MA geometry, presented in Fig. 3.2, is considered here. For this geometry (3.66) is simplified and becomes in the $(k_y,\, \omega)$ domain:

$$\begin{cases} P_x(k_y,\, \omega) = P'_x(k_y,\, \omega) - \dfrac{ik_y}{2} g_y d_x(k_y,\, \omega) \\ Q_{xy}\,(k_y,\, \omega) = Q'_{xy}\,(k_y,\, \omega) + \dfrac{1}{2} g_y d_x(k_y,\, \omega) \\ M_z(k_y,\, \omega) = M'_z(k_y,\, \omega) + \dfrac{i\omega}{2c} g_y d_x(k_y,\, \omega) \end{cases} \tag{3.67}$$

Substituting (3.67) into (3.58) we arrive to the conclusion that the dispersion relation depends on the origin in calculations of the multipoles:

$$\begin{aligned} k_y^2 E_x(k_y, \omega) &= \frac{\omega^2}{c^2}\left(E_x(k_y, \omega) + 4\pi\, P_x(k_y, \omega)\right) - \frac{4\pi\, k_y \omega}{c} M_z(k_y, \omega) \\ &= \frac{\omega^2}{c^2}\left(E_x(k_y, \omega) + 4\pi\, P'_x(k_y, \omega)\right) - \frac{4\pi\, k_y \omega}{c} M'_z(k_y, \omega) - \frac{ik_y \omega^2}{c^2} g_y d_x \end{aligned} \tag{3.68}$$

The result of the origin dependence of the wave vector looks unacceptable, because of the wave vector obviously should not depend on the voluntary choice of the origin of the multipole calculations. It has to be emphasized, that anyway the origin dependence is a natural consequence of the elaborated in [21] multipole model. Equations (3.66) and in our particular case of the double wires (3.67) are direct consequences of the recipe presented in [21]. The origin-dependence of “microscopic” moments does not automatically disappear in the process of averaging, leading to macroscopic models. Additional physical requirements need to be introduced, so that the resulting macroscopic model contains only origin-independent terms. Usually these requirements are the reciprocity of the medium and symmetry considerations applied to averaging formulas.

Let us take a look closer to this problem. The first row of (3.68) is rigorous in terms of phenomenological approach: if we know exactly $P_x(k_y, \omega)$ and $M_z(k_y, \omega)$, then the wave vector can be found unambiguously (nevertheless, several solutions for the wave vector can exist). The question about origin dependence does not appear, because of both functions are assumed to be known without any approximations and irrespective to the multipole expansion. The question about origin dependence appears only when we start to apply the multipole expansion in order to elaborate an analytical form of the $P_x(k_y, \omega)$ and $M_z(k_y, \omega)$. The multipole expansion, nevertheless, is an approximation (especially if we restrict our consideration by quadrupole/magnetic dipole only) and gives the result with some limited accuracy: calculation of the multipole moments gives different results for the different origins, which is obvious. Hence, when we consider the problem of origin dependence, we always have to refer to the accuracy of the approximation we are working with.

The problem of the origin dependence roots to the problem of the field calculation from the known charge densities/dynamics at large enough distance—see [30]. The initial expression for the potential calculation is obviously origin invariant:

$$\varphi(\vec{R}, t) = \sum_{k=1}^{N_{\max}} \frac{q_k}{|\vec{R} - \vec{r}_k|} \tag{3.69}$$

Here q_k, $\vec{r}_k$, $\vec{R}$ are the charge, the radius vector of the charge, and the radius vector of the observation point respectively, summation is going over the charge cloud up to the last charge with number $N_{\max}$.

Assuming that $\vec{R} \gg \vec{r}_k$ we arrive to the standard multipole expansion, which is already not origin invariant:

$$\varphi(\vec{R}, t) = \frac{1}{|\vec{R}|} \sum_{k=1}^{N_{\max}} q_k + \sum_{k=1}^{N_{\max}} q_k \vec{r}_{k,\alpha} \left(\frac{\partial}{\partial \vec{r}_\alpha} \frac{1}{|\vec{R}|} \right) + \sum_{k=1}^{N_{\max}} q_k \vec{r}_{k,\alpha} \vec{r}_{k,\beta} \left(\frac{\partial^2}{\partial \vec{r}_\alpha \vec{r}_\beta} \frac{1}{|\vec{R}|} \right) \tag{3.70}$$

Here, as usual, the first term is proportional to the total charge, and the second and third are the dipole and quadrupole approximation respectively. It is clear, that the origin invariance is lost due to the truncation of the expansion by the quadrupole approximation; in case of further expansion the origin invariance has to be recovered. The truncation at the quadrupole term is stipulated by a reasonability of this approximation: the quadrupole (and respectively magnetic dipole) term is the minimum required order of expansion which describes the magnetic effect due to current distribution. From the other side, this approximation is analytically treatable, and taking into account the next orders (octupoles etc.) makes the problem solvable only numerically; in this case introduction of the multipole approximation does not make too much sense. One can show that approximation (3.70) gives the result with the accuracy described by two scales. The first one is the relation between typical system size a (size of the charge cloud) and the wavelength $\frac{\delta\varphi}{\varphi} \sim \left(\frac{a}{\lambda}\right)^2$; the second estimation is connected with the freedom of origin choice and can be estimated as $\frac{\delta\varphi}{\varphi} \sim \frac{a}{|R|}$; for the case of the MM, considered here, a is the typical size of the MA and $|R|$ is of the order of distance between the MAs. It has to be admitted, that the substitution of exact expression (3.69) by approximation (3.70) leads to the loss of the origin invariance, which in turn leads to the limited accuracy of (3.70). In the frame of the developed here model there is no rigorously justified methodology which could keep the origin invariance for the multipole (up to quadrupole/magnetic dipole order) approximation.

Moreover, it is easy to see, that the voluntarily choice of the origin leads to evidently non-physical results. Consider three systems of charges depicted in Fig. 3.3. For the symmetric charge distribution the origin has to be chosen in the middle due to evident symmetry considerations. From the other side, if the origin is

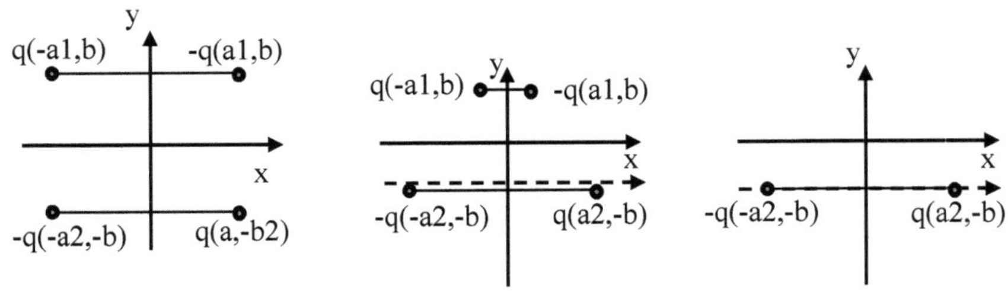

Fig. 3.3 Schematic representation of three charge distribution with the same position of the origin (solid axes) and variable x axe (dashed x axe). For the symmetric charge distribution (left picture) the origin has to appear in the middle due to the symmetry consideration; the same origin position (solid x axe) for a single dipole (right picture) leads to unphysical appearance of magnetic response, while the origin for the dipole has to lay between the charges (dashed x axe). This consideration gives a hint to the choice for the origin position for the intermediate charge system (middle picture)

kept in the same point and the top dipole becomes shorter, then in the extreme case of negligible top dipole the quadrupole moment has to be zero, which takes place only if the origin appears between the charges (dashed $\boldsymbol{x}$ axe in Fig. 3.3 right picture).

It becomes intuitively clear that in the middle case the origin has to be chosen lower towards to the bigger dipole (dashed $\boldsymbol{x}$ axe), which allows us to suggest new approach to the problem of origin for the multipole expansion—instead of require rather questionable condition of the origin independence (which does not appear as a natural consequence of the developed model) we apply an extra condition, which fixes the origin for any system based on its charge distribution.

The origin can be fixed by rather physically evident requirement of zero quadrupole and magnetic dipole moments in the case when only the symmetric oscillation mode is excited (or, better to say, the anti-symmetric mode is not excited). In fact, in case of symmetric oscillations there are no circular currents und hence there is no reason for the appearance of the magnetic response; from the other side, non-zero quadrupole and magnetic dipole moments would mean that this magnetic response takes place. It is enough to consider the quadrupole moment only (the magnetic moment can be considered in the same way). Let us assume (see Fig. 3.2) that the quadrupole moment is calculated in some coordinate system:

$$Q_{xy}\left(k_y,\,\omega\right) = \frac{1}{2}\eta\left[q_1 y_1 x_1\left(k_y,\,\omega\right) - q_2 y_2 x_2\left(k_y,\,\omega\right)\right] \tag{3.71}$$

here $q_{1,2}$, $y_{1,2}$, $x_{1,2}$ are the charges, the positions over y, and the deviation from an equilibrium position of the charges in the upper and lower wires, respectively. In another system of coordinates the new quadrupole moment is given by:

$$Q'_{xy}(k_y,\,\omega) = Q_{xy}(k_y,\,\omega) + \frac{1}{2}g_y d_x = \frac{1}{2}\eta\left[q_1 y_1 x_1(k_y,\,\omega) - q_2 y_2 x_2(k_y,\,\omega)\right] + \frac{1}{2}g_y\eta\left[q_1 x_1(k_y,\,\omega) + q_2 x_2(k_y,\,\omega)\right] \tag{3.72}$$

In case of the symmetric mode $x_1(k_y,\,\omega_{\mathrm{sym}}) = x_2(k_y,\,\omega_{\mathrm{sym}})$ and expression for the quadrupole moment becomes:

$$Q'_{xy}(k_y,\,\omega) = \frac{1}{2}\eta\left([q_1 y_1 - q_2 y_2] + g_y[q_1 + q_2]\right)x_2(k_y,\,\omega_{\mathrm{sym}}) = 0 \tag{3.73}$$

which leads to the fixed position for the origin of the coordinates, at which the quadrupole moment (and, consequently, the magnetic response) is zero:

$$g_y = \frac{q_2 y_2 - q_1 y_1}{q_1 + q_2} \tag{3.74}$$

Hence, at least for the considered here particular case, the origin of the system of coordinates can be unambiguously fixed. Note that for the case of identical wires $q_1 = q_2$ and the origin has to be placed in the middle point $y_1 = y_2$; for the absence of one of the wires (right picture in Fig. 3.3) transformation (3.74) shifts the origin to another one and places the origin such way that the quadrupole/magnetic dipole moments disappear.

In the presented here model the averaged current (3.56) and final expression for the wave vector (3.68) contain the same combination of $P_x(k_y,\omega)$ and $M_z(k_y,\omega)$, namely $\frac{\omega}{c}P_x(k_y,\omega) - k_y M_z(k_y,\omega)$. It means, that the origin independence of the wave vector is equivalent to the origin independence for the averaged current, which cannot be logically justified in case of multipole expansion; the only requirement which could be utilized is that the accuracy of the accepted approximation should not degrade at the choice of different origins.

The extra terms for the quadrupole and magnetic dipoles, appearing due to the origin shift [see (3.67)] cannot cancel each other due to the relation between them $Q_x(k_y,\omega) = \frac{i\omega}{c}M_z(k_y,\omega)$ and the fact that both $Q_x(k_y,\omega)$ and $M_z(k_y,\omega)$ appear in both averaged current and wave vector expression, as it was already mentioned, in the same combination $-\frac{ik_y\omega}{c}Q_x(k_y,\omega) - k_y M_z(k_y,\omega) = -2\frac{ik_y\omega}{c}Q_x(k_y,\omega)$.

One more question to be considered here is the relation of the origin dependence and the SFT. According to the general rules, the SFT (2.37) for the considered here double wires can be written as [compare with (3.67)]:

$$\begin{cases} P_x(k_y,\,\omega) = P'_x(k_y,\,\omega) - \dfrac{i\omega}{4\pi c}T_{1,x} + ik_y T_{2,z} \\ M_z(k_y,\,\omega) = M'_z(k_y,\,\omega) + \dfrac{ik_y}{4\pi}T_{1,z} + \dfrac{i\omega}{c}T_{2,z} \end{cases} \tag{3.75}$$

In order to compensate for the shift, both transformations have to be equivalent, namely:

$$\begin{cases} -\dfrac{ik_y}{2}g_y d_x(k_y,\,\omega) = -\dfrac{i\omega}{4\pi c}T_{1,x} + ik_y T_{2,z} \\ +\dfrac{i\omega}{2c}g_y d_x(k_y,\,\omega) = \dfrac{ik_y}{4\pi}T_{1,z} + \dfrac{i\omega}{c}T_{2,z} \end{cases} \tag{3.76}$$

The last requirement (3.76) can be satisfied by unlimited number of variants for $T_{1,x}$ and $T_{1,z}$. From the other side, (3.76) cannot be satisfied if both $T_{1,x}$ and $T_{1,z}$ are zeros (which is required if we consider the same physical realization), in this case:

$$\begin{cases} P_x(k_y,\,\omega) = P'_x(k_y,\,\omega) + ik_y T_{2,z} \\ M_z(k_y,\,\omega) = M'_z(k_y,\,\omega) + \dfrac{i\omega}{c}T_{2,z} \end{cases} \tag{3.77}$$

$$\begin{cases} -\dfrac{ik_y}{2}g_y d_x(k_y,\,\omega) = ik_y T_{2,z} \\ +\dfrac{i\omega}{2c}g_y d_x(k_y,\,\omega) = \dfrac{i\omega}{c}T_{2,z} \end{cases} \tag{3.78}$$

and there is no nontrivial solution for $T_{2,z}$. Transformations (3.78) do not change dispersion relation (3.59), and cannot compensate for the origin shift (3.67). Using transformation (3.76) it is possible either compensate for the origin shift in quadrupole or in magnetic dipole contribution, but not for both of them simultaneously.

An ability to compensate for the origin shift, given by nonzero solutions $T_{1,x}$ and $T_{1,z}$ of (3.77) means that final field solution has to be transformed as well according to (2.37). Basically, it does not bring any news for the origin dependence problem because of it does not fix the origin based on some extra conditions. As it was shown above (see Sect. 2.5), the nonzero $\vec{T}_1$ means solution for some other charge/current distributions; in the case of origin shift the charge/current are formally changed (due to the limited accuracy of the multipole approximation), and the Serdyukov-Fedorov transformation compensates for this change.

One can finally conclude, that:

1. In the frame of the developed here approach the origin dependence appears as a consequence of the limited accuracy of the multipole approximation, which is estimated to be $\sim \frac{a}{|R|}$, here a is the typical size of the MA and $|R|$ is of the order of distance between the MAs.
2. The origin independence cannot be automatically ensured in the frame of the developed here model.
3. In order to fix the problem, we have introduced another requirement, that fixes the origin for each MA and which clearly leads to reasonable limiting cases.
4. Consideration of this problem in the frame of the presented approach requires further investigation.

The conclusions of this part evidently contradict to the accepted in the literature ones that the multipole model can be constructed such way, that the requirement of the origin independency can be satisfied [23]. Note that in the developed in [23] approach initial expressions for the multipoles differ from ones, elaborated in [21] and which our consideration is based on. The main difference between our consideration and the theory in [23] is in following. In [23] the multipole expansion (i.e. expansion of the charge dynamics) is mixed with the expansion of the local field; in our model the field expansion and consequent consideration of the local/averaged fields and the consideration of their mutual relations is not necessary for the presented above conclusions. Nevertheless, the relation of the developed here model and one in [23] requires further investigation.

3.2.5 Toroidal/Anapole Metamaterials

In order to analyze an appearance of the magnetic response of the MAs, it is enough to extend the multipole expansion above the dipole level and take into account magnetic and quadrupole moments. It is commonly assumed, that the next, third level of expansion containing magnetic quadrupole, octupole, and toroidal moments can be neglected if the previous second one moments (magnetic dipole and quadrupole) has non zero contribution. It turns out, that this statement can be violated, and third order multipole contribution can interfere destructively with the first one—the toroidal moment (third level of multipole expansion) can compensate the dipole one (first level of multipole expansion), which results in non-radiative structure, provided other moments are zeros. In order to understand this phenomena, one must refer to the general multipole expansion expressions. For the sake of generality, assume a nontrivial current distribution $\vec{j}(\vec{r},t)$ producing an electromagnetic field $E(\vec{r})$ [31]:

$$\begin{aligned} \vec{j}(\vec{r},t) &= \sum_{l=0} \frac{(-1)^l}{l!} B^{(l)}_{i\ldots k} \partial_i \ldots \partial_k \delta(\vec{r}) \\ B^{(l)}_{i\ldots k} &= \int \vec{j}(\vec{r},t) \vec{r}_i \ldots \vec{r}_k \mathrm{d}^3\vec{r} \end{aligned} \tag{3.79}$$

Here $B^{(l)}_{i\ldots k}$ is a tensor of lth rank. From these tensors various Cartesian multipoles in form of sums of the unreducible tensors can be obtained. For example, $B^{(1)}_i \sim d^{(1)}_i$ determines the electric dipole moment $d^{(1)}_i$, $B^{(2)}_{ij} \sim Q^{(2)}_{ij} + \mu^{(1)}_i$ consists of electric quadrupole $Q^{(2)}_{ij}$ (symmetric) and magnetic dipole $\mu^{(1)}_i$ (anti-symmetric) moments, and $B^{(3)}_{ijk} \sim O^{(3)}_{ijk} + \mu^{(2)}_{ij} + T^{(1)}_i$ gives rise to electric octupole $Q^{(3)}_{ijk}$, magnetic quadrupole $\mu^{(2)}_{ij}$, and toroidal dipole moments $T^{(1)}_i$. On the other hand, the radiation properties can by described by using the total scattering cross-section in Canonical basis and can be written as a sum of intensities of spherical electric $a_{\mathrm{E}}(l,m)$ and magnetic $a_{\mathrm{E}}(l,m)$ scattering amplitudes [32]:

$$C_{\mathrm{sca}} = \frac{\pi}{k^2}\sum_{l=1}^{\infty}\sum_{m=-l}^{l}(2l+1)\left[|a_{\mathrm{E}}(l,m)|^2 + |a_M(l,m)|^2\right] \tag{3.80}$$

coefficients $a_{\mathrm{E}}(l,m)$ and $a_M(l,m)$ which can be unambiguously determined by multipole coefficients $B^{(l)}_{i\ldots k}$. We are focusing on the situation when a spherical electric dipole mode is the dominant one. In this case the total scattering cross-section is determined solely by the electric dipole scattering coefficient $C_{\mathrm{sca}} \propto |a_{\mathrm{E}}(1,\pm 1)|^2$. In this case, for the first order expansion [32] the relation between Cartesian and spherical multipoles can be written as:

$$\begin{aligned} a_{\mathrm{E}}(1,\pm 1) = {} & C_1\left[\pm B^{(1)}_x + iB^{(1)}_y\right] + 7C_3\left[\pm B^{(3)}_{xxx} + 2B^{(3)}_{xyy} + 2B^{(3)}_{xzz} - B^{(3)}_{yyx} - B^{(3)}_{zzx}\right] \\ & - i\left[B^{(3)}_{yyy} + 2B^{(3)}_{yxx} + 2B^{(3)}_{yzz} - B^{(3)}_{xxy} - B^{(3)}_{zzy}\right] \end{aligned} \tag{3.81}$$

Thus, the total scattering can vanish if the spherical electric dipole scattering coefficient becomes zero, $a_{\mathrm{E}}(1,\pm 1) \approx 0$, provided all higher order scattering amplitudes are also close to zero. In order to zero spherical electric dipole, $a_{\mathrm{E}}(1,\pm 1) = 0$, the first order (electric dipole) coefficients $B^{(1)}_i$ has to be compensated by the third order coefficients $B^{(3)}_{ijk}$, which contains the toroidal dipole moments $T^{(1)}_i$. This simple consideration creates basis for understanding of physics of the anapole mode, namely mutual compensation of $B^{(1)}_i$ and $B^{(3)}_{ijk}$. The other two moments of the third order—octupole $O^{(3)}_{ijk}$ and magnetic quadrupole $\mu^{(2)}_{ij}$—are assumed to be negligible, which in fact takes place for the toroidal structure.

Typically, the toroidal moment is ignored as it appears in the third order of expansion, which is expected to be negligible. The results in Fig. 3.4 indicate that it is necessary to introduce toroidal moments for optically large particles to accurately describe the total scattered field, in particular for a dielectric nanodisk with high refractive index such as the one of silicon. The total scattering cancellation is possible due to the fact that the radiation patterns of the electric and toroidal dipoles are equivalent.

Anapole has been introduced in physics of elementary particles [33]. The electrodynamic analog of a stationary anapole is well known toroid with a constant poloidal surface current, which is also associated with a toroidal dipole moment. It generates no field outside, but possible nonzero potential, which might lead to a violation of the reciprocity theorem and Aharonov-Bohm like phenomena [34, 35].

In the dynamic case, the oscillating toroidal dipole moment produces nonzero electromagnetic radiation with pattern fully repeating one from that of the electric dipole moment, but scaled by a factor of ω^2. For oscillating surface current, the

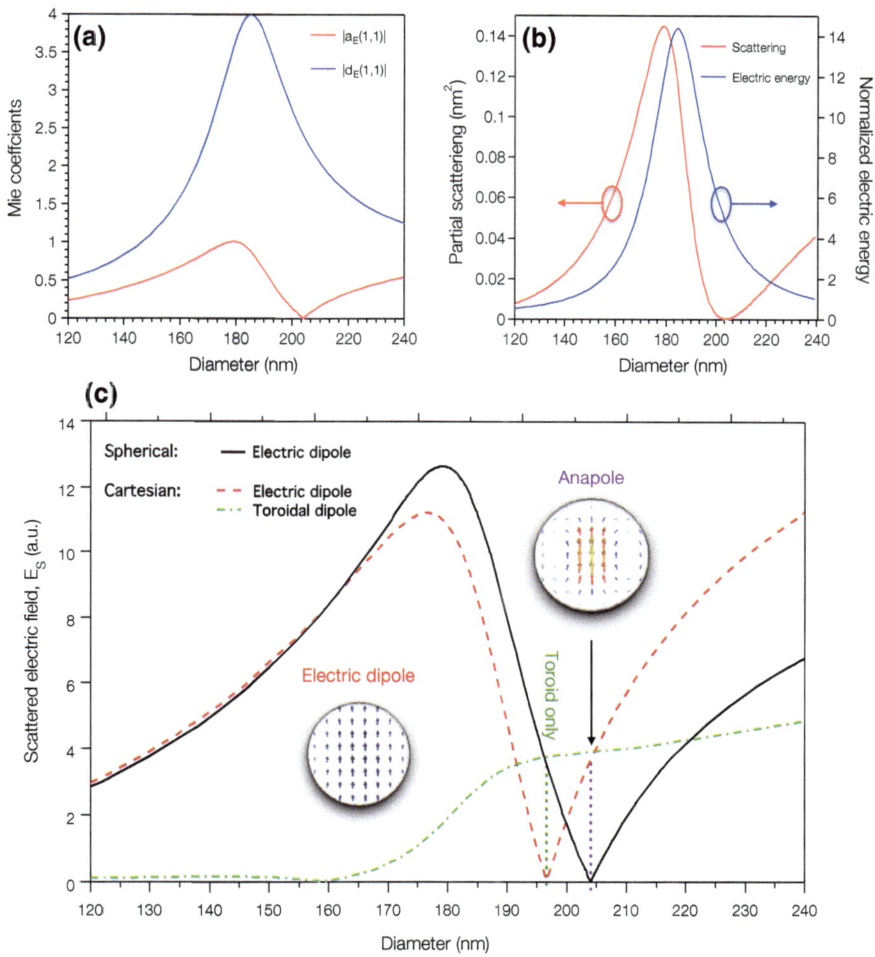

Fig. 3.4 Miroshnichenko [2, Fig. 2]: Decomposition of the contribution to the far-field scattering with electric dipole symmetry in terms of spherical and Cartesian multipoles. We consider scattering by a dielectric spherical particle inside as a function of diameter for refractive index $n = 4$ and wavelength 550 nm: **a** Scattering $|\boldsymbol{a}_{\boldsymbol{E}}(1,1)|$ and internal $|\boldsymbol{d}_{\boldsymbol{E}}(1,1)|$ Mie coefficients; **b** partial scattering cross-section and energy density of the electric dipole; **c** calculated spherical electric dipole $|\mathbf{P}_{\mathrm{sph}}|$ (black), Cartesian electric $|\mathbf{P}_{\mathrm{Car}}|$ (red) and toroidal $|\mathbf{T}_{\mathrm{Car}}|$ (green) dipole moments contributions to the partial scattering. These figures demonstrate that for small particles both contributions of the spherical and Cartesian electric dipoles are identical and the toroidal moment is negligible. For larger sizes, the contribution of the toroidal dipole moments to the total scattered field has to be taken into account. The anapole excitation is associated with the vanishing of the spherical electric dipole $\mathbf{P}_{\mathrm{sph}} = 0$, when the Cartesian electric and toroidal dipoles cancel each other

radiationless properties can be kept by adding another dipole oscillating in anti-phase with the toroid, resulting in complete destructive interference of their radiation due to similar far-field scattering patterns [36]. This radiationless electric and toroid dipoles nontrivial current configuration has also been named “anapole” [36], which is Greek for ‘without poles’. Nevertheless, this compensation is not complete: the compensated toroidal dipole moment is a part of the third order expansion, and all higher order expansions remain radiative.

The anapole concept has attracted considerable attention in metamaterial community as a possible realization of these radiationless objects [37]. The toroidal moment itself and the respective effects (including toroidal metamaterials [38]) have been deeply investigated theoretically [35, 39]. Several experimental verifications in microwave [40] and optical [41] domains for toroidal moments confirm theoretical conclusions. Anapole mode in optical domain using a simple silicon structure has been experimentally demonstrated for the first time in [1].

Anapole being a rediationless structure (in the frame of all mentioned above limitations) is also obviously (with the same approximations) is not sensitive to the external radiation. An anapole-like qubit design that is naturally insensitive to low-frequency noise and is well protected from other ambient noise sources, and therefore could be a good candidate for a superconducting qubit, has been proposed and theoretically investigated in [42].

It is worth to remind again about special form of material equations, introduced in Chap. 2, namely Toroidal (“T”).

It is interesting to compare the microscopically introduced toroidal moment with the phenomenologically elaborated equations for “T” form—see Chap. 2, (2.24)–(2.26). Phenomenologically elaborated $\langle\vec{j}\rangle = -i\omega\vec{P} - i\omega\,\mathrm{rot}[\vec{F}_1] + \mathrm{rot}[\vec{F}_2]$ (see (2.12) in Chap. 2) allows two different allocation of term $i\omega\,\mathrm{rot}[\vec{F}_1]$, namely it can be packed in polarizability $\langle\vec{j}\rangle = -i\omega(\vec{P} + \mathrm{rot}[\vec{F}_1]) + \mathrm{rot}[\vec{F}_2]$ or in magnetization $\langle\vec{j}\rangle = -i\omega\vec{P} + \mathrm{rot}[-i\omega\vec{F}_1 + \vec{F}_2]$. Final solution of MEs of course does not depend on this relocation. This possibility of the different allocations corresponds to SFT between the different representations [see (2.48) and (2.51)].

Let us consider the case of material consisting of toroidal structures only (it means, that the MAs have only toroidal moments and no other ones). The current in the frames of the developed phenomenological scheme is given by (taking into account that $\vec{F}_2$ is responsible for a non-toroidal magnetization and can be set to zero):

$$\langle\vec{j}\rangle = \mathrm{rot}\left[-i\omega\vec{F}_1\right] \tag{3.82}$$

From the other side, referring to [31], the current is written through the toroidal moment (see (2.12) in [31]):

$$\langle\vec{j}\rangle = c\,\mathrm{rot}\left[\mathrm{rot}\left[\vec{T}\delta(r)\right]\right] \tag{3.83}$$

Alternatively, it has been shown [43], that the averaged current can be expressed in form (see (2.15) in [43] without non toroidal contributions):

$$\langle\vec{j}\rangle_{\text{total}} = \text{rot}\left[-i\omega\vec{T^e} + c\,\text{rot}[\vec{T^m}]\right] \tag{3.84}$$

here $\vec{T^e}$ and $\vec{T^m}$ are the contributions appearing due to the presence of structures with the toroidal moments in the frame of alternative formulation of MEs with magnetic charges [44]. From the comparison of (3.82) with (3.83) and (3.84) it is seen that the introduced phenomenologically function $\vec{F}_1$ is fixed by the presence of the toroidal moments. From the other side, the exact functional form depends on the ME initial formulation. In the case of traditional MEs it becomes:

$$\vec{F}_1 = \frac{ic}{\omega}\,\text{rot}\left[\vec{T}\delta(r)\right] \tag{3.85}$$

while in the frame of alternative representation of [44] it reads:

$$\vec{F}_1 = \vec{T^e} \tag{3.86}$$

The toroidal moment determines function $\vec{F}_1$ introduced phenomenologically.

3.3 Introducing of Effective Parameters

3.3.1 Elaboration of Effective Parameters

The introduced above scheme allows us to determine unambiguously the terms "effective parameter" and some other notations, which appeared in the literature in the contents of the homogenization procedure. Most generally, the effective parameters are functions which connect introduced at the first step of the homogenization procedure $\langle\rho\rangle$ and $\langle\vec{j}\rangle$ with the averaged electric and magnetic fields $\vec{E}$ and $\vec{B}$. Usually the relations between $\langle\rho\rangle$ and $\langle\vec{j}\rangle$ and $\vec{E}$ and $\vec{B}$ are not used; instead the relations between $\vec{P}$ and $\vec{M}$ (in any representations) and $\vec{E}$ and $\vec{B}$ are under consideration. It has been shown above, that in the case of the "L&L" representation it is possible to introduce the effective parameter (effective permittivity) $\varepsilon_{\text{LL},\alpha\beta}(\vec{k},\omega) = \delta_{\alpha\beta} + 4\pi R_{\text{LL},\alpha\beta}(\vec{k},\omega)$ (3.5), which connects $\vec{D}$ and $\vec{E}$. The effective permittivity is a function of material properties and is spatially dispersive (depends on wave vector); in other words, is nonlocal.

The non-locality (spatial dispersion) caused a discussion in literature whether it makes sense to call $\varepsilon_{\text{LL},\alpha\beta}(\vec{k},\omega)$ an effective *material* parameter. Following [45], the effective permittivity $\varepsilon_{\text{LL},\alpha\beta}(\vec{k},\omega)$ is an effective parameter, but is not an effective *material* parameter because it includes information not only about material, but also depends on the wave vector, in other words contains information about external electromagnetic fields. Similar considerations have been published in [46] where also two different notations—Characteristic Material Parameters (CMP) and Effective Material Parameters (EMP)—have been suggested.

Here the question about necessity of introducing of new paradigm for the effective parameters is considered in view of the presented consideration based on the types of modes excited in MAs.

In order to create a consistent terminological basis, it has to be mentioned, that the introduced effective parameters in case of strong spatial dispersion (for example, the permittivity in case of the "L&L" representation) solves the problem of homogenization, and to this extend does not require any other comments or discussions. If the functional form of $\varepsilon_{\mathrm{LL},\alpha\beta}(\vec{k},\omega)$ is found, then this function contains information about material properties (for example, eigenmodes of the electron oscillations in the metal nanoresonators), and the excitation conditions (wave-vector dependence). In general, both properties—eigenmodes and excitation conditions—are not separable.

The physical meaning of the effective constants, introduced in the homogenization procedure, should be properly appreciated. The effective constants appear in the model (phenomenological or multipole) as the functions describing charge dynamics. It is necessary to recall that the microscopic MEs consist of not only field equations, but also include equations for the charge dynamics—see the last equation in (2.1). This equation (and all information about charge dynamics) is lost in the homogenization procedure; in other words, this equation is not averaged in the homogenization. It is clear, that the information contained in this equation is also lost and has to be somehow compensated, which is the basic reason for the multiple forms of various homogenization models (it has to be noted, that this equation has been basically kept in [20], which is the reason for pretty straightforward elaboration of the model). Instead of the systematic consideration of the charge dynamics, the phenomenological approach just gives some frames for the homogenization model, resulting in some expressions connecting averaged charge density and current and averaged fields. The coefficients between them (effective parameters) should contain information about charge dynamics caused by the averaged fields.

At this stage it has to be clearly realized, that these coefficients contain information about *charge dynamics* in a particular excitation situation, and *not only about the material properties*. The difference becomes clear if we again recall the symmetric and anti-symmetric modes in, for example, double wires. The existence of the modes itself is the characteristics of the system and does not depend on the external fields. But the type of the excited mode can depend (but not necessarily!) on the external field structure—see Table 3.1 where it is summarized that the anti-symmetric mode can be excited due to both asymmetry of the structure and inhomogeneity of the field (asymmetry of the field distribution). In general, the modes which are excited and which determine the material response are functions of both material properties and properties of the external fields. It is possible to create MAs with modes determined by the properties of the MAs only—asymmetric structures (for example, double wires with different lengths closely placed to each other—see [47]) with the sizes much smaller than the wavelength. In this case the coefficients $\varepsilon^{(0)}_{\mathrm{C},\alpha\beta}(\omega)$, $\varepsilon^{(1)}_{\mathrm{C},\alpha\beta\gamma}(\omega)$ and $\phi^{(0)}_{\mathrm{C},\alpha\beta}(\omega)$ form material equations, which nevertheless describe spatial dispersion of the first order. This clearly shows that the presence/absence of spatial dispersion in material response is not an indicator of the physical properties of the media, but rather a manifestation of interactions of charges with the fields and the charge dynamics. If all five coefficients in Table 3.1 are not zero (the anti-symmetric mode is excited due to both the structure and field asymmetries), the situation is basically the same, but the material equations contain spatial dispersion of the second order as well.

The presented above consideration shows that the spatial dispersion itself can hardly be considered as a criterion for introduction of some new notations. It would be more physically justified to consider the effective parameters based on the types of the modes in particular structures under particular excitation conditions. For example, following the results summarized in Table 3.1, one can subdivide the magnetic response in MAs (and the respective MM) into two categories: the first one where the anti-symmetric modes appear as a result of the asymmetry of the structure itself, and the second one, where the anti-symmetric modes appear as a result of the asymmetric excitation. For both types the coefficients in Table 3.1 can be considered as material effective parameters, because they all depend on the material properties and do not depend on the wave vector.

The MM of the first and second types can be rather easily distinguished experimentally.

If the retrieval procedure based on (3.48) will be applied, it will be clear which type of structure has been tested: for the first type the coefficients $\varepsilon^{(2)}_{\mathrm{C},\alpha\beta\gamma\delta}(\omega)$, $\phi^{(1)}_{\mathrm{C},\alpha\beta\gamma}(\omega)$ are expected to be close to zero, and for the second type the coefficients $\varepsilon^{(1)}_{\mathrm{C},\alpha\beta\gamma}(\omega)$, $\phi^{(0)}_{\mathrm{C},\alpha\beta}(\omega)$ should be about zero. The information about relative contribution of the coefficients would reveal information about basic physical processes in the MAs and undoubtedly enhance our insight about electrodynamics of the MM. Moreover, this method allows us to extract *microscopic* information based on *macroscopic* measurements of the effective response of the MM, which makes this approach an extremely useful tool in the lab.

3.3.2 Impossibility of Unambiguous Effective Parameters Determination for Bulk Materials

It has to be realized that the problem of determination of Effective Parameters for bulk materials does not make much practical sense. From the theoretical point of view, the only problem which can be stated and (potentially) solved for the bulk materials is finding the dispersion relation, which does not assume even introduction of the effective parameters. The effective parameters can be introduced as some coefficients between polarizability and magnetization and the electric and magnetic fields, but it can NOT be done unambiguously—see again SFT (2.37), which give birth of unlimited number of different effective parameters, each set of them nevertheless satisfy MEs.

In fact, most authors do not look for the step by step elaboration of the model, describing the averaged characteristics, but just postulate the relations between polarizability, magnetization and the electric and magnetic fields and then find, for example, dispersion relation—see Fig. 3.4, which illustrates different ways of possible elaboration of the homogenization models. It is clear, that the chain "Microscopic Maxwell equations" to "Bulk material—introduction of ε, μ", shown by the red vertical arrow, is not a natural way of the elaboration of the homogenization model, but rather an attempt to avoid the detailed and consequent consideration.

A functional form of the material equations has to be elaborated based on the phenomenological and/or multipole approaches, which has been done for the bulk materials in this work. It is argued, that the presented here consideration covers all possible forms of the material equations in the frame of the commonly used paradigm of spatial dispersion.

3.3.3 *Effective Retrieved Parameters and Their Relation to the Effective Parameters*

The problem of the effective parameters retrieval is to some extend out of the scope of the presented consideration. The reason is that in the optical domain due to the high level of losses the vast majority of tests have been performed not with bulk MMs (which is the main object of the presented consideration) but rather with meta-surfaces [48, 49], where only several layers of MAs are stacked together. In this case the presented above consideration is not directly applicable. For example, expressions (3.2), (3.3), and (3.10) could not form the basis for further consideration (nevertheless, all the conclusions made in Sect. 2.4 about different representations and their mutual transformations remain valid). This problem of fundamental inapplicability of the "bulk expressions" to meta-surfaces are frequently mentioned, but very rarely properly analyzed; anyway, all the presently accepted approaches for the effective parameter retrieval are reduced to the "bulk expressions" in one or another form—see, for example, [50]. The theoretical background for the retrieval procedures is based on some hypothesis about relations between polarizability and magnetization on the averaged field. In its original form the retrieval procedure assumed very simple forms of the relations for polarizability and magnetization [51–54] which did not include bianisotropy and corresponded in terms of (3.38) to:

$$\begin{pmatrix} \varepsilon^{(0)}_{\mathrm{C},\alpha\beta}(\omega) & 0 & 0 \\ 0 & \phi^{(1)}_{\mathrm{C},\alpha\beta\gamma\delta}(\omega) & 0 \end{pmatrix} \tag{3.87}$$

Bianisotropy has been included into consideration in [55] and in the retrieval algorithm of [56], which corresponds in terms of the developed in here approach to the form:

$$\begin{pmatrix} \varepsilon^{(0)}_{\mathrm{C},\alpha\beta}(\omega) & \varepsilon^{(1)}_{\mathrm{C},\alpha\beta\gamma\delta}(\omega) & 0 \\ \phi^{(0)}_{\mathrm{C},\alpha\beta}(\omega) & \phi^{(1)}_{\mathrm{C},\alpha\beta\gamma\delta}(\omega) & 0 \end{pmatrix} \tag{3.88}$$

It is worth noticing again, that the developed in this publication approach gives a different representation for the retrieval procedure (3.47) which includes not only bianisotropy, but also excludes the influence of the magnetic field on the effective parameters.

The retrieved from the reflection and transmission measurements parameters (whatever representation it is based on) are the so called "effective refractive index" and the "effective surface impedance". The main goal of the theoretical models is to establish a reasonable correspondence between the retrieved from the measured data effective refractive index and impedance and parameters of the developed model. Often it is assumed that the effective parameters of the meta-surfaces can be described by the same type of relations as for bulk MMs—for example, coefficients in (3.47) or (3.49), or the respective coefficients in some other representation.

3.4 Conclusion

The developed in this chapter scheme is summarized in Fig. 3.4. Developing of the averaging procedure results in roughly three levels of MEs: the first level is the microscopic MEs (the starting level), the second level are the different representations for macroscopic MEs, where polarizability and magnetization $\vec{P}$ and $\vec{M}$ can be introduced, and the third level where the homogenization problem results in the dispersion relation for bulk materials and in a surface impedance and an effective refractive index for layered materials (Fig. 3.5).

Possible transitions between the first and second levels are represented by the phenomenological route and the multipole model; it is worth reminding that the multipole model arrives to the same "C" form of MEs representation.

The effective parameters in general turn out to be functions of the wave vector and hence cannot be called the effective material parameters. In order to separate the effect of spatial dispersion, the effective parameters can be expanded into series over the wave vector, and the appearing coefficients can be accepted as effective material parameters.

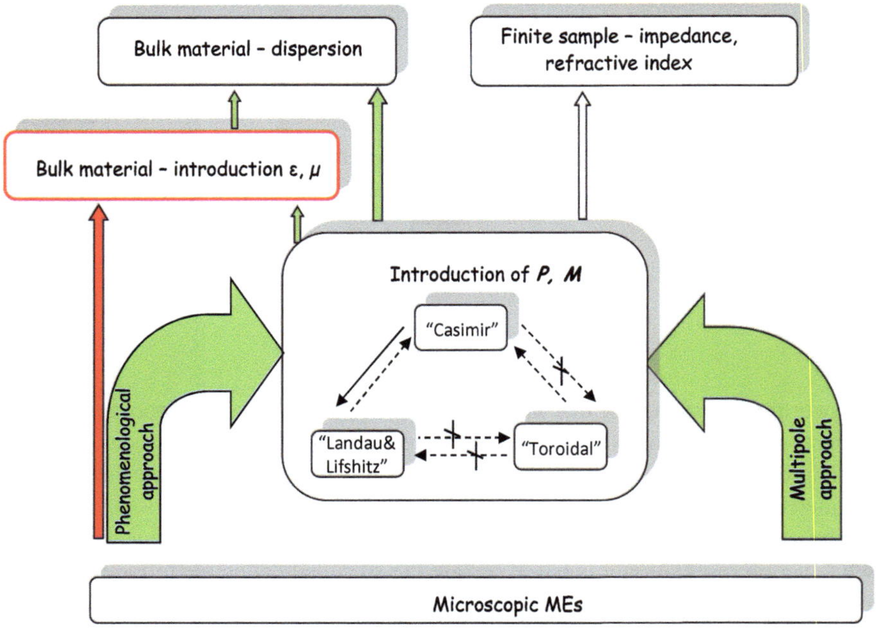

Fig. 3.5 Schematic representation of different levels of the averaging procedure. "Red way"—it's NOT a homogenization, but rather a way to avoid elaboration of the homogenization model

A new form of material equations has been suggested in the frame of the phenomenological approach in the case of weak (up to the second order) dispersion. Comparison between the phenomenological and multipole approaches based on the symmetry consideration allowed us to prescribe a clear physical meaning to all introduced material parameters. The use of the retrieval algorithm based on the suggested form of the material equations would allow us to determine the types of the internal charge mode dynamics, on which the response of MMs is based.

References

1. A. Chipouline, C. Simovski, S. Tretyakov, Basics of averaging of the Maxwell equations for bulk materials. Metamaterials **6**, 77 (2012)
2. A. Miroshnichenko, A. Evlyukhin, Y.F. Yu, R. Bakker, A. Chipouline, A. Kuznetsov, B. Luk'yanchuk, B. Chichkov, Y. Kivshar, Observation of an anapole with dielectric nanoparticles. Nat. Commun. **6**, 8069 (2015)
3. A. Serdyukov, I. Semchenko, S. Tretyakov, A. Sihvola, *Electromagnetics of Bi-Anisotropic Materials—Theory and Applications* (Gordon and Breach, Amsterdam, 2001)
4. S. Pekar, *Crystal Optics and Additional Light Waves* (Naukova Dumka, Kiev, 1982)
5. V. Agranovich, V. Ginzburg, Kristallooptika s Uchetom Prostranstvennoi Dispersii i Teoriya Eksitonov (Crystal Optics with Spatial Dispersion, and Excitons) (Nauka, Moscow, 1965) [Translated into English (Springer, Berlin, 1984)]

6. J.D. Jackson, *Classical Electrodynamics*, 3rd edn. (Wiley, New York, 1999)
7. L.D. Landau, E.L. Lifshitz, *Electrodynamics of Continuous Media*, 2nd edn. (Pergamon Press, New York, 1960) (Chapter IX)
8. A. Vinogradov, *Electrodynamics of Compound Media* (Scientific and Educational Literature Publisher, Russian Federation, 2001). ISBN 5-8360-0283-5 (in Russian)
9. E. Turov, *Material Equations of Electrodynamics* (Nauka, 1983) (in Russian)
10. A. Vinogradov, A. Aivazyan, Scaling theory of homogenization of the Maxwell equations. Phys. Rev. E **60**, 987 (1999)
11. G. Bosi, F. Girouard, V. Truong, J. Appl. Phys. **53**, 648 (1982)
12. E. Graham, R. Raab, JOSA **6**, 1239 (1996)
13. E. Raab, J. Cloete, JEWA **8**, 1073 (1994)
14. C. Simovski, S. Tretyakov, On effective electromagnetic parameters of artificial nanostructured magnetic materials. Photonics Nanostruct. Fundam. Appl. **8**, 254 (2010)
15. C. Simovski, Weak spatial dispersion in composite media Polytechnika (St. Petersburg, 2003) (in Russian)
16. S. Tretyakov, A. Sihvola, A. Sochava, C. Simovski, Magnetoelectric interactions in bi-anisotropic media. J. Electromagn. Wave Appl. **12**, 481 (1998)
17. C. Kriegler, M. Rill, S. Linden, M. Wegener, Bianisotropic photonic metamaterials. IEEE J. Sel. Top. Quantum Electron. **16**, 367–375 (2010)
18. I. Lindell, A. Sihvola, S. Tretyakov, A. Vitanen, *Electromagnetic Waves in Chiral and Bi-Isotropic Media* (Artech House, Boston; London, 1994)
19. B. Tellegen, The gyrator: a newelectric network element. Philips Res. Rep. **3**, 81 (1948)
20. J. Petschulat, C. Menzel, A. Chipouline, C. Rockstuhl, A. Tünnermann, F. Lederer, T. Pertsch, Multipole approach to metamaterials. Phys. Rev. B **78**, 043811 (2008)
21. P. Mazur, B. Nijboer, On the statistical mechanics of matter in an electromagnetic field. I. Physica **XIX**, 971 (1953)
22. G. Rusakoff, A derivation of the macroscopic Maxwell equations. Am. J. Phys. **38**(10), 1188 (1970)
23. R. Raab, O. De Lange, *Multipole Theory in Electromagnetism* (Clarendon, Oxford, 2005)
24. T.W. Ebbesen, H.J. Lezec, H.F. Ghaemi, T. Thio, P.A. Wolff, Nature (London) **391**, 667 (1998)
25. E. Pshenay-Severin, U. Hübner, C. Menzel, C. Helgert, A. Chipouline, C. Rockstuhl, A. Tünnermann, F. Lederer, T. Pertsch, Double-element metamaterial with negative index at near-infrared wavelengths. Opt. Lett. **34**, 1678 (2009)
26. A. Chipouline, J. Petschulat, A. Tuennermann, T. Pertsch, C. Menzel, C. Rockstuhl, F. Lederer, Multipole approach in electrodynamics of metamaterials. Appl. Phys. A **103**, 899–904 (2011)
27. J. Petschulat, A. Chipouline, A. Tünnermann, T. Pertsch, C. Menzel, C. Rockstuhl, T. Paul, F. Lederer, Simple and versatile analytical approach for planar metamaterials. Phys. Rev. B **82**, 075102 (2010)
28. E. Pshenay-Severin, A. Chipouline, J. Petschulat, U. Huebner, A. Tuennermann, T. Pertsch, Optical properties of metamaterials based on asymmetric double-wire structures. Opt. Express **19**, 6269 (2011)
29. J. Petschulat, A. Chipouline, A. Tüunnermann, T. Pertsch, C. Menzel, C. Rockstuhl, F. Lederer, Phys. Rev. A **80**, 063828 (2009)
30. L.D. Landau, E.L. Lifshitz, *Field Theory*, 2nd edn. (Pergamon Press, New York, 1960)
31. V. Dubovik, V. Tugushev, Toroid moments in electrodynamics and solid-state physics. Phys. Rep. **187**(4), 145 (1990)
32. P. Grahn, A. Shevchenko, M. Kaivola, Electromagnetic multipole theory for optical nanomaterials. New J. Phys. **14**, 093033 (2012)
33. I.B. Zeldovich, Electromagnetic interaction with parity violation. JETP **33**, 1531 (1957)
34. G. Afanasiev, Simplest source of electromagnetic fields as a tool for testing the reciprocity-like theorems. J. Phys. D: Appl. Phys. **34**, 539 (2001)

35. G. Afanasiev, Vector solutions of the Laplace equation and the influence of helicity on Aharonov-Bohm scattering. J. Phys. A: Math. Gen. **27**, 2143 (1994)
36. G.N. Afanasiev, Y.P. Stepanovsky, J. Phys. A: Math. Gen. **8**, 4565 (1995)
37. V. Dubovik, L. Tosunyan, V. Tugushev, Axial toroidal moments in electrodynamics and solid-state physics. Zh. Eksp. Teor. Fiz. **90**, 590 (1986)
38. T. Kaelberer, V.A. Fedotov, N. Papasimakis, D.P. Tsai, N.I. Zheludev, Science **330**, 1510 (2010)
39. K. Marinov, A.D. Boardman, V.A. Fedotov, N. Zheludev, Toroidal metamaterial. New J. Phys. **9**, 324 (2007)
40. V.A. Fedotov, A. Rogacheva, V. Savinov, D. Tsai, N.I. Zheludev, Resonant transparency and non-trivial non-radiating excitations in toroidal metamaterials. Sci. Rep. **3**, 2967 (2013)
41. B. Ögüt, N. Talebi, R. Vogelgesang, W. Sigle, P.A. van Aken, Toroidal plasmonic eigenmodes in oligomer nanocavities for the visible. Nano Lett. **12**, 5239 (2012)
42. A.M. Zagoskin, A. Chipouline, E. Il'ichev, J.R. Johansson, F. Nori, Toroidal qubits: naturally decoupled quiet artificial atoms. Sci. Rep. **5**, 16934 (2015). https://doi.org/10.1038/srep16934
43. V. Dubovik, M.A. Martsenyuk, B. Saha, Material equations for electromagnetism with toroidal polarizations. Phys. Rev. E **61**(6), 7087 (2000)
44. D. Singleton, Am. J. Phys. **64**, 452 (1996)
45. C. Menzel, C. Rockstuhl, T. Paul, F. Lederer, T. Pertsch, Retrieving effective parameters for metamaterials at oblique incidence. Phys. Rev. B **77**, 195328 (2008)
46. C. Simovski, On electromagnetic characterization and homogenization of nanostructured metamaterials. J. Opt. **13**, 013001 (2011)
47. C. Menzel, R. Alaee, E. Pshenay-Severin, C. Helgert, A. Chipouline, C. Rockstuhl, T. Pertsch, F. Lederer, Genuine effectively biaxial left-handed metamaterials due to extreme coupling. Opt. Lett. **37**, 596 (2012)
48. A. Vinogradov, A. Ignatov, A. Merzlikin, S. Tretyakov, C. Simovski, Additional effective medium parameters for composite materials (excess surface current). Opt. Express **19**, 6699 (2011)
49. D. Morits, C. Simovski, Electromagnetic characterization of planar and bulk metamaterials: a theoretical study. Phys. Rev. B **82**, 165114 (2010)
50. M. Albooyeh, D. Morits, C. Simovski, Electromagnetic characterization of substrated metasurfaces. Metamaterials **5**, 178 (2011)
51. W.B. Weir, Proc. IEEE **62**, 33 (1974)
52. A.M. Nicholson, G.F. Ross, IEEE Trans. Instrum. Meas. **IM-19**, 377 (1970)
53. D. Smith, S. Schultz, P. Markos, C. Soukoulis, Determination of effective permittivity and permeability of metamaterials from reflection and transmission coefficients. Phys. Rev. B **65**, 195104 (2002)
54. D.R. Smith, D.C. Vier, T. Koschny, C.M. Soukoulis, Electromagnetic parameter retrieval from inhomogeneous metamaterials. Phys. Rev. E **71**, 036617 (2005)
55. M. Silveirinha, Metamaterial homogenization approach with application to the characterization of microstructured composites with negative parameters. Phys. Rev. B **75**, 115104 (2007)
56. X. Chen, B.-I. Wu, J. Kong, T. Grzegorczyk, Retrieval of the effective constitutive parameters of bianisotropic metamaterials. Phys. Rev. E **71**, 046610 (2005)

Chapter 4
Multipole Approach for Homogenization of Metamaterials: "Classical" Metamaterials

4.1 Charge Dynamics in Isolated Plasmonic Metaatoms: Antisymmetric Modes as a Source for Magnetization

Now the multipole expansion developed in Chap. 3 is applied to describe the widely used double-wire geometry [1]. In what follows we assume again the geometry shown in Fig. 3.2 with the electric field $\vec{E}_x$ polarized along the long axis of the wires and propagation along the y axis $(0, \vec{k}_y, 0)$. Our goal is to elaborate a dispersion relation as a function of the particular parameters of the MAs using the general expressions obtained in chapter (3.61). In order to find the relations for the dipole, quadrupole, and magnetic dipole moments (3.57) it is necessary to express charge dynamics in the MAs as the functions of the averaged fields. As it was mentioned above, the microscopic interaction between charges and the electromagnetic wave is determined by the interaction with the electric field. The interaction with the magnetic field becomes significant only for relativistic velocities or extremely large magnetic fields [2] which are irrelevant in the present study. A rigorous description of the charge dynamics in terms of eigenmodes can be rather straightforwardly done for the nanospheres [3] and can be easily adopted for the elliptical particles of various eccentricities [4], but the physical picture qualitatively remains unchanged—the dynamics is described to the first approximation by the harmonic oscillation equation(s) with the eigenfrequencies, damping rates, and mutual coupling coefficients. The coefficients depend on particular geometry of the MAs and can be found analytically or semi-analytically with some accuracy, which in most cases does not allow direct comparison with the experiments and a fitting with the experimental data is required in one or another form anyway. For the developed in the frame of our approach model we have adopted another way. We assume that the MA consist of several "harmonic oscillators" coupled to each other —for example, for the double wires in Fig. 3.2 the system dynamics is supposed to be modeled by two coupled harmonic oscillators. The oscillator parameters (the

A. Chipouline and F. Küppers, *Optical Metamaterials: Qualitative Models*, Springer Series in Optical Sciences 211, https://doi.org/10.1007/978-3-319-77520-3_4

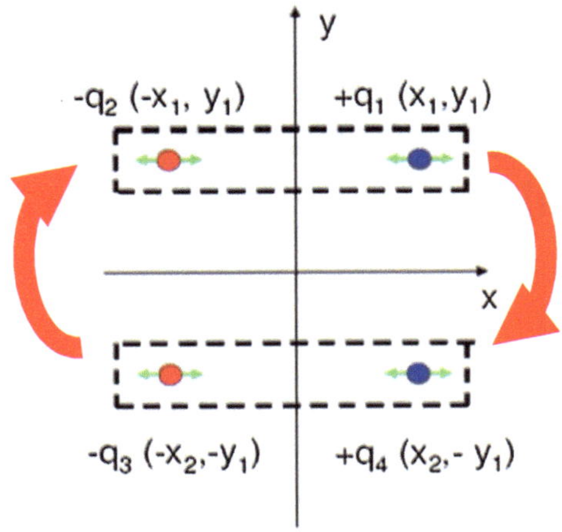

Fig. 4.1 Double-wire MM geometry and corresponding suitable charge distributions that support electric dipole, electric quadrupole, and magnetic dipole moments. The dynamics including interactions between the top and the bottom wires is described by a coupled harmonic oscillator model which is indicated by the red (gray) arrows [6]

eigenmodes, damping constants, and coupling coefficients) are introduced phenomenological and are supposed to be found from the fitting with the experimental data or with the data from rigorous numerical simulation. This approach allows us to keep basic physics and at the same time do not overload the model with unreasonably complicated math. It is worth noticing that the same approach is adopted for the natural materials, for example for the analytical description of the frequency dependence of the dielectric constant and does not seem to be a significant drawback of the model. Under these restrictions the damped and driven harmonic oscillator equation, describing the dynamics of the charge qk, takes the following form [5]:

$$\begin{aligned} &\frac{\partial^2 \vec{r}_k(t)}{\partial t^2} + \gamma_k \frac{\partial \vec{r}_k(t)}{\partial t} + \omega_k^2 \vec{r}_k(t) + \sigma_{ki} \vec{r}_i(t) \\ &= \frac{q_k}{m_k} \left[\vec{E}_{k,\text{loc}}(\vec{R}, t) + \left(\frac{\partial \vec{r}_k(t)}{\partial t} \times \vec{B}_{k,\text{loc}}(\vec{R}, t) \right) \right] \approx \frac{q_k}{m_k} \vec{E}_{k,\text{loc}}(\vec{R}, t) \end{aligned} \tag{4.1}$$

In (4.1) γ_k represents the damping constant, σ_{ki} the coupling between oscillators, and ω_k the eigenfrequency of the charge in the microscopic coordinates $\vec{r}(t)_k$. The oscillators are driven by local fields $\vec{E}_{k,\text{loc}}(\vec{R}, t)$ at the point of the oscillator locations; the relation between the local and averaged fields will be discussed later. This set of equations of motion can be analytically solved and the system parameters ω_k^2, γ_k, and σ_{ki} can be evaluated in phenomenological way by comparison with experimental or numerical data. To apply this model to the double-wire geometry we propose the charge arrangement as shown in Fig. 4.1:

$$\vec{r}_1(t) = \begin{pmatrix} x_1 \\ y_1 \\ 0 \end{pmatrix}, \quad \vec{r}_2(t) = \begin{pmatrix} -x_1 \\ y_1 \\ 0 \end{pmatrix},$$
$$\vec{r}_3(t) = \begin{pmatrix} -x_2 \\ -y_1 \\ 0 \end{pmatrix}, \quad \vec{r}_4(t) = \begin{pmatrix} x_2 \\ -y_1 \\ 0 \end{pmatrix} \tag{4.2}$$

$$q_1 = q_4 = q, \quad q_2 = q_3 = -q$$

Thus the double-wire geometry is modeled by four charges where the two upper and the two lower ones represent electric dipoles. One important detail of the charge arrangement has to be mentioned explicitly, namely, the introduced asymmetry of the charge dynamics in the top and bottom wires (Fig. 4.1). The asymmetry has its origin in the finite size of the charge distribution and the spatial retardation of the exciting electric field, i.e., the field is different at both dipole sites, which will be considered in the coupled-oscillator approach. This difference occurs due to the phase and amplitude variation of the electric field propagating along the y direction between the wires. The interplay of the external field with the induced local field excited in the wires results in the excitation of symmetric and anti-symmetric dipole moments in the upper and lower wires. It is worth noting that the required asymmetry to excite second-order multipoles can also be modeled by different oscillator properties (e.g., the damping constant) which will be discussed in detail in the next paragraph. By substituting (4.2) into (3.56), the electric dipole, the electric quadrupole, and the magnetic dipole moment can be calculated straightforwardly (in frequency domain):

$$\begin{cases} \vec{\mathrm{P}}(\vec{R}, \omega) = 2\eta q \begin{pmatrix} x_1 + x_2 \\ 0 \\ 0 \end{pmatrix} - \nabla \cdot Q(\vec{R}, \omega) \\ Q_{ij}(\vec{R}, \omega) = \eta q y_1 \begin{pmatrix} 0 & x_1 - x_2 & 0 \\ x_1 - x_2 & 0 & 0 \\ 0 & 0 & 0 \end{pmatrix} \\ \vec{\mathrm{M}}(\vec{R}, \omega) = \frac{i\omega\eta q y_1}{c} \begin{pmatrix} 0 \\ 0 \\ x_1 - x_2 \end{pmatrix} \end{cases} \tag{4.3}$$

From (4.3) one can recognize that all second-order expansion moments (the electric quadrupole and the magnetic dipole moment) vanish for a symmetric charge configuration $x_1 = x_2$, while the polarization (electric dipole) is still nonzero. The symmetric system $x_1 = x_2$ would consist of two classical dipoles with no influence of higher-order multipoles, and hence with no magnetization.

It has to be realized that the anti-symmetric modes (and in turn magnetization) appear due to two factors only, namely:

(a) asymmetric nature of the system of coupled oscillators (for example, not equal parameters of the oscillators), or
(b) asymmetric driving fields $\vec{E}_{k,\mathrm{loc}}(\vec{R}, t)$ (e.g. in (4.1) $\vec{E}_{1,\mathrm{loc}}(\vec{R}, t) \neq \vec{E}_{2,\mathrm{loc}}(\vec{R}, t)$).

Both options and consequent effective parameters will be considered below.

4.2 Dispersion Relations and Effective Parameters for Metamaterials: Asymmetric Structures

First, we consider the case when the anti-symmetric modes appear due to asymmetry of the structure of the MA itself, which can be achieved by not equal damping constants γ_k and/or not equal eigenfrequencies ω_k of the oscillators (4.1). At the same time, it is assumed that the local field is equal for both oscillators and equal to the averaged field $\vec{E}_{1,\mathrm{loc}}(\vec{R}, t) = \vec{E}_{2,\mathrm{loc}}(\vec{R}, t) = \vec{E}(\vec{R}, t)$. Physically, this corresponds to a very small in compare with the wavelength MAs, which normally does not take place in optical region for MMs based on plasmonic structures like shown in Fig. 4.1. The system of equations in this case becomes:

$$\begin{cases} \dfrac{\partial^2 x_1(t)}{\partial t^2} + \gamma_1 \dfrac{\partial x_1(t)}{\partial t} + \omega_1^2 x_1(t) - \sigma x_2(t) = \dfrac{q}{m} E_x \\ \dfrac{\partial^2 x_2(t)}{\partial t^2} + \gamma_2 \dfrac{\partial x_2(t)}{\partial t} + \omega_2^2 x_2(t) - \sigma x_1(t) = \dfrac{q}{m} E_x \end{cases} \tag{4.4}$$

System (4.4) can be solved in Fourier domain:

$$\begin{cases} x_1(\omega) = \alpha_1(\omega) E_x \\ x_2(\omega) = \alpha_2(\omega) E_x \\ \alpha_1(\omega) = \dfrac{q}{2m} \dfrac{\left(i\gamma_2\omega + \omega^2 - \omega_2^2 + \sigma\right)}{\left(\sigma^2 - \omega_1^2\omega_2^2 - \omega^4 - i\omega^3(\gamma_1 + \gamma_2) + \omega^2\left(\omega_1^2 + \omega_2^2 + \gamma_1\gamma_2\right) + i\omega\left(\omega_1^2\gamma_1 + \omega_2^2\gamma_2\right)\right)} \\ \alpha_2(\omega) = \dfrac{q}{2m} \dfrac{\left(i\gamma_1\omega + \omega^2 - \omega_1^2 + \sigma\right)}{\left(\sigma^2 - \omega_1^2\omega_2^2 - \omega^4 - i\omega^3(\gamma_1 + \gamma_2) + \omega^2\left(\omega_1^2 + \omega_2^2 + \gamma_1\gamma_2\right) + i\omega\left(\omega_1^2\gamma_1 + \omega_2^2\gamma_2\right)\right)} \end{cases} \tag{4.5}$$

According to the arguments presented in Sect. 3.24, the origin for the asymmetric structure has to be shifted by an amount given by (3.72) which after substitution (4.5) becomes:

$$g_y = \left[\frac{\alpha_2(\omega_{\mathrm{sym}}) N_2 - \alpha_1(\omega_{\mathrm{sym}}) N_1}{\alpha_1(\omega_{\mathrm{sym}}) N_1 + \alpha_2(\omega_{\mathrm{sym}}) N_2}\right] \frac{y}{2} \tag{4.6}$$

here $y = y_{1,\mathrm{init}} + y_{2,\mathrm{init}}$, and N_1, N_2 are the numbers proportional to the number of electrons in the top and bottom wires, see Fig. 3.2. We have chosen middle point

$y_{1,\mathrm{init}} = y_{2,\mathrm{init}}$ as an initial origin; it is easy to see, that the final origin position given by (3.27) does not depend on the initial choice $y_{1,\mathrm{init}}, y_{2,\mathrm{init}}$ for the origin. The eigen frequency for the symmetric mode is obtained under requirement of equal phases for the oscillation of the both wires, namely:

$$\arg\left[x_1(\omega_{\mathrm{sym}})\right] = \arg\left[x_2(\omega_{\mathrm{sym}})\right] \tag{4.7}$$

As a result, the final position of the origin is:

$$\begin{cases} y_{1,\mathrm{final}} = \dfrac{y}{2} + g \\ y_{2,\mathrm{final}} = \dfrac{y}{2} - g \end{cases} \tag{4.8}$$

The multipole moments are:

$$\begin{cases} P_x(k_y,\omega) = \left(p_x(k_y,\omega) - ik_y u_{xy}(k_y,\omega)\right)E_x(k_y,\omega) \\ Q_{xy}(k_y,\omega) = u_{xy}(k_y,\omega)E_x(k_y,\omega) \\ M_z(k_y,\omega) = m_z(k_y,\omega)E_x(k_y,\omega) \\ p_x(k_y,\omega) = 2\eta q(\alpha_1(\omega)N_1 + \alpha_2(\omega)N_2) \\ u_{xy}(k_y,\omega) = \dfrac{\eta q}{4}(\alpha_1(\omega)N_1y_1 - \alpha_2(\omega)N_2y_2) \\ m_z(k_y,\omega) = \dfrac{i\omega\eta q}{4c}(\alpha_1(\omega)N_1y_1 - \alpha_2(\omega)N_2y_2) \end{cases} \tag{4.9}$$

which finally gives according to general expression (3.61a) the dispersion relation:

$$k_y^2 = \frac{\omega^2}{c^2}\left(1 + 8\pi\eta q(\alpha_1(\omega)N_1 + \alpha_2(\omega)N_2) - i2\pi\eta q k_y(\alpha_1(\omega)N_1y_1 - \alpha_2(\omega)N_2y_2)\right) \tag{4.10}$$

Material equations (3.61a) and (3.61b) become:

$$\begin{cases} D_x(k_y,\omega) = \left(1 + 8\pi\eta q(\alpha_1(\omega)N_1 + \alpha_2(\omega)N_2) - i\pi\eta q k_y(\alpha_1(\omega)N_1y_1 - \alpha_2(\omega)N_2y_2)\right)E_x(k_y,\omega) \\ H_z(k_y,\omega) = \left(-\dfrac{k_y c}{\omega} - i\dfrac{\pi\omega\eta q}{c}(\alpha_1(\omega)N_1y_1 - \alpha_2(\omega)N_2y_2)\right)E_x(k_y,\omega) \end{cases} \tag{4.11a}$$

$$\begin{cases} \varepsilon_x(k_y,\omega) = 1 + 8\pi\eta q(\alpha_1(\omega)N_1 + \alpha_2(\omega)N_2) - i\pi\eta q k_y(\alpha_1(\omega)N_1y_1 - \alpha_2(\omega)N_2y_2) \\ \mu_z(k_y,\omega) = \left(1 + \frac{i\pi\omega^2\eta q}{k_y c^2}(\alpha_1(\omega)N_1y_1 - \alpha_2(\omega)N_2y_2)\right)^{-1} \\ \xi_{zx}(k_y,\omega) = -\left(\frac{k_y c}{\omega} + \frac{i\pi\omega\eta q}{c}(\alpha_1(\omega)N_1y_1 - \alpha_2(\omega)N_2y_2)\right) \end{cases} \tag{4.11b}$$

Reminding phenomenological form of the material equations (3.47), we write down straightforwardly the expressions for the material parameters introduced in (3.64), (3.65):

$$\begin{cases} \varepsilon_x^{(0)}(\omega) = 1 + 8\pi\eta q(\alpha_1(\omega)N_1 + \alpha_2(\omega)N_2) \\ \varepsilon_x^{(1)}(\omega) = -i\pi\eta q(\alpha_1(\omega)N_1 y_1 - \alpha_2(\omega)N_2 y_2) \\ \varepsilon_x^{(2)}(\omega) = 0 \\ \phi_x^{(0)}(\omega) = -i\frac{\pi\omega\eta q}{c}(\alpha_1(\omega)N_1 y_1 - \alpha_2(\omega)N_2 y_2) \\ \phi_x^{(1)}(\omega) = -\frac{c}{\omega} \end{cases} \tag{4.12}$$

The analysis above demonstrates that the coefficient in (3.47) have clear physical mean and indicate the reason of the asymmetric modes appearing in the MAs.

Dispersion relation (4.7) can be solved for k_y:

$$k_y = \frac{-i\frac{\omega^2}{c^2}4\pi\eta q y_1(\alpha_1(\omega)N_1 y_1 - \alpha_2(\omega)N_2 y_2)}{2} \pm \\ \pm \frac{\sqrt{-\left(\frac{\omega^2}{c^2}4\pi\eta q y_1(\alpha_1(\omega)N_1 y_1 - \alpha_2(\omega)N_2 y_2)\right)^2 + \frac{4\omega^2}{c^2}(1 + 8\pi\eta q(\alpha_1(\omega)N_1 + \alpha_2(\omega)N_2))}}{2} \tag{4.13}$$

The coefficients α_1, α_2 are complex and imaginary part of k_y is always positive, which corresponds to the energy dissipation from the propagating wave. The two solutions of (4.10) manifest, as usual, propagation in opposite directions. In fact, if we consider solution of (4.10) for the k_y for two "inverted" structures ($\alpha_1(\omega)N_1 - \alpha_2(\omega)N_2 = \Delta(\omega)$ and $\alpha_1(\omega)N_1 - \alpha_2(\omega)N_2 = -\Delta(\omega)$) we see, that (4.10) remains the same in case of simultaneous transformations $k_y \to -k_y$ and $\Delta(\omega) \to -\Delta(\omega)$.

The detailed consideration of the asymmetric media can be found in [7].

4.3 Dispersion Relations and Effective Parameters for Metamaterials: Symmetric Structures (Retarded Field)

For a polarization along the wire axis (x direction) the complete equations describing the charge dynamics in the two wires is given by (4.1) and in case of equal oscillators (symmetric structure) and not equal local fields are reduced to:

$$\begin{cases} \frac{\partial^2 x_1(t)}{\partial t^2} + \gamma\frac{\partial x_1(t)}{\partial t} + \omega_0^2 x_1(t) + \sigma x_2(t) = \frac{q}{m}E_{x1,\mathrm{loc}} \\ \frac{\partial^2 x_2(t)}{\partial t^2} + \gamma\frac{\partial x_2(t)}{\partial t} + \omega_0^2 x_2(t) + \sigma x_1(t) = \frac{q}{m}E_{x2,\mathrm{loc}} \end{cases} \tag{4.14}$$

The inhomogeneous solution of the system (4.14) can be obtained in the Fourier domain:

$$\begin{cases} x_1(\omega) = \frac{q}{m}\frac{E_{x1,\mathrm{loc}}\left(i\gamma\omega+\omega^2-\omega_0^2\right)-\sigma E_{x2,\mathrm{loc}}}{\left(\sigma^2-\left(i\gamma\omega+\omega^2-\omega_0^2\right)^2\right)} \\ x_2(\omega) = \frac{q}{m}\frac{E_{x2,\mathrm{loc}}\left(i\gamma\omega+\omega^2-\omega_0^2\right)-\sigma E_{x1,\mathrm{loc}}}{\left(\sigma^2-\left(i\gamma\omega+\omega^2-\omega_0^2\right)^2\right)} \end{cases} \tag{4.15}$$

For the symmetric and asymmetric oscillations, entering the formalism:

$$\begin{cases} x_1(\omega)+x_2(\omega) = \left(E_{x1,\mathrm{loc}}+E_{x2,\mathrm{loc}}\right)\chi^+(\omega) \\ x_1(\omega)-x_2(\omega) = \left(E_{x1,\mathrm{loc}}-E_{x2,\mathrm{loc}}\right)\chi^-(\omega) \\ \chi^\pm(\omega) = \frac{q}{m}\frac{1}{\left(\omega_0^2-\omega^2-i\gamma\omega\pm\sigma\right)} \end{cases} \tag{4.16}$$

The relations between local amplitudes on the electric field in the upper and bottom wires $E_{x1,\mathrm{loc}}(y, \omega)$, $E_{x2,\mathrm{loc}}(y, \omega)$ are assumed to be caused by the retardation and are taken into account by following expressions:

$$\begin{cases} E_{x1,\mathrm{loc}}(y,\omega) = E_x(\omega)\exp\left(ik_y y_1\right) \\ E_{x2,\mathrm{loc}}(y,\omega) = E_x(\omega)\exp\left(-ik_y y_1\right) \end{cases} \tag{4.17}$$

Here $E_x(y, \omega) = E_x(\omega)\exp(ik_y y)$ is the macroscopic (averaged) field in Maxwell equations. At this point we explicitly mention that in our model the excitation via plane waves is described in an approximate manner. The electric field propagating from the first to the second wire is determined by k_y, the complex wave number. In addition to this external electric field evolution, the excitation process is also governed by the near-field coupling of the two wires. This mechanism is taken into account by the empirical coupling constant σ between the two wires and not by the additional electric fields on the right side of (4.14). This approximation provides an analytical solution for the equations of motion and prevents us from regarding the complex near-field interactions between the wires by introducing that coupling constant. Therefore k_y represents the propagation vector of the corresponding effective medium, not the free-space wave vector. Upon substitution into (4.16), the oscillation amplitudes depend only on the electric field at a single site required for performing the multipole expansion:

$$\begin{cases} x_1(\omega)+x_2(\omega) = 2E_x(\omega)\cos\left(k_y y_1\right)\chi^+(\omega) \\ x_1(\omega)-x_2(\omega) = 2iE_x(\omega)\sin\left(k_y y_1\right)\chi^-(\omega) \end{cases} \tag{4.18}$$

The quantities $\chi^\pm(\omega)$ introduced above represent the polarizabilities related to the eigenmodes of the coupled system, where the (+) and (−) the signs indicate the symmetric and the antisymmetric modes, respectively. It turns out that the antisymmetric mode induces both the magnetic dipole and the electric quadrupole moments, whereas the symmetric mode is related to the electric dipole moment. Now the calculated multipole moments (4.3) can be rewritten:

$$\begin{cases} P_x(k_y,\omega) = p_x(k_y,\omega)E_x(k_y,\omega) \\ Q_{xy}(k_y,\omega) = u_{xy}(k_y,\omega)E_x(k_y,\omega) \\ M_z(k_y,\omega) = m_z(k_y,\omega)E_x(k_y,\omega) \\ p_x(k_y,\omega) = 4\eta q\chi^+(\omega)\cos(k_y y_1) \\ u_{xy}(k_y,\omega) = 2i\eta q y_1\chi^-(\omega)\sin(k_y y_1) \\ m_z(k_y,\omega) = -\frac{2\omega\eta q y_1}{c}\chi^-(\omega)\sin(k_y y_1) \end{cases} \tag{4.19}$$

The dispersion relation for the plane wave can be found straightforwardly by plugging (4.19) into (3.61a):

$$k_y^2 = \frac{\omega^2}{c^2}\left(1 + 16\pi\eta q\chi^+(\omega)\cos(k_y y_1) + 8\pi\eta q k_y y_1\chi^-(\omega)\sin(k_y y_1)\right) \tag{4.20}$$

The implicit dispersion relation obtained can be solved only numerically. To keep the model analytical, we can approximate the trigonometric functions $\cos(k_y y_1)$ and $\sin(k_y y_1)$ for small arguments $k_y y_1 \ll 1$:

$$\begin{cases} \cos(k_y y_1) \approx 1 - \frac{(k_y y_1)^2}{2} \\ \sin(k_y y_1) \approx k_y y_1 \end{cases} \tag{4.21}$$

and:

$$k_y^2 = \frac{\omega^2}{c^2}\frac{1 + 16\pi\eta q\chi^+(\omega)}{1 + \frac{\omega^2}{c^2}16\pi\eta q y_1^2\left(\frac{1}{2}\chi^+(\omega) - \chi^-(\omega)\right)} \tag{4.22}$$

Now with (4.22) the dispersion relation is explicit in k_y and mirror symmetric with respect to $k_y = 0$, as in (4.20). The above approximation is justified because $y_1 \ll \lambda$ holds in the considered cut-wire geometry.

Material equations (3.61a, b) are:

$$\begin{cases} D_x(k_y,\omega) = \left(1 + 16\pi\eta q\chi^+(\omega) + 8\pi\eta q(\chi^-(\omega) - \chi^+(\omega))(k_y y_1)^2\right)E_x(k_y,\omega) \\ H_z(k_y,\omega) = \left(\left(-\frac{c}{\omega} + 4\pi\frac{\omega\eta q y_1^2}{c}\chi^-(\omega)\right)k_y\right)E_x(k_y,\omega) \end{cases} \tag{4.23a}$$

$$\begin{cases} \varepsilon_x(k_y,\omega) = 1 + 16\pi\eta q\chi^+(\omega) + 8\pi\eta q(\chi^-(\omega) - \chi^+(\omega))(k_y y_1)^2 \\ \mu_z(k_y,\omega) = \left(1 - \frac{8\pi\omega^2\eta q y_1^2}{c^2}\chi^-(\omega)\right)^{-1} \\ \xi_{zx}(k_y,\omega) = -\left(\frac{c}{\omega} - \frac{8\pi\omega\eta q y_1^2}{c}\chi^-(\omega)\right)k_y \end{cases} \tag{4.23b}$$

Reminding phenomenological form of the material equations (3.47), we write down the expressions for the material parameters introduced in (3.64), (3.65):

$$\begin{cases} \varepsilon_x^{(0)}(\omega) = 1 + 1 + 16\pi\eta q\chi^+(\omega) \\ \varepsilon_x^{(1)}(\omega) = 0 \\ \varepsilon_x^{(2)}(\omega) = 8\pi\eta q y_1^2(\chi^-(\omega) - \chi^+(\omega)) \\ \phi_x^{(0)}(\omega) = 0 \\ \phi_x^{(1)}(\omega) = -\frac{c}{\omega} + 8\pi\frac{\omega\eta q y_1^2}{c}\chi^-(\omega) \end{cases} \tag{4.24}$$

In this section the model will be validated by the comparison with the results of the rigorous numerical calculations.

The multipole expansion leads to a coupling between dielectric and magnetic responses that would be completely decoupled in a purely dipole interaction regime. This coupling can be interpreted as follows. In our model the interaction of the incoming plane wave with matter is determined only by the electric field (interaction with the magnetic field is negligible). This interaction with the electric field can be expressed in terms of electric and magnetic multipole responses that consequently determine the dependence of all quantities on the electric field. The excitation of coupled-charge oscillations leads to a magnetic response that can be described by an effective magnetic permeability, which is again a consequence of the interaction with the electric (not the magnetic) field. It should be emphasized again that the physical picture of the magnetic response differs basically from that taking place in solid state physics. The magnetic response in the latter case is caused by a magnetic field which induces or aligns existing magnetic moments of atoms or molecules (the free-electron magnetism effect again is caused by interaction with the magnetic components of the field). In the case of MMs the electric field excites localized or surface (like in case of fish-net structure) plasmon-polaritons which contribute to both electric and magnetic responses, while the microscopic magnetic component does not participate in the light-matter interaction.

The introduction of effective parameters might look artificial because the electric as well as the magnetic response are caused by an interaction of carriers confined in the nanostructure with the electric field [8, 9], and are mutually related to each other. However, the decoupling of these two responses, e.g., by introducing permittivity and permeability, might be necessary, for example, for comparing them with numerically determined electric and magnetic properties in terms of ε and μ.

4.4 Validation of the Model

In this section, a quantitative comparison of results obtained by the outlined analytical approach and rigorous numerical calculations is performed. First it is necessary to summarize that the wave vector (4.22) depends on frequency ω, on the

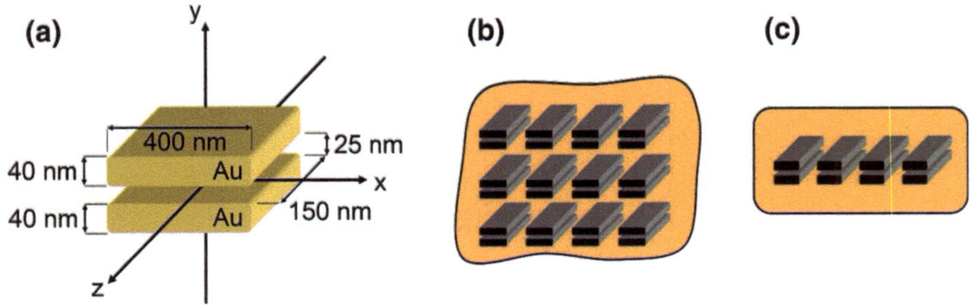

Fig. 4.2 **a** Geometry of the simulated double wire meta-atom is shown. **b** Three-dimensional (3D) bulk MM alignment to calculate the dispersion relation of a bulk MM and **c** the slab arrangement which allows additionally the calculation of effective parameters [6]

product of the carrier density η with the charge q, on the geometrical parameter y_1, on the two quantities $\chi^+(\omega)$ and $\chi^-(\omega)$ (which are in turn functions of the eigenfrequency ω_0 on the damping constant γ, and on the coupling constant σ of the carrier oscillations. In order to determine rigorously the dispersion relation of the geometry shown in Fig. 4.2 the Fourier modal method [10] has been used. To describe the propagation of electromagnetic waves in 3D bulk media (see Fig. 4.2b) the calculation has been examined with periodic boundary conditions in all three space dimensions. The periods used in the x, y, and z directions are $\Lambda_x = 600\,\text{nm}$, $\Lambda_y = 500\,\text{nm}$, and $\Lambda_z = 150\,\text{nm}$, respectively. The double wires are formed from gold with the sizes shown in Fig. 4.2), and are separated by a thin glass layer with $n = 1.44$. The corresponding effective material parameters of the same geometry have been obtained by FMM calculations for one layer (see Fig. 4.2c).

The effective permeability and permittivity are calculated from the complex reflected and transmitted amplitudes which are used in inverted expressions for the reflection and transmission of light at a homogeneous slab [11]. First the dispersion relation of the 3D infinite MM was calculated numerically and the analytical version was fitted to it (Fig. 4.3).

The comparison shows very good agreement in the small-frequency domain, while for larger frequencies the analytical dispersion relation differs from the numerical one. This can be explained by the violation of the sub-wavelength criterion for larger frequencies. Since no numerical data for the corresponding effective parameters are available due to the absence of boundaries, a finite geometry has been simulated. Therefore Fig. 4.4 shows the effective refractive index that corresponds to the dispersion relation of the slab arrangement as well as the retrieved effective parameters (see Fig. 4.4d, g). The second column contains the fitted effective refractive index (Fig. 4.4b) and the following effective parameters (see Fig. 4.4e, h). Since only the dispersion relation in terms of the refractive index has been fitted, the coincidence of the effective parameters is significant. Another remarkable property of the analytically determined effective parameters is the vanishing of the anti-resonances. This feature appears in Fig. 4.4g in the

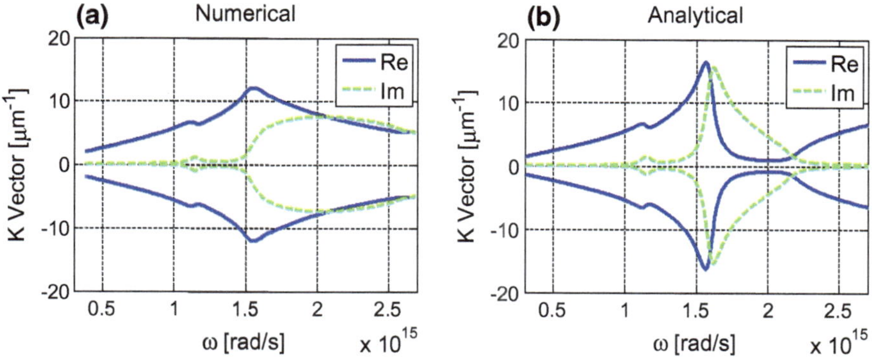

Fig. 4.3 **a** Dispersion relation obtained from the numerical calculations in an infinite 3D MM as shown in Fig. 4.2b and the corresponding fitted analytical version (**b**) [6]

effective permeability exactly at the resonance position of the effective permittivity. The same anti-resonance appears in the effective permittivity at the resonance position of the effective permeability but it is much weaker and cannot be observed in Fig. 4.4d. This discrepancy reflects the described in Chap. 2 unambiguity in "C" representation, where the electric and magnetic responses can be redistributed; the math behind this is given by SFT. The last column in Fig. 4.4 shows the effective parameters that are a direct result of the fit of the dispersion relation of the bulk MM, presented in Fig. 4.3. It can be seen that the resonance positions as well as their absolute values differ from the slab parameters [12]. This can be explained by the coupling of neighboring MM layers, which leads to slightly different effective material parameters from those for a decoupled MM, e.g., a MM slab. In the following we describe the fitting procedure in detail. First, the locations of both resonances and their bandwidths were fitted. The resonance positions in the analytical model have been tuned by selecting the eigenfrequency ω_0 in between the two resonance frequencies of the numerical dispersion relation. We notice that the eigenfrequency ω_0 corresponds to the localized plasmon-polariton frequency of an isolated wire.

The eigen frequency of such a wire can be estimated with an ellipsoidal particle in the quasi-static regime and beyond [13] with the same dimensions shown in Fig. 4.2a. The calculated value coincides approximately with the eigen frequency used for the fit (see the value in parentheses in Table 4.1). To realize the presence of two resonances, the coupling constant σ was increased until the two resonance positions coincide with the numerical ones. The resonance width, which represents the damping, has been retrieved from the full width at half maximum value of the imaginary part of the peak in the dispersion relation for the 3D geometry and the refractive index for the 2D arrangement. The unknown term $\frac{16\pi\eta q^2}{m}$ this term appears in the expressions for $k(\omega)$, $\varepsilon(\omega)$, and $\mu(\omega)$ represents the only remaining free fitting parameter. The values found for both scenarios are listed in Table 4.1.

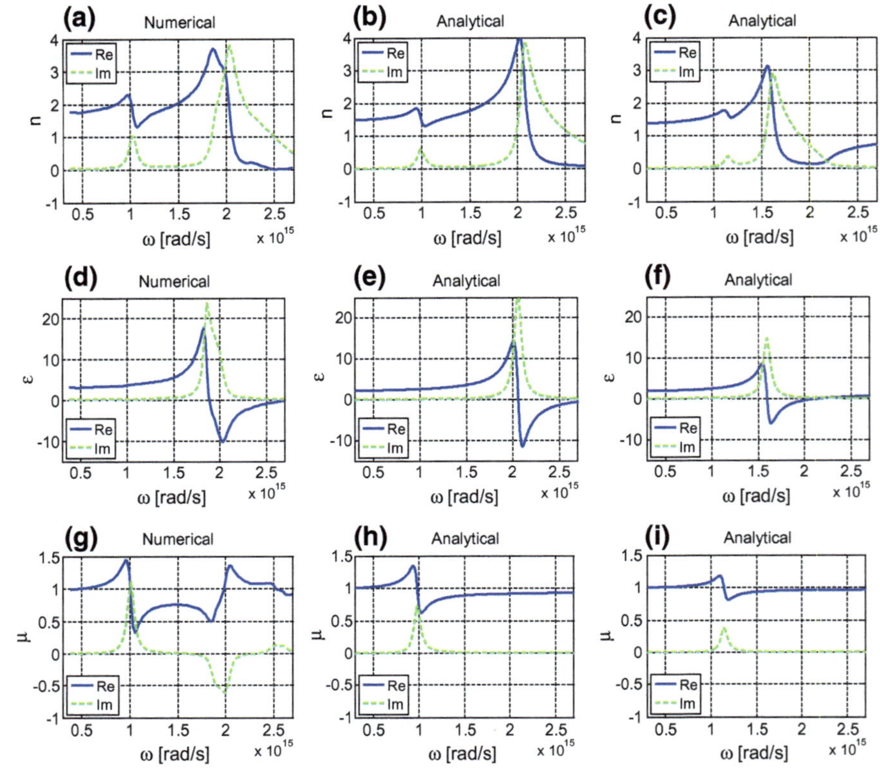

Fig. 4.4 Retrieved effective index (**a**), effective permittivity (**d**), and effective permeability (**g**) from FMM simulations of a 2D single-layer MM. Based on the coefficients from fitting the dispersion relation that is equivalent to the refractive index in the 2D case, the resulting effective index (**b**), the effective permittivity (**e**), and the effective permeability (**h**) resulting from the analytical model are shown. **c**, **f**, and **i** correspond to the analytically determined effective parameters for the 3D infinite bulk MM. Therefore the dispersion relation (Fig. 4.3) has been fitted by the analytical one [6]

Table 4.1 The fitting parameters that have been applied to match the dispersion relation for the slab as well as for the bulk arrangement

Fitting parameter	Slab MM	Bulk MM
$\omega_0\,(\mathrm{rad/s})$	1.62×10^{15} (1.85×10^{15})	1.39×10^{15} (1.85×10^{15})
$\gamma\,(\mathrm{rad/s})$	9.42×10^{30}	9.42×10^{30}
$\sigma\,(\mathrm{rad/s})^2$	1.60×10^{30}	0.60×10^{30}
$\frac{16\pi\eta q^2}{m}\,(\mathrm{AsV/m^2Kg})$	5.00×10^{30}	2.20×10^{30}

Together with (4.23b) the effective parameters can be calculated as shown in Fig. 4.4. Additionally, the eigen frequencies of the single wire calculated separately in the quasistatic regime are listed in parentheses

4.5 Conclusion

The multipole approach to the homogenization of the MM has been formulated in case of classical charge dynamics of the MAs, and the dispersion relation has been elaborated. While the dispersion relation can be fitted, the permittivity and the permeability are not direct outcomes of the fitting procedure. Having the dispersion relation (and respective constants) fitted, the assigned effective parameters can be calculated from (4.23b) without any further adaptation. The comparison of these functions (refractive index, permittivity, and permeability) validates the procedure introduced for the modeling of effective parameters as well as the application of the multipole expansion (see Fig. 4.4). Nevertheless, one can conclude that for such a rather primitive model the analytical results are in quite good agreement with the rigorous simulations, especially for small frequencies (long wavelengths) where the approach based on the averaged Maxwell's equations is supposedly more appropriate.

References

1. V. Shalaev, W. Cai, U. Chettiar, H.-K. Yuan, A. Sarychev, V. Drachev, A. Kildishev, Negative index of refraction in optical metamaterials. Opt. Lett. **30**, 3356 (2005)
2. M. Wegener, *Extreme Nonlinear Optics* (Springer, Berlin, 2005)
3. G. Sun, J.B. Khurgin, C.C. Yang, APL **95**, 171103 (2009)
4. T. Kalkbrenner, U. Hakanson, V. Sandoghdar, Nano Lett. **4**, 2309 (2004)
5. C. Kittel, *Introduction to Solid State Physics* (Wiley, New, York, 1986)
6. J. Petschulat, C. Menzel, A. Chipouline, C. Rockstuhl, A. Tünnermann, F. Lederer, T. Pertsch, Multipole approach to metamaterials. Phys. Rev. B **78**, 043811 (2008)
7. E. Pshenay-Severin, A. Chipouline, J. Petschulat, U. Huebner, A. Tuennermann, T. Pertsch, Optical properties of metamaterials based on asymmetric double-wire structures. Opt. Express **19**, 6269 (2011)
8. S. Tretyakov, *Analytical Modeling in Applied Electromagnetics* (Artech House, Boston, 2003)
9. C. Rockstuhl, T. Zentgraf, E. Pshenay-Severin, J. Petschulat, A. Chipouline, J. Kuhl, T. Pertsch, H. Giessen, F. Lederer, Opt. Express **15**, 8871 (2007)
10. B. Lee, H. Kim, Hwi Kim (Autor) › Entdecken Sie Hwi Kim bei Amazon Finden Sie alle Bücher, Informationen zum Autor und mehr. Suchergebnisse für diesen Autor Sind Sie ein Autor? Erfahren Sie mehr über Author Central
11. D.R. Smith, D.C. Vier, Th Koschny, C.M. Soukoulis, Electromagnetic parameter retrieval from inhomogeneous metamaterials. Phys. Rev. E **71**, 036617 (2005)
12. C. Rockstuhl, T. Paul, F. Lederer, T. Pertsch, T. Zentgraf, T. Meyrath, H. Giessen, Phys. Rev. B **77**, 035126 (2008)
13. V.V. Klimov, Nanoplasmonika. ISBN 978-5-9221-1205-5 (2010) (in Russian)

Chapter 5
Applications of the "Classical" Metamaterial Model—Optical Activity and Electromagnetically Induced Transparency

5.1 Review of Optical Activity with Metamaterials

In the initial stage of research on MMs emphasis was put on exploring materials that potentially lead to a biaxial anisotropic (linear dichroism) effective material response [1–6]. Recently research was also extended toward the exploration of meta-atoms that affect off-diagonal elements of the effective material tensors (elliptical dichroism). It expands the number of observable optical phenomena, leading to, e.g., optical activity [7–11] bidirectional and asymmetric transmission [12–14] or chirality-induced negative refraction [15–17]. In general, investigating the geometry of the MM (the meta-atoms geometry and their arrangement) allows us to determine the form of the effective material tensors in the quasistatic limit as extensively discussed in [8]. From such considerations it is possible to conclude on the symmetry of the plasmonic eigenmodes sustained by the MAs and on the polarization of the eigenmodes allowed to propagate in the effective medium [12]. But in order to determine the actual frequency dependence of the tensor elements, more extended models are needed which start in their description of the MA properties from scratch [14]. Such models are required to be universal, simple and assumption free to the largest possible extent. Here, an approach which meets these requirements is outlined. It is based on developed in previous chapters conceptually decomposing the complex MAs into a set of coupled plasmonic entities that sustain the excitation of dipolar resonances [18]. The knowledge of the plasmonic properties of these dipoles and their coupling suffices to derive the material response and the symmetry of the eigenmodes. This, in turn, permits to predict the observable quantities in the far-field, such as the polarization and frequency-dependent reflected and transmitted complex amplitudes. Since the major focus is devoted to investigate structural modifications on planar MAs the here considerations will be

A. Chipouline and F. Küppers, *Optical Metamaterials: Qualitative Models*, Springer Series in Optical Sciences 211, https://doi.org/10.1007/978-3-319-77520-3_5

restricted to normal incidence. In principle, the introduced formalism can be similarly extended to investigate different illumination scenarios. The most appealing aspect of the model is that once the plasmonic entities and their coupling strengths are characterized, far-field properties remain predictable by the model even if substantial modifications of the MA geometry have been made. Even a modification that leads to a different symmetry of the material tensor does not prevent a quantitative description of the dispersive behavior of the tensor elements. To become specific, the investigations start with an optical inactive, biaxial anisotropic MM, namely, the split-ring resonator (SRR). From the observable far-field quantities the properties of the dipolar oscillators will be derived. The properties govern the plasmonic response of the three wires forming the SRR, namely, the individual eigenfrequencies, oscillator strengths and damping constants as well as their mutual coupling strengths. By relying on these quantities, the effective properties of MMs consisting of modified MAs with respect to the initial SRR for normal incidence will be determined. An attention will be focused on two modifications that evoke elliptical dichroism enabling asymmetric transmission for circular polarized light, as reported in structures with comparable symmetry [14]. With this model the effective properties of these MMs are predicted. The optical coefficients will be then computed for a slab made of these MMs and compare it to rigorous simulations. Excellent agreement between the introduced model and rigorous calculations for the optical coefficients is observed throughout the work. Therefore, the proposed method can be used for the parameter retrieval without resorting to rigorous simulations. And since it is based on simple analytical calculations the approach potentially allows a large variety of meta-atom modifications and to systematically tailor its effective properties. Hence, the approach provides a powerful and versatile tool for a systematic analysis of achievable material properties by varying only a few constituents that may couple in some well-defined ways. Furthermore, it will also be shown that such a parametrical treatment provides further insight into MM properties. Specifically, it is shown that it is possible to directly infer that the model’s predictions are valid in terms of the Casimir-Onsager relations [19, 20] the requirement for time reversal and reciprocity in linear media [21]. Thus, the main aspects of the chapter can be summarized as follows. At first, the localized carrier dynamics occurring in MAs may be properly described by a set of coupled oscillators, representing the decomposition of the MA in nanowire pieces. Second, the dynamics of these oscillators, determined by the shape of the nanowires and their coupling, result in electric dipole polarizabilities that permit the calculation of the effective permittivity tensor. The main advantage of this simple approach is that modifications of a MA, for which the oscillator dynamics and parameters have been found, leave the effective permittivity tensors predictable. Hence, the far-field reflectance and transmittance can be calculated. It will be shown that this holds also for modifications changing the character of the eigenstates from linear to non-linear polarized. Moreover, the approach can be useful to determine the effective permittivity tensor for MMs whose eigenstates are no longer linear, but elliptically polarized, by intensity measurements of linearly polarized light only.

5.2 Example of Calculation Procedure for SRR Metaatoms

To reveal the versatile character of this approach the consideration will be directly started by conceptually replacing the planar SRR geometry, shown in Fig. 5.1b, by a set of coupled oscillators. Each of the oscillators introduced here represents a MAs piece, i.e., a straight nanowire that is coupled to its electrically conductive neighbors. The oscillators accounting for the isolated nanowire are associated with the excitation of carriers representing the free-electron gas of the metal. Hereby, the carrier dynamics is solely influenced by the external electromagnetic field, including contributions from adjacent nanowires.

The dynamics becomes resonant for the eigenfrequency of the localized plasmon polariton resonance of the individual nanowire. This localized resonance described by the individual oscillator corresponds to the fundamental electric dipole mode since the dimensions of the nanowires forming the MA investigated here are small compared with the wavelength. It will be shown later that the carrier oscillations of MAs assembled from several coupled wires still correspond to electric dipole polarizabilities. This holds as long as the wires are assembled in-plane. For out-of-plane structures higher order multipoles come generally into play [18, 22]. Here, the spatial coordinate represents the elongation of a negatively charged carrier density driven by an external electromagnetic field, which is the usual assumption in plasmonics [23]. It will be shown below that this assumption is sufficient to entirely predict the optical response. The equations for the coupled oscillators are:

$$\begin{cases} \frac{\partial^2 x_1(t)}{\partial t^2} + \gamma_1 \frac{\partial x_1(t)}{\partial t} + \omega_{01}^2 x_1(t) + \sigma_{21} z_2(t) = -\frac{q_1}{m} E_x \\ \frac{\partial^2 z_2(t)}{\partial t^2} + \gamma_2 \frac{\partial z_2(t)}{\partial t} + \omega_{02}^2 z_2(t) + \sigma_{21} x_1(t) - \sigma_{23} x_3(t) = -\frac{q_2}{m} E_z \\ \frac{\partial^2 x_3(t)}{\partial t^2} + \gamma_3 \frac{\partial x_3(t)}{\partial t} + \omega_{03}^2 x_3(t) - \sigma_{23} z_2(t) = -\frac{q_3}{m} E_x \end{cases} \tag{5.1}$$

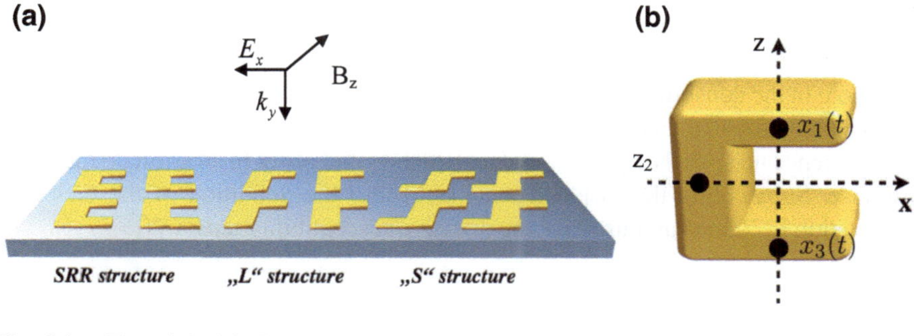

Fig. 5.1 **a** The original SRR structure (left), the first modification, namely, the L (center) and the second modification, the S structure (right); **b** the SRR together with the carrier oscillators, marked by black dots, which are used to phenomenologically replace the SRR [50]

where it was assumed nearest-neighbor coupling between adjacent, i.e., conductively coupled dipolar oscillators. In passing it was mentioned that in general also capacitive, inductive or in general free-space coupling appears, e.g., between the facing SRR wires. This coupling compared with the conductive coupling is much smaller for the present configuration. In general this free-space coupling could be easily taken into account by introducing yet another coupling constant between the coupled oscillators. The oscillators are driven by the electric field component of an external illuminating field propagating in y direction (normal incidence is assumed throughout the entire consideration). Excitation of the oscillators by magnetic field components can be safely neglected [11, 18, 21]. In (5.1) γ_i is the damping accounting for radiative and non radiative losses, ω_{0i} the eigenfrequency, σ_{ij} the coupling constant and q_i is the charge for the three oscillators $(i, j) \in (1, 2, 3)$, respectively, similarly to those introduced in [18]. The coordinates (x_1, z_2, x_3) themselves are understood as the displacement of the negatively charged carriers representing oscillating currents. Here, the multipole approach is applied to map the displacement onto electric dipole moments, since this is consistent with the previous works [18, 22]. Accounting for the contributions of the electric dipole and quadrupole as well as the magnetic dipole moment the wave equation is:

$$\frac{\partial^2 E_{x,z}(y,\omega)}{\partial y^2} + \frac{\omega^2}{c^2}\left(E_{x,z}(y,\omega) + 4\pi P_{x,z}(y,\omega)\right) + \frac{i4\pi\omega}{c}\frac{\partial M_{z,x}(y,\omega)}{\partial y} = 0 \qquad (5.2)$$

Emphasis is putted on the fact that the higher order multipole moments, leading to Q, M, appear in general [18], but do not provide any contribution to the far-field in the present configuration, as mentioned above. Thus, there is no effective magnetic response (effective permeability tensor components equal to 1) since the incoming magnetic field and the magnetic dipole moment of the SRR are perpendicular. A magnetic response as well as bi-anisotropic effects would emerge on the base of measurable quantities in the far-field if the SRRs would be oriented uprightly or the angle of incidence would be different [24, 25]. Thus it is sufficient to consider the dispersion in the electric susceptibility, resulting in a linear effective permittivity tensor which has dispersive entries only in the x–z components. Please note once again, that the observation of bi-anisotropic effects and dispersion in the permeability is possible using such split rings; but it is not possible to probe for the respective entries in the constitutive relations at normal incidence relative to the SRR plane as done here. Therefore in the constitutive tensors these coefficients are deliberately set to zero [26]. This will hold for all planar configurations with an illuminating field invariant in the x–z plane, hence no spatial dispersion occurs which would result in an artificial magnetic response. Substituting the displacements (5.1) into the definition of the dielectric polarization [27] one can introduce the effective susceptibility tensor $\overline{\chi}(\omega)$:

$$P_i(\vec{r}, t) = \eta \sum_{l=1}^{N} q_l r_{l,i} = \overline{\chi}_{ij}(\omega) E_j(\omega) \qquad (5.3)$$

where η accounts for the carrier density. This effective susceptibility can be easily calculated. As usual the effective permittivity tensor is defined as:

$$\varepsilon_{ij} = 1 + 4\pi \overline{\chi}_{ij}(\omega) \tag{5.4}$$

that governs the wave propagation of an incident plane wave in an effective medium composed of SRR meta-atoms. Next the possible eigenmodes of (5.1) are considered for the two polarization directions (x or z). It suffices to investigate these two polarizations as long as the system under consideration is linear. In order to describe a SRR with two identical side arms one can set $\omega_{01} = \omega_{02} = \omega_{0x}$, $\gamma_1 = \gamma_3 \equiv \gamma_x$, $\sigma_{21} = \sigma_{23} \equiv \sigma$, and $q_1 = q_3 = -q$. For the oscillator associated with the SRR base having a different geometry it is assumed $\omega_{02} = \omega_{0z}$, $\gamma_2 \equiv \gamma_z$, $q_2 = -q$. It has to be noted that the latter distinction could have been dropped if the geometry of all constituents of the SRR would have been the same. For the reasons obvious from the consideration below it is refrained from doing so. For a polarization of the incoming plane wave parallel to the x-axis one can solve (5.1) and obtain the following displacements in Fourier domain:

$$\begin{aligned} x_1(\omega) = x_3(\omega) &= -\frac{q_x}{m}\frac{1}{A_x(\omega)}E_x(y,\omega) \\ z_2(\omega) &= 0 \end{aligned} \tag{5.5}$$

where $A_x(\omega) = \omega_{0x}^2 - \omega^2 - i\omega\gamma_x$ has been introduced. For the polarization in z direction one can obtain:

$$\begin{aligned} x_1(\omega) = -x_3(\omega) &= -\frac{q_x}{m}\frac{\sigma}{A_x(\omega)A_z(\omega) - 2\sigma^2}E_z(y,\omega) \\ z_2(\omega) &= -\frac{q_z}{m}\frac{A_x(\omega)}{A_x(\omega)A_z(\omega) - 2\sigma^2}E_z(y,\omega) \end{aligned} \tag{5.6}$$

Considering the eigen modes for x polarization, it is observed that two parallel dipoles ($x_1 = x_3$) are induced, while due to symmetry constraints no dipole is induced in y direction ($z_2 = 0$), see Fig. 5.2c. By contrast, besides a dipole in z direction the z-polarized illumination induces oscillating dipoles in x direction in both arms. But due to the anti-symmetric oscillation $x_3 = -x_1$ (5.6), the dipoles in the SRR arms (Fig. 5.1b) do not radiate into the far-field because they oscillate out-of-phase and interfere destructively. Hence, no cross-polarization is observed and the far-field polarization equals that of the illumination. Later it will be proven that any radiation emerging from cross-polarized dipole moments will result in elliptical dichroism, as expected. By substituting (5.5) and (5.6) in (5.3) one gets the susceptibility tensor for the pertinent SRR configuration:

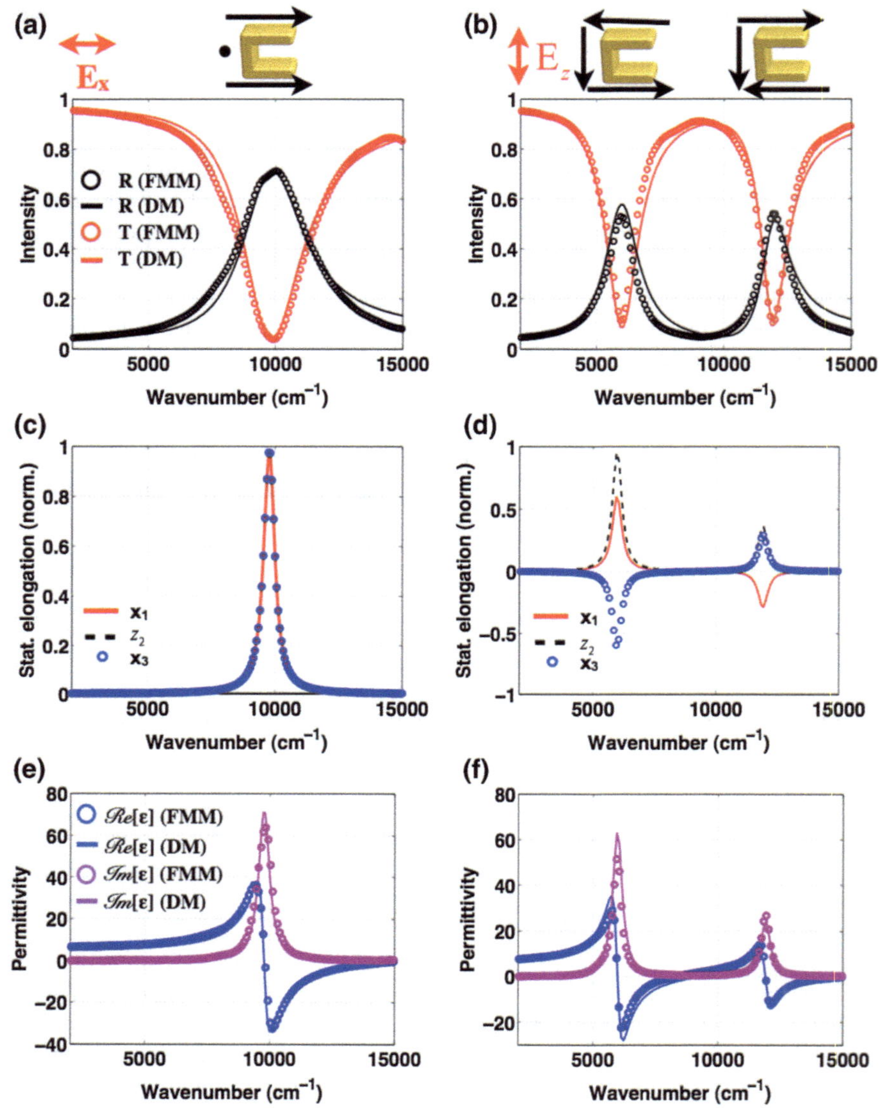

Fig. 5.2 The rigorously calculated (FMM) far-field transmission/reflection spectra compared with those obtained by the coupled dipole model (DM) for the two indicated polarization directions [**a** x polarization and **b** z polarization]. The stationary carrier elongation (normalized imaginary part of $x_{1,3}$ and z_2) for the two corresponding polarizations [(**c**) and (**d**)], **e** and **f** are the exactly retrieved parameters compared to calculations performed with the model [22]

$$\overline{\chi}_{ij}(\omega) = \begin{pmatrix} \chi_{xx}(\omega) & & \\ & 0 & \\ & & \chi_{zz}(\omega) \end{pmatrix}$$
$$\chi_{xx}(\omega) = \frac{q_x^2 \eta}{m} \frac{2}{A_x(\omega)} \tag{5.7}$$
$$\chi_{zz}(\omega) = \frac{q_z^2 \eta}{m} \frac{A_x(\omega)}{A_x(\omega) A_z(\omega) - 2\sigma^2}$$

with the polarization $P_i(\vec{r}, t) = -\eta(q_x(x_1 + x_2), 0, q_z z_2)$, according to (5.4).

As expected, the susceptibility tensor is diagonal. Hence, the eigenmodes of the effectively homogenous medium are linearly polarized and orthogonal. Due to the polarization dependent carrier dynamics the SRR shows a linear dichroitic behavior. Hence, our model correctly predicts the linear polarization eigenstates as required by the mirror symmetry with respect to the x–y plane. The only unknown parameters are those describing the oscillators and their coupling strengths. They can be determined by matching the optical coefficients of an effective medium whose permittivity is described by (5.7) to the spectra obtained by rigorous simulations or far-field measurements of the structure. In order to find the oscillator parameters, numerical Fourier modal method (FMM) calculations [28] of aperiodic array of gold SRR's[1] have been performed similar to those reported in [2]. These numerical far-field observables (reflected and transmitted intensities) were fitted by using the effective permittivity tensor (5.4) for both polarization directions in a conventional transfer matrix formalism that computes reflection and transmission from a slab of equal thickness [30]. Results are shown in Fig. 5.2a, b. Once the unknown parameters have been found the frequency dependent stationary elongations of the oscillators can be determined as shown in Fig. 5.2c, d. They can be used to identify the carrier oscillations of the different plasmonic eigenmodes. These carrier oscillations are shown on top of Fig. 5.2 and their horizontal position relates to the respective resonance frequency. The observed eigenmodes are documented in literature [26]. Figure 5.2e, f show the rigorously retrieved effective material properties together with the ones of the model.

Excellent agreement is observed. The rigorous results were obtained by applying a common parameter retrieval based on an inversion of the matrix formalism to calculate the effective parameters on the basis of complex reflected and transmitted amplitudes for a certain slab thickness. It has to be underlined that the unknown oscillator parameters can be obtained by comparing the far-field intensity only.

[1]The period in x and y direction is 0.4 μm, the SRR arm length 0.2 μm, the base width 0.08 μm, the arm width 0.04 μm, and the metal film thickness 0.025 μm. Gold material parameters were taken from literature [29]. As a substrate index we used $n_{\text{sub}} = 1.5$ and for the ambient material $n_{\text{amb}} = 1$.

The derived effective material parameters are in excellent agreement with the rigorous results, but they were derived without the necessity of knowing the complex-valued fields. This will be advantageous for MAs with nonlinear polarized eigenstates in the following sections where the experimental determination of these parameters is in general complicated (e.g., phase resolved measurements).

5.3 Results for L-Type of Metaatoms

As outlined in the introduction low-symmetry MAs are investigated in order to reveal new optical phenomena such as asymmetric transmission. It will be therefore extended the previous considerations toward MAs exhibiting elliptical dichroism by rearranging the SRR constituents. In the following the consideration will be relied on the oscillator parameters obtained by the fitting procedure above. Using these parameters in conjunction with the analytical expressions for the rearranged constituents as derived below, the optical response of the modified meta-atoms can be predicted. The first investigated structure is the L-MA [31, 32]. In this geometry one of the SRR arms is omitted in order to prevent the cancellation of the far-field that originated from the antiparallel dipole moments. Hence, it is expected to obtain polarization rotation. To get specific in (5.1) one horizontal oscillating dipole, e.g., dipole "3," has to be dropped. Hence, one obtains for x polarization:

$$\begin{aligned} x_1^x(\omega) &= -\frac{q_x}{m}\frac{A_z(\omega)}{A_x(\omega)A_z(\omega)-\sigma^2}E_x(y,\omega) \\ z_2^x(\omega) &= -\frac{q_x}{m}\frac{\sigma}{A_x(\omega)A_z(\omega)-\sigma^2}E_x(y,\omega) \end{aligned} \tag{5.8}$$

and for z polarization:

$$\begin{aligned} x_1^z(\omega) &= -\frac{q_z}{m}\frac{\sigma}{A_x(\omega)A_z(\omega)-\sigma^2}E_z(y,\omega) \\ z_2^z(\omega) &= -\frac{q_z}{m}\frac{A_x(\omega)}{A_x(\omega)A_z(\omega)-\sigma^2}E_z(y,\omega) \end{aligned} \tag{5.9}$$

From the resulting polarization (5.3):

$$P^j(y,t) = -\eta\begin{pmatrix} q_x x_1^j \\ 0 \\ q_z z_2^j \end{pmatrix}, \quad j \in [x, z] \tag{5.10}$$

the susceptibility reads as:

$$
\overline{\chi}_{ij}(\omega) = \begin{pmatrix} \chi_{xx}(\omega) & 0 & \chi_{xz}(\omega) \\ 0 & 0 & 0 \\ \chi_{zx}(\omega) & 0 & \chi_{zz}(\omega) \end{pmatrix}
$$

$$
\begin{aligned}
\chi_{xx}(\omega) &= \frac{q_x^2\eta}{m}\frac{A_z(\omega)}{A_x(\omega)A_z(\omega)-\sigma^2} \\
\chi_{zz}(\omega) &= \frac{q_z^2\eta}{m}\frac{A_x(\omega)}{A_x(\omega)A_z(\omega)-\sigma^2} \\
\chi_{xz}(\omega) = \chi_{zx}(\omega) &= \frac{q_x q_z\eta}{m}\frac{\sigma}{A_x(\omega)A_z(\omega)-\sigma^2}
\end{aligned}
\tag{5.11}
$$

The most significant change compared to the SRR is the appearance of off-diagonal elements in the susceptibility tensor $\overline{\chi}_{ij}(\omega)$ which is, however, symmetric leading also to $\varepsilon_{ij}(\omega) = \varepsilon_{ji}(\omega)$.

This symmetry relation is important because it is required for time reversal, known as the Onsager-Casimir principle [19, 20]. As expected the tensor of the effective permittivity has the same form as that for planar optical active media [7, 14] resulting in asymmetric transmission due to elliptical dichroism. Furthermore one can mention that for all considerations performed here lossy meta-atoms, i.e., effective permittivity tensor entries that are complex valued, are considered. Note that the optical response would change dramatically if both arms would be identical. In this case the diagonal elements $\overline{\chi}_{ii}(\omega)$ are identical too and the tensor can be diagonalized by a rotation of $\pi/4$. So the polarization eigenstates would be linear and the effective medium would be linearly dichroitic. This is clear since the meta-atom would have an additional mirror symmetry with respect to the plane defined by the surface normal and the line $x = z$, see [33]. Another difference while comparing the L to the SRR meta-atom is that the splitting σ between both resonances is reduced (by a factor of $\sqrt{2}$), which follows from dropping one of the SRR arms. In order to check whether this simple description is valid and to reveal the relation between the SRR and the L-structure eigen modes, numerical FMM simulations has been performed for the L-meta-atom and compared the results to the model in Fig. 5.3. Since we deal now with more complex effective media where the full tensorial nature of the permittivity tensor has to be taken into account the standard transfer matrix algorithm cannot be applied. Hence, it is challenging to determine the scattering coefficients analytically [34] or even to invert them to retrieve effective parameters directly.

Therefore an adapted Fourier modal method was used to determine the transmitted and reflected intensities [35]. In a first approximation the parameters determined for the SRR were used. Based on these parameters $\overline{\chi}_{ij}(\omega)$ was calculated and the far-field intensities as shown in Fig. 5.3a, b for the L-structure. The associated eigenmodes for

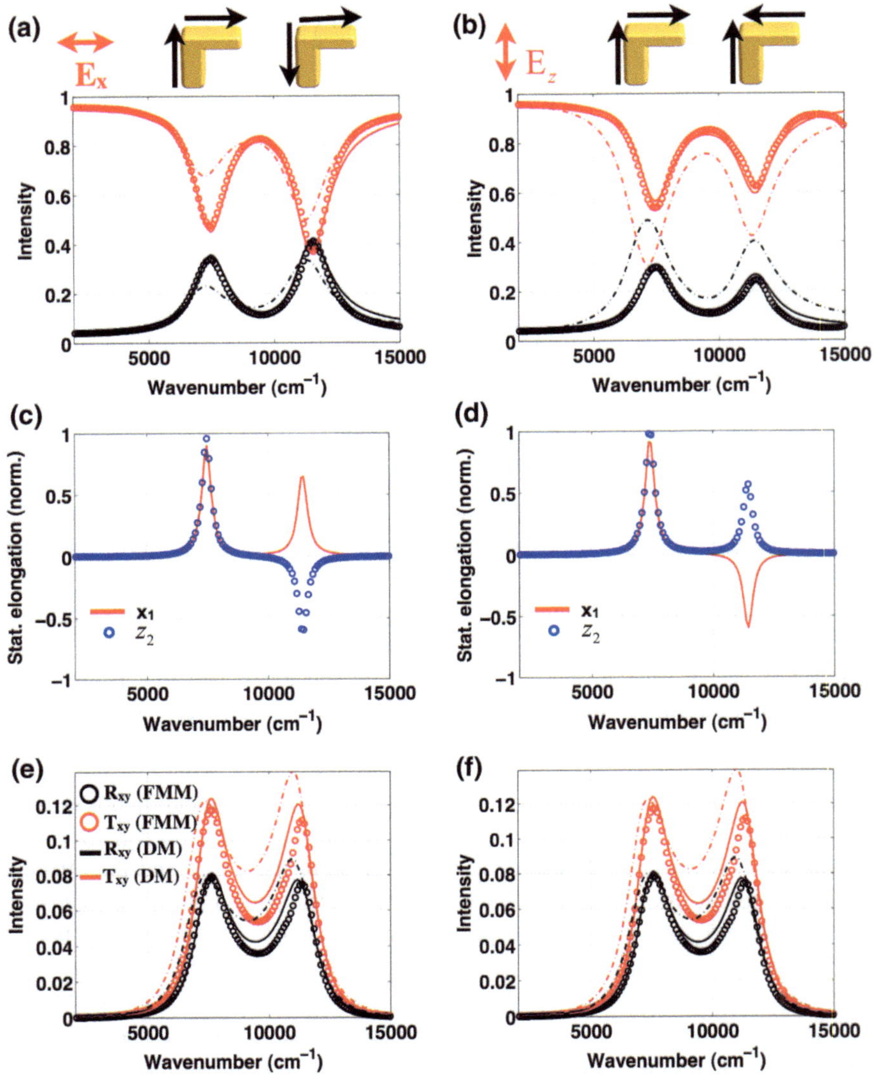

Fig. 5.3 The far-field response of the L structure for **a** x and **b** z polarization. In addition to the numerical (FMM, spheres) and the fitted data (DM, solid lines) the predicted spectra incorporating the SRR parameters (dashed-dotted lines) are plotted. **c** and **d** In contrast to the SRR both eigenmodes can be excited for each polarization direction. The respective numerical cross polarization contributions (FMM, circles) compared with the model predicted (dashed-dotted lines) and the fitted (solid lines) values are shown in (**e**) and (**f**). Note that figures (**e**) and (**f**) are identical as required for such kind of effective media and are only shown for completeness [22]

the carriers are shown in Fig. 5.3c, d. The curves for both the co-polarized Fig. 5.3a, b and the cross-polarized intensities Fig. 5.3e, f are in good qualitative agreement for the rigorous FMM results (dotted line) and the model (dashed line) based on the previously derived SRR oscillator parameters. Although there are deviations between the actual resonance strength, the agreement, e.g., for the resonance positions, is obvious. Note that the intensities, in particular for the cross-polarized fields, can be solely predicted by the coefficients obtained from the SRR. There, an almost perfect agreement is observed. In a second step the oscillator parameters have been adapted in order to fit the exact calculations (solid lines in Fig. 5.3a, b, e, f providing an almost perfect coincidence with the numerical values. Note that the fitting is done only for the co-polarized intensities from which the cross polarized intensities follow. A last step yields the effective permittivity tensor that is inherently accessible and already applied in order to fit the spectra in Fig. 5.4. It can be seen that for the two polarization directions two eigenmodes appear as Lorentz-shaped resonances for the effective permittivity. They differ in strength, due to the different geometrical parameters of the L-arms, Fig. 5.4a, b. The off-diagonal elements $\varepsilon_{xz}(\omega)$ are identical as discussed before, (Fig. 5.3c). Considering especially the second resonance for the off-diagonal tensor elements near $\nu = 11000\,\mathrm{cm}^{-1}$, we observe a Lorentzian anti-resonance that might suggest gain within the system due to the negative imaginary part. However, the nature of this resonance can be explained by the introduced formalism as well. Since the permittivity is proportional to the susceptibility (5.4) and hence also to the carrier displacements (5.11), a negative sign corresponds to a phase difference of π between both oscillating carrier densities (antiparallel oscillations), while for the positive Lorentz resonance at around $\nu = 7000\,\mathrm{cm}^{-1}$, both are oscillating in-phase (sketched by the arrows in Fig. 5.3a). Thus, the negative sign in the imaginary part of the permittivity can by fully explained by means of the mutual interplay of the coupled oscillators.

5.4 Results for S-Type of Metaatoms

In order to verify the model another modification of the SRR has been investigated; namely, the S structure [36, 37] Fig. 5.1a. To observe elliptical dichroism with the same number of coupled entities as for the SRR its mirror symmetry has to be broken. Therefore one of the SRR arms (e.g., x_3) is turned with respect to the SRR

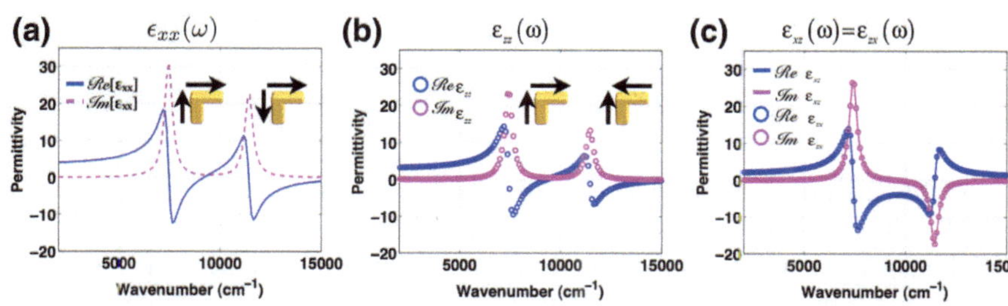

Fig. 5.4 The diagonal effective permittivity tensor elements of the L meta-atom: **a** $\varepsilon_{xx}(\omega)$ and **b** $\varepsilon_{zz}(\omega)$. The arrows indicate the current flow, given by direction of the carrier oscillation for the particular eigen mode. **c** The off-diagonal elements $\varepsilon_{xz}(\omega)$ comprising a Lorentz resonance around $\nu = 7000\,\mathrm{cm}^{-1}$ and an anti-Lorentz resonance at $\nu = 11000\,\mathrm{cm}^{-1}$ [22]

base. This *opening* of the SRR structure is expected to enable the observation of two modes for x polarization, since the oscillator in the base z_2 can now oscillate in-phase or out of-phase with the two remaining x-oriented oscillators of the S structure that are excited by the x-polarized electric field. For z polarization the situation is similar, but now the two oscillators in the horizontal arms (x_1, x_3) are allowed to oscillated in-phase or out-of-phase to the excited oscillator in the base (z_2). Mathematically this modification can be considered by setting $\sigma_{23} = -\sigma_{12} \equiv \sigma$ (5.1), while all other parameters appear similar to the ones applied for the SRR. With these initial assumptions, which reflect all modifications to the geometry, the calculations can be repeated in analogy to those for the SRR and the L structure. Thus, one obtains the elongations for x-polarized excitation:

$$\begin{aligned} x_1^x(\omega) = x_3^x(\omega) &= -\frac{q_x}{m} \frac{A_z(\omega)}{A_x(\omega)A_z(\omega) - 2\sigma^2} E_x(y, \omega) \\ z_2^x(\omega) &= -\frac{q_z}{m} \frac{2\sigma}{A_x(\omega)A_z(\omega) - 2\sigma^2} E_x(y, \omega) \end{aligned} \tag{5.12}$$

and for z-polarized excitation:

$$\begin{aligned} x_1^z(\omega) = x_3^z(\omega) &= -\frac{q_x}{m} \frac{\sigma}{A_x(\omega)A_z(\omega) - 2\sigma^2} E_z(y, \omega) \\ z_2^z(\omega) &= -\frac{q_z}{m} \frac{A_x(\omega)}{A_x(\omega)A_z(\omega) - 2\sigma^2} E_z(y, \omega) \end{aligned} \tag{5.13}$$

as well as the respective effective susceptibility tensor:

$$\begin{aligned} \overline{\chi}_{ij}(\omega) &= \begin{pmatrix} \chi_{xx}(\omega) & 0 & \chi_{xz}(\omega) \\ 0 & 0 & 0 \\ \chi_{zx}(\omega) & 0 & \chi_{zz}(\omega) \end{pmatrix} \\ \chi_{xx}(\omega) &= \frac{q_x^2\eta}{m} \frac{2A_z(\omega)}{A_x(\omega)A_z(\omega) - 2\sigma^2} \\ \chi_{zz}(\omega) &= \frac{q_z^2\eta}{m} \frac{A_x(\omega)}{A_x(\omega)A_z(\omega) - 2\sigma^2} \\ \chi_{xz}(\omega) = \chi_{zx}(\omega) &= \frac{q_x q_z \eta}{m} \frac{2\sigma}{A_x(\omega)A_z(\omega) - 2\sigma^2} \end{aligned} \tag{5.14}$$

In (5.14) the polarization (5.4) is used, which is found to coincide with that of the SRR structure, whereas the elongations are different.

With respect to the splitting of the resonances, it is expected the same resonance positions as for the SRR for z polarization, since 2σ appears in the denominator of the SRR oscillation amplitudes (5.8) for z polarization as well as in all elongations in (5.12), (5.13). Performing the respective numerical and analytical calculations as before for the L structure, one can predict, based on the plasmonic eigenmodes, the

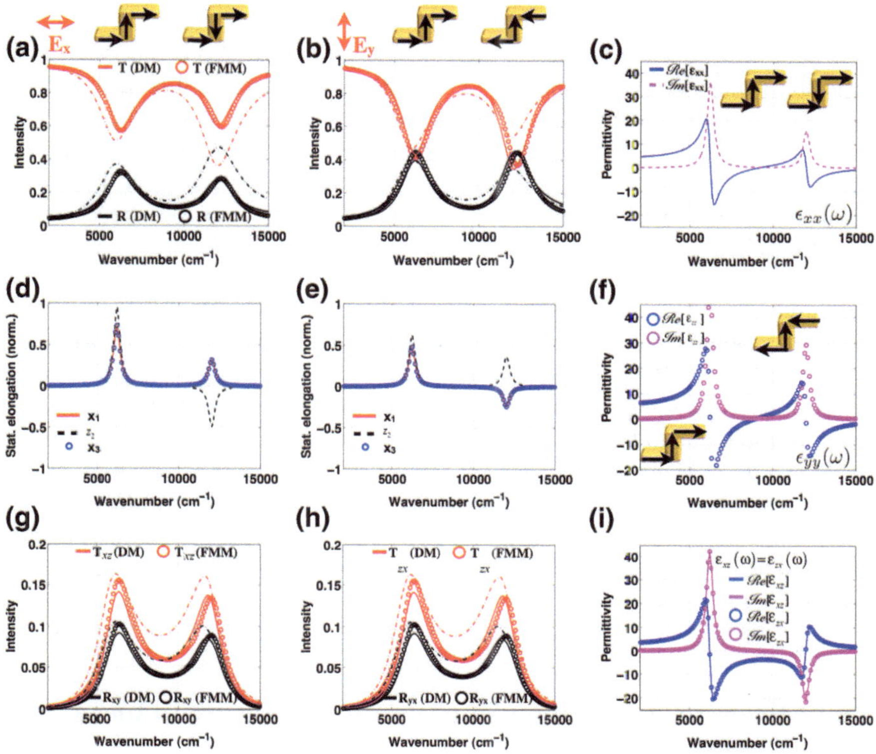

Fig. 5.5 The far-field spectra of the S-structure **a** x polarization and **b** z polarization obtained by numerical simulations (circles), predictions based on the SRR structure parameters (dashed-dotted lines), and adapting the parameters to fit the numerical values (solid lines). The carrier eigenmodes, i.e., an in line current over the entire structure and antiparallel currents (normalized imaginary part of $x_1(\omega)$, $z_2(\omega)$, $x_3(\omega)$ with respect to the center part of the S-structure are observed for the two polarization directions **d** x and **e** z. The comparison between the cross-polarization contributions (T_{ij}, R_{ij}) for the numerical simulations (circles), the predicted lines from the SRR parameters (dashed-dotted lines) and the fitted parameter spectra (solid lines) for the found parameters deduced for fitting the co-polarized response (T_{ij}, R_{ij}). The effective permittivity tensor (**c**), **f** diagonal and **i** the off-diagonal components [22]

spectral response as well as the effective material properties. The results are shown in Fig. 5.5. Considering these eigenmodes in Fig. 5.5d, e, one may distinguish two situations. At first the eigenmode, represented by three dipoles, being in-phase along the entire structure. The second eigenmode is characterized by two dipoles being in-phase in the arms and out-of-phase in the base. Both eigenmodes are excited for x and z polarization, respectively, and will lead to spectral resonances appear in gas dips and peaks in transmission or reflection, respectively (Fig. 5.5a, b).

The spectral positions of the resonances are in agreement with the expectations from the SRR structure resonances. As expected, for both the S and the L meta-atom the in phase eigenmodes appear at smaller wave numbers (larger wavelengths) compared to the out-of-phase ones. This is completely in agreement with arguments from plasmon hybridization theory [38]. Again the use of the oscillator parameters as optimized for the SRR structure (dash-dotted lines in Fig. 5.5a, b, g, h) reveals the relationship between both structures, since the numerically (circles) calculated spectra agree with respect to the overall shape and the resonance positions Fig. 5.5d, e very well with the parametrical predictions (solid and dashed lines). A subsequent fitting again improves the results toward almost excellent agreement, which can also be observed for the cross-polarization observables Fig. 5.5g, h. Considering the tensor components of the effective permittivity Fig. 5.5c, f, i, one observes a difference between the diagonal entries due to the geometrical differences in the S-structure center and arms, while the anti-resonance for the out-of-phase eigenmode is observed in the off-diagonal elements with the same origin as discussed for the L structure. In passing, it was mentioned that due to the presence of the substrate the planar meta-atoms forming the S and L MM are chiral as well. However, chiral effects induced by the substrate can be neglected as compared to the effects of the anisotropy of our planar meta-atoms itself [12, 13]. All parameters required for the calculations presented here for the SRR as well as the two presented modifications are provided in [22].

5.5 Metamaterial Analogy of Electromagnetically Induced Transparency

Electromagnetic Induced Transparency is a quantum effect, which appears for example in 3-level system under appropriate conditions and special requirements for coherence of the fields involved—see Fig. 5.6, where two transitions 1–3 and 2–3 are supposed to be dipole allowed, and transition 1–2 is forbidden in dipole approximation. Without the control external field at the frequency ω_c the measured transmission spectrum around the transition resonance 13 is shown in Fig. 5.7 by dotted line.

In case of presence of the second field at frequency ω_{23} the spectrum exhibits an extra transmission peak at frequency ω_{13}. This effect can be explained by

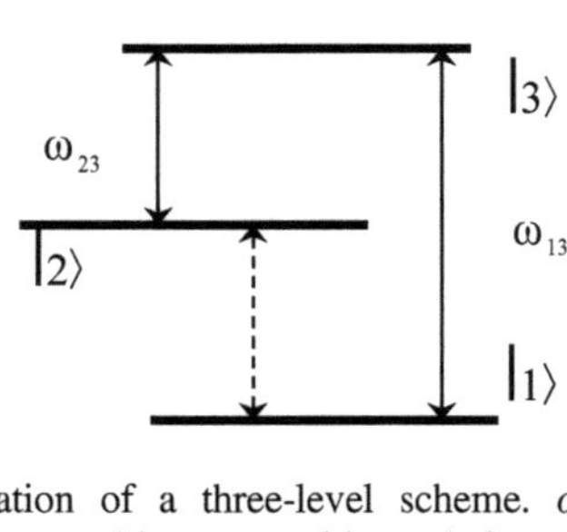

Fig. 5.6 Schematic representation of a three-level scheme. ω_{13} and ω_{23} are the resonant frequencies of the respective transitions; transitions 1–3 and 2–3 are allowed in dipole approximation, transition 1–3 is forbidden in dipole approximation

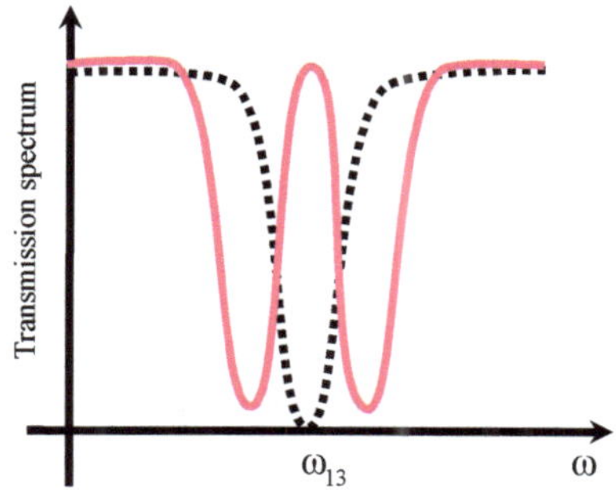

Fig. 5.7 Transmission spectrum of a three-level scheme: dotted line—transmission spectrum without control field at frequency ω_{23} and solid magenta line—transmission spectrum with the control field at frequency ω_{23}

interference between different transition pathways in a three-level system under simultaneous action of two fields at frequencies ω_{23} and ω_{13} [39–41]. There are several classic systems exhibiting similar properties (transmission spectrum behavior shown in Fig. 5.7), namely coupled microresonators [42], resonators coupled with waveguides [43, 44], metallic structures [45], and in particular plasmonic induced transparency effect in MMs, which has been theoretically described [46] and investigated experimentally [47, 48]. Another one similarity (with Fano resonances) has been both theoretically predicted and experimentally observed [49].

It has to be mentioned that originally quantum effect of the EIT has not so much to do physically with its classical analog in MMs. In quantum system the interference appears in purely quantum description of the polarizability dynamics, and the respective interference between two fields exhibits itself in two respective terms in full Hamiltonian, which could really cancel each other under appropriate conditions. In case of the plasmonic structures (for example, in case of the Split Ring Resonators—see Fig. 5.8) the respective effect can be considered as a just one of particular representation of the modes of this structure and has, of course, nothing to do with the quantum mechanical interference effect; in fact, in case of plasmonic EIT the second field (analog of the field at frequency ω_{23}) is not required.

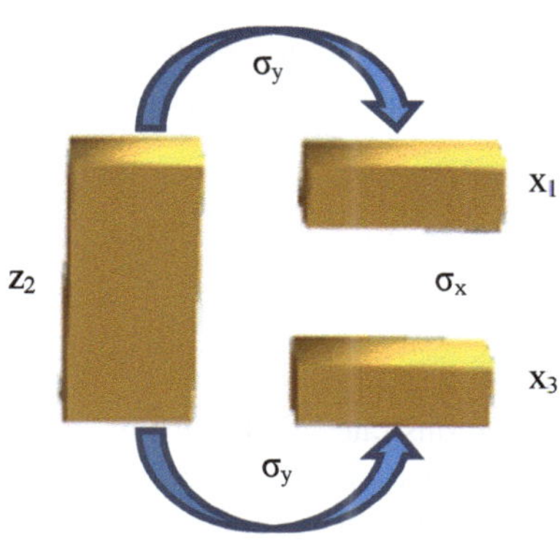

Fig. 5.8 Partially coupled SRR structure which exhibits quasi EIT optical properties

The set of dynamical equations, describing the optical properties of the structure in Fig. 5.8 in the case of z-polarization of the external electric field is:

$$\begin{cases} \dfrac{\partial^2 x_1(t)}{\partial t^2} + \gamma_x \dfrac{\partial x_1(t)}{\partial t} + \omega_x^2 x_1(t) + \sigma_z z_2(t) = 0 \\ \dfrac{\partial^2 z_2(t)}{\partial t^2} + \gamma_z \dfrac{\partial z_2(t)}{\partial t} + \omega_z^2 z_2(t) + \sigma_z (x_1(t) - x_3(t)) = -\dfrac{q}{m} E_z \\ \dfrac{\partial^2 x_3(t)}{\partial t^2} + \gamma_x \dfrac{\partial x_3(t)}{\partial t} + \omega_x^2 x_3(t) - \sigma_z z_2(t) = 0 \end{cases} \quad (5.15)$$

In the frequency domain (5.15) is:

$$\begin{cases} x_1(\omega)\left(\omega_x^2 - \omega^2 - i\omega\gamma_x\right) + \sigma_x x_3(\omega) + \sigma_z z_2(\omega) = 0 \\ z_2(\omega)\left(\omega_y^2 - \omega^2 - i\omega\gamma_y\right) + \sigma_x (x_1(\omega) - x_3(\omega)) = -\dfrac{q}{m} E_z \\ x_2(\omega)\left(\omega_x^2 - \omega^2 - i\omega\gamma_x\right) + \sigma_x x_1(\omega) - \sigma_z z_2(\omega) = 0 \end{cases} \quad (5.16)$$

The corresponding solution for y dynamics, polarization, and respective permittivity for the MMs consisting of the structures shown in Fig. 5.8 are:

$$\begin{cases} x_1(\omega) = -x_3(\omega) & z_2(\omega) = -\dfrac{q}{m} \dfrac{E_z(\omega)}{A_z(\omega) + \frac{2\sigma_z^2}{\sigma_x - A_x(\omega)}} \\ P_z(\omega) = E_z(\omega) + 4\pi q z_2(\omega) & \varepsilon(\omega) = 1 + 4\pi \dfrac{q^2}{m} \dfrac{1}{A_z(\omega) + \frac{2\sigma_z^2}{\sigma_x - A_x(\omega)}} \\ A_z(\omega) = \omega_z^2 - \omega^2 - i\omega\gamma_z & A_x(\omega) = \omega_x^2 - \omega^2 - i\omega\gamma_x \end{cases} \quad (5.17)$$

Using expression for the dielectric permittivity, one can calculate transmission and reflection spectra for the respective homogenized MM; the results are presented in Fig. 5.9. The results, obtained here based on the developed model have been compared with ones obtained in [46] from more rigorous calculations of the same structure—see Fig. 5.9c, d. It is clearly seen, that the developed simple model, based on the coupled harmonic oscillator dynamics, gives very good results and can be undoubtedly used for estimations of the optical properties of MMs.

5.6 Conclusion

In summary, it has been demonstrated that the developed model permits the calculation of effective material parameters for planar MMs consisting of variable meta-atoms formed by a few straight wire sections of potentially different shape.

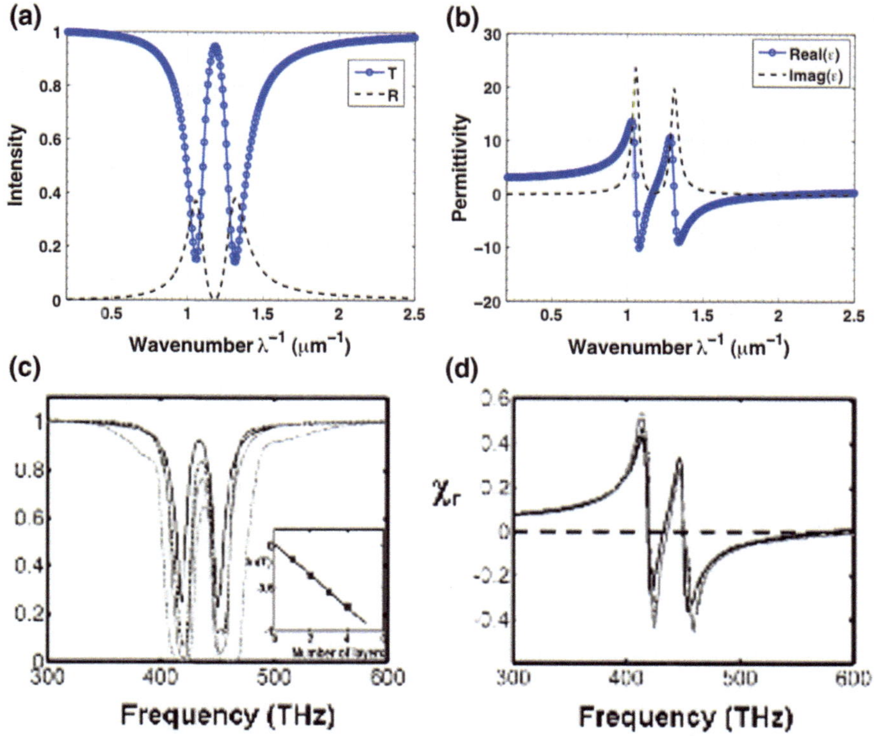

Fig. 5.9 Transmission/reflection spectra and respective permittivity of the MM with the MAs shown in Fig. 5.8. **a**, **b**—Calculated data, **c**, **d**—data from [46], Fig. 3, presented here for comparison

The model takes advantage of resonant electric dipole oscillations in the wires and their mutual coupling. The pertinent form of the meta-atom determines the actual coupling features. Although the model is parametric since it requires at least one rigorous simulation to fix all parameters, the determination of modified MAs can be easily performed analytically. Here the SRR has been used as a particular reference meta-atom, whereas two modified meta-atoms, the L and the S meta-atom have been modeled on the basis of the initial SRR parameters. Thus, this model represents a kind of building block approach for quite different meta-atoms. Since in particular the effect of asymmetric transmission for circular polarized light attracted a lot of research interest recently, an attention here is focused on planar meta-atoms that are optically active in the same manner due to elliptical dichroism. Within our model all properties of the effective permittivity tensors for such kind of media are correctly predicted and the corresponding scattering characteristics are in very good agreement with the rigorous numerical results. Moreover, some effects which are of interest now in terms of mimicking quantum mechanical ones can be successfully described using the developed approach.

References

1. U.K. Chettiar, A.V. Kildishev, H.-K. Yuan, W. Cai, S. Xiao, V.P. Drachev, V.M. Shalaev, Opt. Lett. **32**, 1671 (2007)
2. G. Dolling, M. Wegener, C. Soukoulis, Opt. Lett. **32**, 53 (2007)
3. J. Valentine, S. Zhang, T. Zentgraf, G. Ulin-Avila, D. Genov, X. Zhang, Nature **455**, 376 (2008)
4. C. Helgert, C. Menzel, C. Rockstuhl, E. Pshenay-Severin, E.B. Kley, A. Chipouline, A. Tunnermann, F. Lederer, T. Pertsch, Opt. Lett. **34**, 704 (2009)
5. C. Garcia-Meca, R. Ortuno, F.J. Rodriguez-Fortuno, J. Marti, A. Martinez, Opt. Lett. **34**, 1603 (2009)
6. M. Rill, C. Kriegler, M. Thiel, G. von Freymann, S. Linden, M. Wegener, Opt. Lett. **34**, 19 (2009)
7. B. Bai, Y. Svirko, J. Turunen, T. Vallius, Phys. Rev. A **76**, 023811 (2007)
8. L. Arnaut, J. Electromagn. Waves Appl. **11**, 1459 (1997)
9. J. Reyes, A. Lakhtakia, Opt. Commun. **266**, 565 (2006)
10. S. Prosvirnin, N. Zheludev, J. Opt. A: Pure Appl. Opt. **11**, 074002 (2009)
11. S. Tretyakov, I. Nefedov, A. Shivola, S. Maslovski, C. Simovski, J. Electromagn. Waves Appl. **17**, 695 (2003)
12. V. Fedotov, P. Mladyonov, S. Prosvirnin, A.V. Rogacheva, Y. Chen, N. Zheludev, PRL **97**, 167401 (2006)
13. V. Fedotov, A. Schwanecke, N. Zheludev, V. Khardikov, S. Prosvirnin, Nano Lett. **7**, 1997 (2007)
14. S. Zhukovsky, A. Novitsky, V. Galynsky, Opt. Lett. **34**, 1988 (2009)
15. J. Pendry, Science **306**, 1353 (2004)
16. S. Tretyakov, A. Sihvola, L. Jylhä, Photonics Nanostruct. Fundam. Appl. **3**, 107 (2005)
17. J. Zhou, J. Dong, B. Wang, T. Koschny, M. Kafesaki, C. Soukoulis, Phys. Rev. B **79**, 1 (2009)
18. J. Petschulat, C. Menzel, A. Chipouline, C. Rockstuhl, A. Tünnermann, F. Lederer, T. Pertsch, Multipole approach to metamaterials. Phys. Rev. B **78**, 043811 (2008)
19. L. Onsager, Phys. Rev. **37**, 405 (1931)
20. H. Casimir, Rev. Mod. Phys. **17**, 343 (1945)
21. S. Tretyakov, A. Sihvola, B. Jancewicz, J. Electromagn. Waves Appl. **16**, 573 (2002)
22. J. Petschulat, A. Chipouline, A. Tüunnermann, T. Pertsch, C. Menzel, C. Rockstuhl, F. Lederer, Phys. Rev. A **80**, 063828 (2009)
23. H. Raether, *Surface Plasmons* (Springer, New York, 1988)
24. C. Enkrich, M. Wegener, S. Linden, S. Burger, L. Zschiedrich, F. Schmidt, J.F. Zhou, T. Koschny, C.M. Soukoulis, PRL **95**, 203901 (2005)
25. R. Marqués, F. Medina, R. Rafii-El-Idrissi, Phys. Rev. B **65**, 144440 (2002)
26. C. Rockstuhl, T. Zentgraf, E. Pshenay-Severin, J. Petschulat, A. Chipouline, J. Kuhl, T. Pertsch, H. Giessen, F. Lederer, Opt. Express **15**, 8871 (2007)
27. R. Raab, O. De Lange, *Multipole Theory in Electromagnetism* (Clarendon, Oxford, 2005)
28. L. Li, J. Opt. Soc. Am. A **14**, 2758 (1997)
29. P.B. Johnson, R.W. Christy, Phys. Rev. B **6**, 4370 (1972)
30. P. Yeh, *Optical Waves in Layered Media* (Wiley, New York, 1998)
31. B. Canfield, S. Kujala, M. Kauranen, K. Jemovs, T. Vallius, J. Turunen, Appl. Phys. Lett. **86**, 183109 (2005)
32. B.K. Canfield, S. Kujala, K. Jefimovs, T. Vallius, J. Turunen, M. Kauranen, J. Opt. A: Pure Appl. Opt. **7**, S110 (2005)
33. M. Decker, S. Linden, M. Wegener, Opt. Lett. **34**, 1579 (2009)
34. G. Borzdov, J. Math. Phys. **38**, 6328 (1997)
35. L. Li, J. Opt. A: Pure Appl. Opt. **5**, 345 (2003)

36. H. Chen, L. Ran, J. Huangfu, X. Zhang, K. Chen, T.M. Grzegorczyk, J.A. Kong, APL **86**, 151909 (2005)
37. H.S. Chen, L.X. Ran, J.T. Huangfu, X.M. Zhang, K.S. Chen, T.M. Grzegorczyk, J.A. Kong, Prog. Electromagn. Res. **51**, 231 (2005)
38. E. Prodan, C. Radloff, N. Halas, P. Nordlander, Science **302**, 419 (2003)
39. K. Boller, A. Imamoglu, S. Harris, Observation of electromagnetically induced transparency. PRL **66**, 2593 (1991)
40. S.E. Harris, Electromagnetically induced transparency. Phys. Today **50**, 36 (1997)
41. M. Fleischhauer, A. Imamoglu, J. Marangos, Electromagnetically induced transparency: optics in coherent media. Rev. Mod. Phys. **77**, 633 (2005)
42. Q. Xu et al., Experimental realization of an on-chip all-optical analogue to electromagnetically induced transparency. PRL **96**, 123901 (2006)
43. E. Waks, J. Vuckovic, Dipole induced transparency in drop-filter cavity-waveguide systems. PRL **96**, 153601 (2006)
44. M. Yanik, W. Suh, Z. Wang, S. Fan, Stopping light in a waveguide with an all-optical analog of electromagnetically induced transparency. PRL **93**, 233903 (2004)
45. N. Papasimakis, V. Fedotov, N. Zheludev, PRL **101**, 253903 (2008)
46. S. Zhang, D. Genov, Y. Wang, M. Liu, X. Zhang, Plasmon-induced transparency in metamaterials. PRL **101**, 047401 (2008)
47. N. Liu, L. Langguth, J.K.T. Weiss, M. Fleischhauer, T. Pfau, H. Giessen, Nat. Mater. **8**, 758 (2009)
48. M. Liu, T.-W. Lee, S. Gray, P. Guyot-Sionnest, M. Pelton, Excitation of dark plasmons in metal nanoparticles by a localized emitter. PRL **102**, 107401 (2009)
49. B. Luk'yanchuk, N. Zheludev, S. Maier, N. Halas, P. Nordlander, H. Giessen, C. Chong, The Fano resonance in plasmonic nanostructures and metamaterials. Nat. Mater. **9**, 707–715 (2010)
50. J. Petschulat, A. Chipouline, A. Tünnermann, T. Pertsch, C. Menzel, C. Rockstuhl, T. Paul, F. Lederer, Simple and versatile analytical approach for planar metamaterials. Phys. Rev. B **82**, 075102 (2010)

Chapter 6
Applications of the "Classical" Metamaterial Model—Metamaterials with Interaction Between Meta-Atoms

6.1 Introduction

The interaction between the small particles (meta-atoms), either dielectric or metallic, and the propagation of an optical excitation in a regular chain of such particles has been extensively investigated [1–6]. Interest in the behavior of chains of metallic nanoparticles was driven mainly by the pursuit of subwavelength guiding structures for a new generation of the optoelectronic components in the area of communication and information processing. Nevertheless, theoretical tools for the modeling of these chains (irrespective to the nature and sizes) remain invariant: the electromagnetic excitation in the particles is supposed to be described by taking into account all possible eigenmodes [1, 3] and interactions between all particles in a chain. There are several approximations which are typically accepted in these kinds of problems. Firstly, depending on the size of the particles, the model can be restricted by consideration of dipole moment only (for metallic nanoparticles) [2, 6]; the higher moments can be taken into consideration and similarly in the case of investigation of magnetic response [5, 7]. Usually, for the problem of only electromagnetic excitation propagation the chain the dipole approximation is enough [8], provided distance between particles is not less than about three times their dimensions. Secondly, the interaction between the particles in a chain can be considered in the frame of the quasi-static approximation, where no retardation between particles is retained; otherwise interaction between dipoles contains terms proportional to the $1/r$ and $1/r^2$ in addition to the quasistatic term of $1/r^3$ (r is the distance between dipoles). The problem possesses an exact solution for the infinite chain in the quasistatic limit, while taking into consideration the retardation leads to known mathematical difficulties and requires continuation into the lower half frequency plane [2]. Consideration of the finite chain is free from these excessive mathematical problems, but can be treated only numerically; the respective solutions for both longitudinal and transverse modes are presented in [2, 9].

A. Chipouline and F. Küppers, *Optical Metamaterials: Qualitative Models*, Springer Series in Optical Sciences 211, https://doi.org/10.1007/978-3-319-77520-3_6

In this chapter the multipole approach for the homogenization of the MM [10, 11] will be extended in the case of regularly placed interacting MAs in the form of a double-wire structure. As in the previous chapters, the MM consisting of the identical layers with the regularly spaced MAs in each is considered; one layer of the MM (the layers repeat themselves in y direction), is presented in Fig. 6.1.

The interaction between the MAs is assumed to be negligible in the longitudinal direction (wave propagation direction, perpendicular to the layer surfaces), in other words, the layers are assumed to be well separated from each other. The effect of interaction in lateral direction (parallel to the layer surface) and its influence on the dispersion relation of the plane waves propagating in the MM is the subject of this chapter.

When there is a coupling between the MAs, the response of the medium is no longer truly local. As a result, depending upon the configuration of the MA, the medium responds differently to electromagnetic waves propagating in different directions. This phenomenon is called spatial dispersion (akin to the phenomenon of temporal dispersion—where the response of a medium at a given time depends on the history of its excitation). The direction dependence of the medium response is not just unique for the spatial dispersion—in the case of anisotropic media this effect appears as well. The qualitative markers for differentiation of these two effects have been considered in details in [12, 13] (see also references therein), and have already been discussed in Sect. 2.3. One of the ways to describe spatial dispersion is to use a model of a chain of coupled harmonic oscillators. Such a model is adequate as a first approximation for the interactions between plasmonic nanoresonators. Eigenmodes of the response are obtained as wave solutions, giving the oscillations of the plasmonic charges in each nanoresonator (see Fig. 6.2).

A similar approach has been used to study the effect of interaction of the MAs on the bulk properties of the MM in [14–17] the microwave frequency region. To better understand the problem, the case of a one dimensional chain of coupled harmonic oscillators (coupled dipoles) is studied, after which the problem is extended to the case of coupled MA in the form of double-wires (coupled quadrupoles). The results of the analysis for both ensembles are presented side by side to enable a comparative study of the dispersion characteristics.

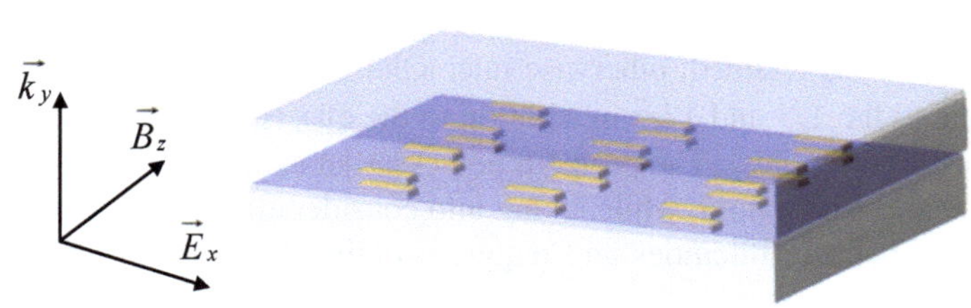

Fig. 6.1 Artificial MAs (plasmonic nanoresonators) embedded in a dielectric matrix form the MM (only one layer is presented). Polarization of the electric and magnetic fields, and direction of the wave vector are shown. The interaction between MAs takes place in z direction; the possible interaction in y direction between the MAs (wave propagation direction) is not taken into account (the interaction between the nanowires in a single MA in y direction is taken into account)

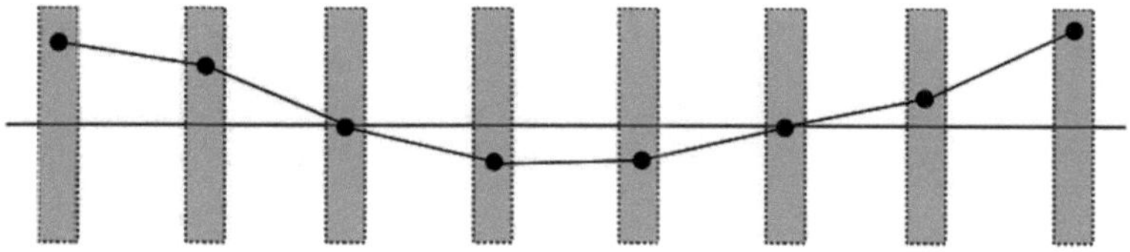

Fig. 6.2 Spatial dispersion viewed as a consequence of a coupling effect in a chain of dipoles. The problem is equivalent to the study of transverse oscillation dynamics in a chain of the coupled harmonic oscillators

Knowledge of the dispersion relation is very important; actually, it is the only relationship that is required to analyze the wave propagation in a media (boundary condition problems are not included in the discussion here). For example, it is known that in order to provide a better resolution (in an optical imaging system) the media has to allow propagation of the lateral components of the wave vector $\vec{k}$ with as high values as possible (as higher as possible k_z components in Fig. 6.3 for the same wavelength). This can be achieved, for example, in a hyperbolic dispersion media [18]. The analysis performed below provides a tool to analyze whether the coupling between MAs can enhance the available spatial spectrum of the propagating waves and hence increase resolution of the optical systems, which use the respective MM.

6.2 Dispersion Relations for Material Eigenwaves

6.2.1 *Periodic Chain of Coupled Dipoles*

A chain of the periodically positioned dipoles (oriented with the long axis along the x direction) is considered (see Fig. 6.3). For clarity, only one row is shown in the figure; it is assumed that rows of dipoles are placed along the y direction. The treatment of the problem remains the same, and coupling between neighboring rows are neglected. The arrangement of the dipoles is along the z direction and the period of spacing is taken to be z_0. The effect of coupling with adjacent oscillators is introduced via a coupling constant σ, which is a function of the distance between the oscillators. Well known solution in form of the transverse spatial modes which can be sustained in such a medium under the above mentioned conditions can be straightforwardly obtained.

The dynamic equation for the nth oscillator is:

$$\frac{\partial^2 x_n}{\partial t^2} + \gamma \frac{\partial x_n}{\partial t} + \omega_0^2 x_n + \sigma(x_{n+1} + x_{n-1}) = \frac{q}{m} E_x \tag{6.1}$$

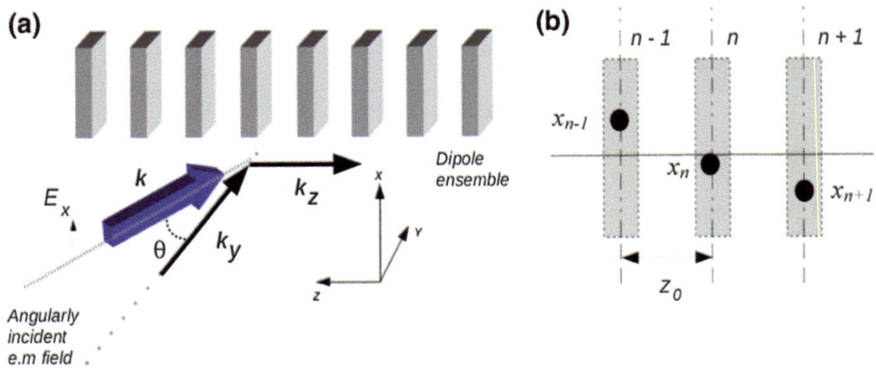

Fig. 6.3 Geometry of propagation: **a** the electric field is polarized along the long axis of the cut-wires, angular incidence gives rise to spatial modes in the ensemble; **b** the dipoles in an arbitrary triplet are labeled as n, $n+1$ and $n-1$. The positional coordinates of the charge clouds x_n within these dipoles are also indexed with these labels

The eigenmodes ($E_x = 0$) can then be obtained by transferring the problem to the Fourier domain using the ansatz:

$$x_n = A_0 \exp(ik_z n z_0 - i\omega t) \tag{6.2}$$

The amplitude of the nth oscillator in terms of the wave vector k_z is thus given by:

$$x_n = A_0 \exp(ik_z n z_0) \tag{6.3}$$

Essentially then, the ansatz describes a system of oscillators vibrating at the same frequency, with k giving the periodicity of the spatial mode. Substitution of this ansatz into (6.1) gives the dispersion relation:

$$\omega_0^2 - \omega^2 - i\omega\gamma + 2\sigma \cos(k_z z_0) = 0 \tag{6.4}$$

The solution of this equation is presented in Figs. 6.8 and 6.9 and the respective discussion is given in Sect. 6.4.

6.2.2 *Periodic Chain of Coupled Quadrupoles*

The above arguments are now extended to the case of the chain of the coupled quadrupoles. The two double-wires forming the quadrupoles are assumed to be oriented with their long axes along the x direction as before, while being separated by a small distance $2y_1$ in the y direction. It is further assumed, that the quadrupoles themselves are periodically spaced along the z direction, with the spatial period z_0. The interaction of a single cut-wire with five of (suggests that there are alternative choices) its nearest neighbors must be considered (see Fig. 6.4) for both cut-wires (that is, both x_n and $x_n^{'}$), as they experience different excitation conditions due to the retardation of the wave propagating into y direction.

As before, we are looking for the transverse spatial modes which can be sustained in such a medium. The dynamic equations for the respective eigenmodes can be written as:

$$\begin{cases} \frac{\partial^2 x_n}{\partial t^2} + \gamma \frac{\partial x_n}{\partial t} + \omega_0^2 x_n + \sigma(x_{n+1} + x_{n-1}) + \sigma^{'}\left(x_{n+1}^{'} + x_{n-1}^{'}\right) + \sigma_0 x_n^{'} = 0 \\ \frac{\partial^2 x_n^{'}}{\partial t^2} + \gamma \frac{\partial x_n^{'}}{\partial t} + \omega_0^2 x_n^{'} + \sigma\left(x_{n+1}^{'} + x_{n-1}^{'}\right) + \sigma^{'}(x_{n+1} + x_{n-1}) + \sigma_0 x_n = 0 \end{cases} \tag{6.5}$$

where x_n and $x_n^{'}$ are the instantaneous coordinates of the plasmonic charge clouds on the cut-wires; σ_0, σ, and $\sigma^{'}$ are the coupling coefficients given by:

$$\sigma = \sigma_0 \frac{(2y_1)^3}{z_0^3}, \quad \sigma^{'} = \sigma_0 \frac{(2y_1)^3}{\left(z_0^2 + (2y_1)^2\right)^{3/2}} \tag{6.6}$$

where $2y_1$ is the spatial period of the quadrupoles in the direction of the wave propagation y. The following ansatz is assumed [compare with (6.2)]:

$$x_n = A_0 \exp(ik_z n z_0 - i\omega t), \quad x_n^{'} = A_0^{'} \exp(ik_z n z_0 - i\omega t) \tag{6.7}$$

Substituting the ansatz in the dynamic equations, we arrive at:

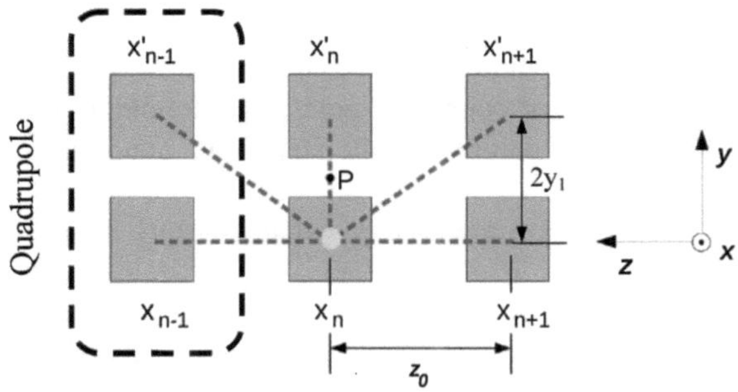

Fig. 6.4 Nearest neighbor interactions—top view of the one dimensional chain of the quadrupoles (two double-wires forming one from the three shown quadrupoles are surrounded by dashed frame). The dashed lines indicate the interactions that have to be taken into account. Point P indicates the center of the nth quadrupole

$$\begin{cases} A_0(R+2\sigma\cos(k_z z_0)) + A_0^{'}\left(2\sigma^{'}\cos(k_z z_0)+\sigma_0\right) = 0 \\ A_0\left(2\sigma^{'}\cos(k_z z_0)+\sigma_0\right) + A_0^{'}(R+2\sigma\cos(k_z z_0)) = 0 \end{cases} \tag{6.8}$$

with $R = \omega_0^2 - \omega - i\gamma\omega$. For the above system of equations to be consistent, the determinant must vanish:

$$\begin{vmatrix} R+2\sigma\cos(k_z z_0) & 2\sigma^{'}\cos(k_z z_0)+\sigma_0 \\ 2\sigma^{'}\cos(k_z z_0)+\sigma_0 & R+2\sigma\cos(k_z z_0) \end{vmatrix} = 0 \tag{6.9}$$

This gives the dispersion relation equation:

$$\left(R+\sigma_0+2\left(\sigma+\sigma^{'}\right)\cos(k_z z_0)\right)\left(R-\sigma_0+2\left(\sigma-\sigma^{'}\right)\cos(k_z z_0)\right) = 0 \tag{6.10}$$

which results in two equations, each one giving the bands of frequencies capable of producing the spatial modes. The equations are:

$$\begin{cases} \omega_0^2 - \omega^2 - i\omega\gamma + \sigma_0 + 2\left(\sigma+\sigma^{'}\right)\cos(k_z z_0) = 0 \\ \omega_0^2 - \omega^2 - i\omega\gamma - \sigma_0 + 2\left(\sigma-\sigma^{'}\right)\cos(k_z z_0) = 0 \end{cases} \tag{6.11}$$

The solutions of the above equations are presented in Figs. 6.8 and 6.9 and the respective discussion is given in Sect. 6.4.

6.3 Dispersion Relations for Electromagnetic Waves

6.3.1 Periodic Chain of Coupled Dipoles

To obtain the dispersion relation for the electromagnetic wave propagating in a media with periodic chains of coupled dipoles, the Helmholtz equation has to be employed. The electric field solution of the Helmholtz equation is of the form:

$$E_x = E_{x,0}\exp\left(ik_y y + ik_z n z_0 - i\omega t\right) \tag{6.12}$$

which is a plane wave polarized along the x axis, with its wave vector in the yz plane. The propagation vector components along y and z directions are denoted by k_y and k_z. In turn, k_y and k_x can be expressed in their polar forms:

$$\begin{aligned} k_y &= k\cos(\theta) \\ k_z &= k\sin(\theta) \end{aligned} \tag{6.13}$$

where k is the magnitude of the propagation vector k, and θ is the angle of incidence measured from the normal to the xz plane (i.e. the plane containing the dipoles)—see Fig. 6.3. The dynamics of the system can then be modeled via (6.1). Substituting the ansatz for x_n and E_x, namely:

$$\begin{cases} E_x = E_{x,0}\exp\left(ik_y y + ik_z n z_0 - i\omega t\right) \\ x_n = A_0 \exp\left(ik_y y + ik_z n z_0 - i\omega t\right) \end{cases} \tag{6.14}$$

and dynamical (6.1) becomes:

$$\left[\omega_0^2 - \omega^2 - i\omega\gamma + 2\sigma\cos(kz_0\sin(\theta))\right] A_0 \exp(ik_z n z_0) = \frac{q}{m} E_{x,0}\exp(ik_z z) \tag{6.15}$$

Considering the field (6.12) at the discrete points:

$$z = nz_0 \tag{6.16}$$

(6.15) becomes an equation in terms of amplitude A_0:

$$\left[\omega_0^2 - \omega^2 - i\omega\gamma + 2\sigma\cos(kz_0\sin(\theta))\right] A_0 = \frac{q}{m} E_{x,0} \tag{6.17}$$

or:

$$A_0 = \frac{q}{m}\frac{E_{x,0}}{\left(\omega_0^2 - \omega^2 - i\omega\gamma + 2\sigma\cos(kz_0\sin(\theta))\right)} \tag{6.18}$$

The polarization of the medium can thus be given by:

$$P_{x,0} = \eta q A_0 = \frac{\eta q^2}{m}\frac{E_{x,0}}{\left(\omega_0^2 - \omega^2 - i\omega\gamma + 2\sigma\cos(kz_0\sin(\theta))\right)} \tag{6.19}$$

where η is the concentration of the dipoles.

An effective susceptibility of the medium is defined via the equation:

$$P_{x,0}(\omega, k) = \eta q \chi(\omega, k) E_{x,0} \tag{6.20}$$

such that:

$$\chi(\omega, k) = \frac{q}{m}\frac{1}{\left(\omega_0^2 - \omega^2 - i\omega\gamma + 2\sigma\cos(kz_0\sin(\theta))\right)} \tag{6.21}$$

The Helmholtz equation for the wave propagation can now be used. For the present case of dipoles $Q = 0$, $M = 0$, and using the plane wave ansatz, the Helmholtz equation is written as:

$$-\left(k_y^2+k_z^2\right)E_{x,0}+\frac{\omega^2}{c^2}\left(E_{x,0}+4\pi P_{x,0}(\omega,k)\right)=0 \tag{6.22}$$

Substituting the polar form for the wave vector components (6.13), the dispersion relation takes the final form:

$$k^2=\frac{\omega^2}{c^2}\left[1+4\pi\frac{q^2\eta}{m}\frac{1}{\left(\omega_0^2-\omega^2-i\omega\gamma+2\sigma\cos(kz_0\sin(\theta))\right)}\right] \tag{6.23}$$

This transcendental equation for k must be solved numerically. The results of the numerical solution of (6.23) are presented in Fig. 6.8.

6.3.2 *Periodic Chain of Coupled Quadrupoles*

The following analysis is carried out using the definitions made in Sect. 6.2 (see also Fig. 6.4). To derive the dispersion relation for electromagnetic waves propagating in a medium with chains of the coupled quadrupoles, the dynamics of system (6.5) has to be investigated under the influence of an electromagnetic plane wave. The dynamic equations can now be written as:

$$\begin{cases}\dfrac{\partial^2 x_n}{\partial t^2}+\gamma\dfrac{\partial x_n}{\partial t}+\omega_0^2 x_n+\sigma(x_{n+1}+x_{n-1})+\sigma'\left(x'_{n+1}+x'_{n-1}\right)+\sigma_0 x'_n=\dfrac{qE_x}{m}\exp(ik_y y_1)\\ \dfrac{\partial^2 x'_n}{\partial t^2}+\gamma\dfrac{\partial x'_n}{\partial t}+\omega_0^2 x'_n+\sigma\left(x'_{n+1}+x'_{n-1}\right)+\sigma'(x_{n+1}+x_{n-1})+\sigma_0 x_n=\dfrac{qE_x}{m}\exp(-ik_y y_1)\end{cases} \tag{6.24}$$

where the phase retardations are measured with respect to the center point P (see Fig. 6.4). Using the same ansatz:

$$\begin{cases}x'_n=A'_0\exp\left(ik_y y+ik_z n z_0-i\omega t\right)\\ x_n=A_0\exp\left(ik_y y+ik_z n z_0-i\omega t\right)\end{cases} \tag{6.25}$$

and taking $z=nz_0$ as before, the dynamic equations are rewritten as:

$$\begin{cases}A_0(R+2\sigma\cos(k_z z_0))+A'_0\left(2\sigma'\cos(k_z z_0)+\sigma_0\right)=\dfrac{qE_x}{m}\exp(ik_y y_1)\\ A_0\left(2\sigma'\cos(k_z z_0)+\sigma_0\right)+A'_0(R+2\sigma\cos(k_z z_0))=\dfrac{qE_x}{m}\exp(-ik_y y_1)\end{cases} \tag{6.26}$$

here $R=\omega_0^2-\omega-i\gamma\omega$. Eventually, position coordinates (6.25) are substituted into the relations for the polarization P, volume averaged quadrupole moment Q, and the magnetization M. Equation (6.26) are then solved for A_0 and A_0'. This can be done by using the Cramer's method, which gives:

$$A_0 = \frac{\begin{vmatrix} \frac{qE_x}{m}\exp(ik_y y_1) & 2\sigma' \cos(k_z z_0) + \sigma_0 \\ \frac{qE_x}{m}\exp(-ik_y y_1) & R + 2\sigma\cos(k_z z_0) \end{vmatrix}}{\begin{vmatrix} R + 2\sigma\cos(k_z z_0) & 2\sigma' \cos(k_z z_0) - \sigma_0 \\ 2\sigma' \cos(k_z z_0) + \sigma_0 & R + 2\sigma\cos(k_z z_0) \end{vmatrix}} \tag{6.27}$$

$$\begin{aligned} A_0 &= \frac{qE_x}{m\Delta}\left[(R + 2\sigma\cos(k_z z_0))\exp(ik_y y_1) - \left(2\sigma' \cos(k_z z_0) + \sigma_0\right)\exp(-ik_y y_1)\right] \\ \Delta &= \left(R + \sigma_0 + 2\left(\sigma + \sigma'\right)\cos(k_z z_0)\right)\left(R - \sigma_0 + 2\left(\sigma - \sigma'\right)\cos(k_z z_0)\right) \end{aligned} \tag{6.28}$$

Similarly:

$$A_0' = \frac{\begin{vmatrix} \frac{qE_x}{m}\exp(ik_y y_1) & 2\sigma' \cos(k_z z_0) + \sigma_0 \\ \frac{qE_x}{m}\exp(-ik_y y_1) & R + 2\sigma\cos(k_z z_0) \end{vmatrix}}{\begin{vmatrix} R + 2\sigma\cos(k_z z_0) & 2\sigma' \cos(k_z z_0) + \sigma_0 \\ 2\sigma' \cos(k_z z_0) + \sigma_0 & R + 2\sigma\cos(k_z z_0) \end{vmatrix}} \tag{6.29}$$

or:

$$\begin{aligned} A_0' &= \frac{qE_x}{m\Delta}\left[(R + 2\sigma\cos(k_z z_0))\exp(-ik_y y_1) - \left(2\sigma' \cos(k_z z_0) + \sigma_0\right)\exp(ik_y y_1)\right] \\ \Delta &= \left(R + \sigma_0 + 2\left(\sigma + \sigma'\right)\cos(k_z z_0)\right)\left(R - \sigma_0 + 2\left(\sigma - \sigma'\right)\cos(k_z z_0)\right) \end{aligned} \tag{6.30}$$

For the symmetric mode:

$$A_0 + A_0' = \frac{qE_{x,0}}{m}\frac{2\cos(k_y y_1)}{(R + \sigma_0 + 2(\sigma + \sigma')\cos(k_z z_0))} \tag{6.31}$$

and for the anti-symmetric mode:

$$A_0 - A_0' = \frac{qE_{x,0}}{m}\frac{2i\sin(k_y y_1)}{(R - \sigma_0 + 2(\sigma - \sigma')\cos(k_z z_0))} \tag{6.32}$$

If the effective susceptibility is defined as:

$$\chi^{\pm}(\omega, k_z) = \frac{q}{m}\frac{1}{(R \pm \sigma_0 + 2(\sigma \pm \sigma')\cos(k_z z_0))} \tag{6.33}$$

then following the multipole approach [10], the averaged multipole moments are:

$$\begin{cases} P_{x,0} = 4q\eta\chi^{+}(\omega,\, k_z)\cos\left(k_y y_1\right)E_{x,0} \\ Q_{x,0} = 2iq\eta y_1\chi^{-}(\omega,\, k_z)\sin\left(k_y y_1\right)E_{x,0} \\ M_{z,0} = (-i\omega)\,2iq\eta y_1\chi^{-}(\omega,\, k_z)\sin\left(k_y y_1\right)E_{x,0} \end{cases} \tag{6.34}$$

It can be shown that substituting the moments into the Helmholtz equation leads to the following dispersion relation:

$$k_y^2 + k_z^2 = \frac{\omega^2}{c^2}\left[1 + \chi^{+}(\omega,\, k_z)\cos\left(k_y y_1\right) + \chi^{-}(\omega,\, k_z)k_y y_1 \sin\left(k_y y_1\right)\right] \tag{6.35}$$

Furthermore, inserting the polar form for the wave vector components (6.13), the dispersion relation takes the final form:

$$k^2 = \frac{\omega^2}{c^2}\left[1 + \chi^{+}(\omega,\, k\sin\theta)\cos(ky_1\cos\theta) + \chi^{-}(\omega,\, k\sin\theta)ky_1\cos\theta\sin(ky_1\cos\theta)\right] \tag{6.36}$$

with:

$$\chi^{\pm}(\omega,\, k\sin\theta) = \frac{q}{m}\frac{1}{(R \pm \sigma_0 + 2(\sigma \pm \sigma')\cos(kz_0\sin\theta))} \tag{6.37}$$

Equation (6.37) reveals the directional dependence of the material susceptibility. The dispersion relation (6.36) represents the basic equation used to study the dispersion characteristics of the spatially dispersive MM with lateral interaction between the constituent MAs. To adhere to the methodology of the previous analysis [10] (i.e. in the study of isolated MAs without any form of interaction), the following approximation can be made:

$$\cos\left(k_y y_1\right) \approx 1 - \left(k_y y_1\right)^2,\quad \sin\left(k_y y_1\right) \approx k_y y_1 \tag{6.38}$$

and the propagation vector may be rewritten under this approximation as:

$$k^2 = \frac{\omega^2}{c^2}\left[\frac{1 + A\chi^{+}(\omega,\, k\sin\theta)}{1 + \frac{\omega^2}{c^2}Ay_1^2\left(\frac{\chi^{+}(\omega, k\sin\theta)}{2} - \chi^{-}(\omega,\, k\sin\theta)\right)}\right] \tag{6.39}$$

This transcendental equation for k has to be solved numerically. The results of the numerical solution of (6.39) are presented in Fig. 6.8.

6.4 Numerical Solution of the Dispersion Relations

In what follows, various methods of analysis of the general dispersion characteristics are jointly developed for both the dipole and quadrupole systems. The dispersion characteristics are presented side by side to facilitate a comparative study of both systems, where (6.23) and (6.39) form the primary working equations for the above presented analysis. A modified version of the Regula Falsi method applicable to complex variables and the commercial software MATLAB was used [19] to implement the numerical routine in order to solve (6.23) and (6.39).

6.4.1 *Verification of the Computer Code*

The normalized forms of (6.23) and (6.39) were used. Specifically, (6.23) was reformulated as:

$$k^2 y_1^2 = \omega_n^2 \left(\frac{\omega_0^2 y_1^2}{c^2} \right) \left[1 + 4\pi \frac{q^2 \eta}{m\omega_0^2} \frac{1}{\left(1 - \omega_n^2 - i\omega_n/Q + 2C\cos(kz_0 \sin(\theta))\right)} \right] \quad (6.40)$$

where $\omega_n = \omega/\omega_0$, $Q = \omega_0/\gamma$, $C = \sigma/\omega_0^2$, $z_n = z_0/(2y_1)$; $2y_1$, the cut wire spacing, was taken as 65 nm. The factor in the numerator $q^2\eta/(m\omega_0^2)$ was taken as $2.20 \times 10^{30}/4$ SI units. The other parameters are: the resonant frequency of a single

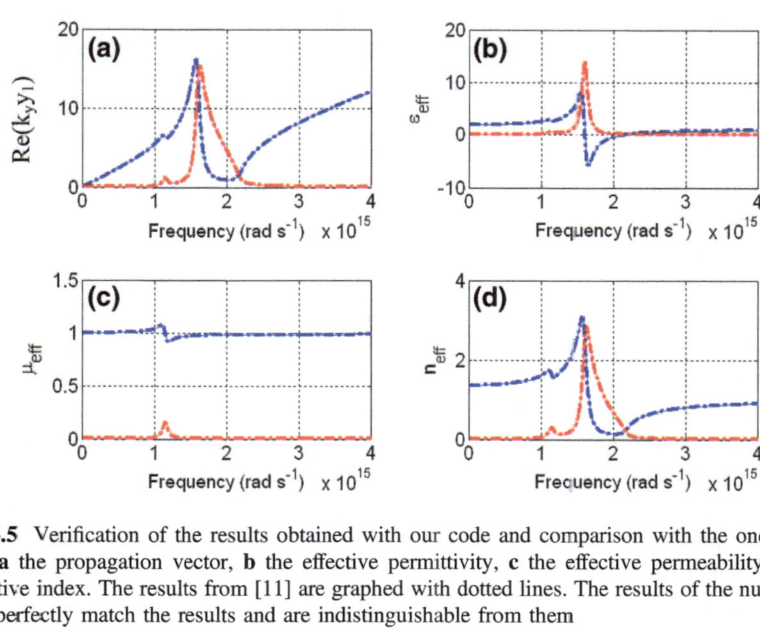

Fig. 6.5 Verification of the results obtained with our code and comparison with the ones from [11]: **a** the propagation vector, **b** the effective permittivity, **c** the effective permeability, **d** the refractive index. The results from [11] are graphed with dotted lines. The results of the numerical code perfectly match the results and are indistinguishable from them

independent resonator is $\omega_0 = 1.39 \times 10^{15}$ rad/s, the damping coefficient is $\gamma = 9.42 \times 10^{13}$ rad/s. The values taken here are the same as in [11], where these parameters were obtained using the fitting with the rigorous computer simulations. To ensure a proper functioning of the computer code, the results of (6.40) at $\sin(\theta) = 0$ were compared with ones presented in [11]. Figure 6.5 shows the wave vector and the effective material parameters obtained with $\sigma = 0.6 \times 10^{30}$ rad/s, $\gamma = 9.42 \times 10^{13}$ rad/s, and $z_n = 4.65$ corresponding to the values used in [11]. The large distance ($z_n = 4.65$) ensures that we are in the weak coupling regime and the values obtained match perfectly, confirming the correct functioning of the code.

6.4.2 Results and Discussions

With the correctness of the numerical procedure ensured, the focus is now shifted to the actual dispersion characteristics. First of all, for the dipole chain the eigenvalue (6.4) in its normalized form is:

$$1 - \omega_n^2 - i\omega_n/Q + 2C\cos(k_z z_0) = 0 \tag{6.41}$$

while the quadrupole chain (6.11) after normalization becomes:

$$\begin{cases} 1 - \omega_n^2 - i\omega_n/Q + C + 2C(b+c)\cos(k_z z_0) = 0 \\ 1 - \omega_n^2 - i\omega_n/Q - C + 2C(b-c)\cos(k_z z_0) = 0 \end{cases} \tag{6.42}$$

The propagation vector k_z is treated as the independent variable, and (6.41), (6.42) are solved for the normalized frequency $\omega_n = \frac{\omega}{\omega_0}$ treating the normalized spatial period $u = k_z z_0$ as a free parameter.

The solutions of (6.41), (6.42) are obtained for different lateral coupling strengths characterized by the normalized lateral distance $z_n = z_0/(2y_1)$ for two values of the damping constant, $\gamma = 0$ (see Fig. 6.6) and $\gamma = 9.42 \times 10^{13}$ rad/s (see Fig. 6.7).

The real part gives the eigenfrequencies, while the imaginary part gives the time decay constant for the particular mode. The values of u are limited to the range 0 to 2π, as the solutions are periodic. A widening of the band of eigenfrequencies is observed as the spatial period decreases. Further, beyond a certain value for the periodicity, the medium starts to exhibit a band gap. The imaginary part of the solution remains independent of the propagation vector until the distance between the dipoles reaches the critical value and becomes k-dependent around the band stop.

For the quadrupoles, the solutions of the above equations are obtained for the symmetric $\omega_{n,\,\mathrm{symm}}$ and anti-symmetric $\omega_{n,\,\mathrm{asymm}}$ modes: the symmetric modes are in the higher band of eigenfrequencies, while the anti-symmetric modes are for frequencies in the lower band. For both types of the modes, there is a band stop accompanied by a non-zero value for the imaginary part of the solution.

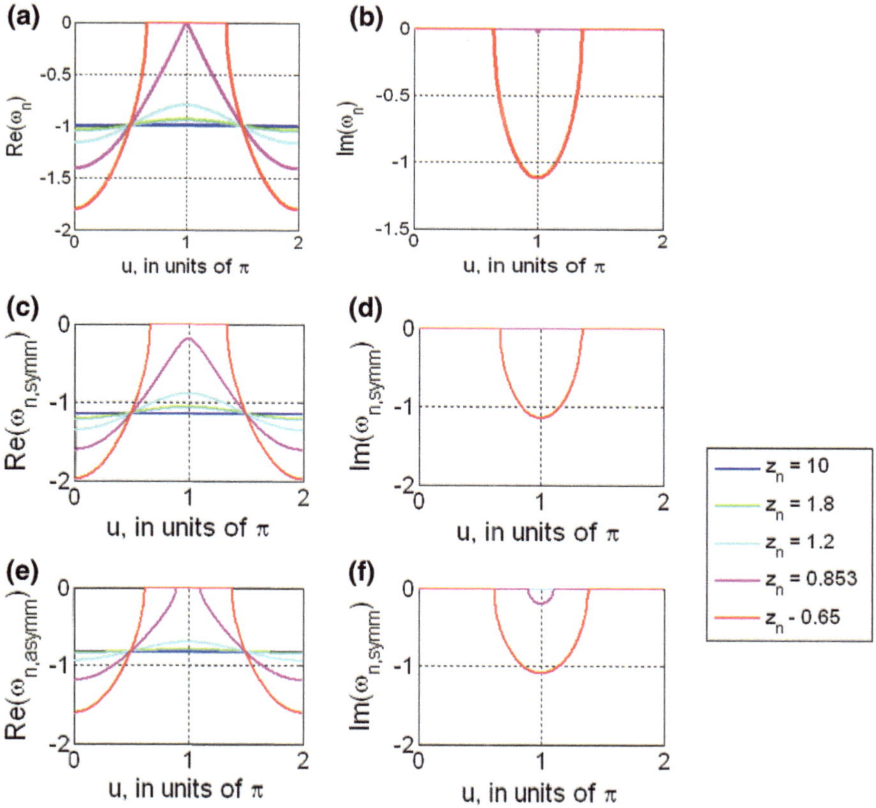

Fig. 6.6 Solution of the material eigenwave dispersion relation in media with chain of coupled dipoles and quadrupoles in the absence of material damping ($\gamma = 0$). The normalized propagation vector $u = kz_0$ is treated as an independent variable, and the solutions are obtained in terms of the normalized eigenfrequency ω_n. The spatial period $z_n = z_0/(2y_1)$ characterizes the coupling strength. The top row gives the real parts of the solutions, while the bottom row gives the imaginary part. **a**, **b**—Solution for dipole chain medium; **c**, **d**—solution for symmetric mode in quadrupole chain medium; **e**, **f**—solution for the anti-symmetric mode in quadrupole chain medium. The curves clearly indicate the onset of a stop band beyond a certain value of the coupling strength (lateral periodic spacing)

The dispersion relations for the electromagnetic wave propagating in a media with the coupled dipoles and quadrupoles (6.23) and (6.39) in their normalized forms are:

$$u^2 = k^2 y_1^2 = \omega_n^2 \left(\frac{\omega^2 y_1^2}{c^2}\right)\left[1 + 4\pi \frac{q^2 \eta}{m\omega_0^2} \frac{1}{\left(1 - \omega_n^2 - i\omega_n/Q + 2C\cos(kz_0 \sin(\theta))\right)}\right] \tag{6.43}$$

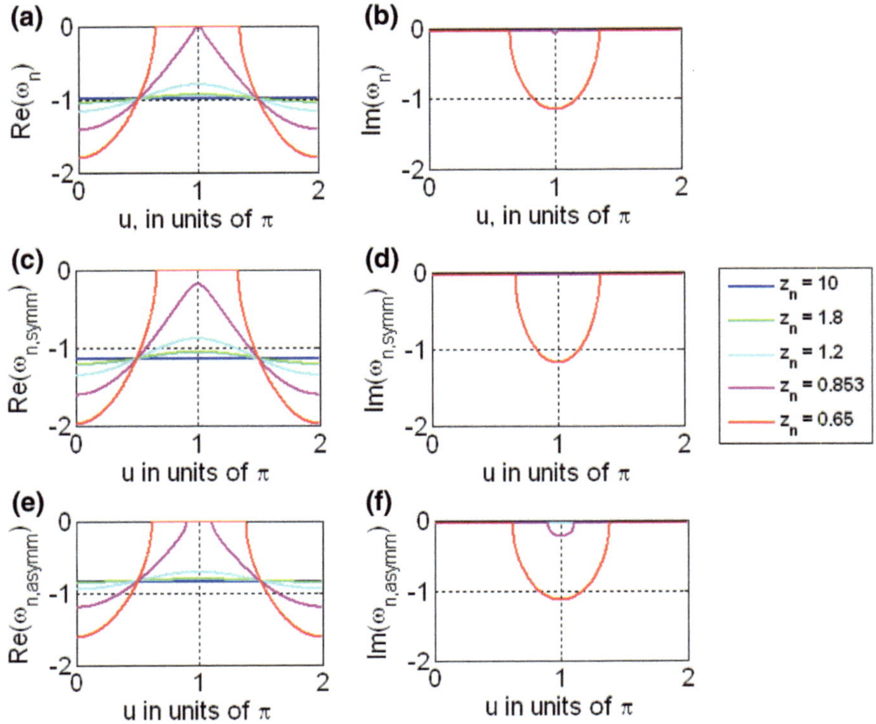

Fig. 6.7 Solution of the material eigenwave dispersion relation in media with chain of coupled dipoles and quadrupoles, in the presence of material damping ($\gamma \neq 0$). The normalized propagation vector $u = kz_0$ is treated as an independent variable, and the solutions are obtained in terms of the normalized eigenfrequency ω_n. The spatial period $z_n = z_0/(2y_1)$ characterizes the coupling strength. The top row gives the real parts of the solutions, while the bottom row gives the imaginary part. **a**, **b**—Solution for dipole chain medium; **c**, **d**—solution for symmetric mode in quadrupole chain medium; **e**, **f**—solution for the anti-symmetric mode in quadrupole chain medium. There are no significant changes in the real part of the solutions, but the imaginary parts are now slightly shifted downwards owing to the presence of the material damping

$$u^2 = k^2 y_1^2 = \frac{\omega^2 y_1^2}{c^2}\left[\frac{1 + A\chi^+(\omega,\, k\sin\theta)}{1 + \frac{\omega^2}{c^2} A y_1^2 \left(\frac{\chi^+(\omega,k\sin\theta)}{2} - \chi^-(\omega,\, k\sin\theta)\right)}\right] \tag{6.44}$$

where:

$$\chi^{\pm}(\omega,\, k\sin\theta) = \frac{q}{m\omega_0^2}\frac{1}{\left(1 - \omega_n^2 - i\omega_n/Q + \pm C + 2C(b \pm c)\cos(kz_0\sin(\theta))\right)} \tag{6.45}$$

with: $z_n = z_0/(2y_1)$, $\omega_n = \omega/\omega_0$, $b = 1/z_n^3$, $c = 1/(1+z_n^2)^{3/2}$, $Q = \omega_0/\gamma$. Solution of dispersion relations (6.43), (6.44) were found numerically for two different periodicities $z_n = 1.8$ and $z_n = 1.2$. The reference spacing was taken to be $2y_1 = 65$ nm. For each of these two periodicities, the dispersion characteristics were calculated for two values of the damping coefficient $\gamma = 9.42 \times 10^{13}$ rad/s and $\gamma = 5 \times 10^{13}$ rad/s. This resulted in overall four sets of data. The results for both dipoles and quadrupoles are given together in Fig. 6.8, which shows a set of dispersion curves obtained for different angles of incidence. The difference between the dispersion curves for dipoles and quadrupoles is in the presence of another set of spatial modes corresponding to quadrupoles, at a frequency lower than the dipole resonance frequency (see for e.g., Fig. 6.8c, d). However, the response from the quadrupoles is much smaller. Also, the response at these lower frequencies completely disappears for incidence at $\theta = \pi/2$—i.e. when light is propagating along the z direction. No quadrupolar moments are excited at this angle as both the cut wires are excited by the same field. The width of the band gap is different in the two systems, which is caused by the fact that in the case of the quadrupoles each meta-atom comprises of two double-wires, giving rise to the enhanced absorption width. It is further observed that the angle of incidence has a clear and pronounced effect on the position of the symmetric response. Specifically, the resonance peak shifts towards the blue part of the spectrum as the angle of incidence increases from 0 to $\pi/2$, while it is red shifted for the imaginary part (see for e.g., Fig. 6.8g, h).

The curves in Figs. 6.9, 6.10 and 6.11 show the variation of the effective material parameters (permittivity, permeability and the refractive index) with the incident angle. The trends as observed for the propagation vector are replicated here as well. The permeability goes to zero at all wavelengths for an incident angle of $\pi/2$ (Fig. 6.10). Figure 6.11 shows the variation of the effective material parameters with the spatial periodicity. In the curves, the frequency of maximum response, and the corresponding value are plotted as a function of the normalized spatial

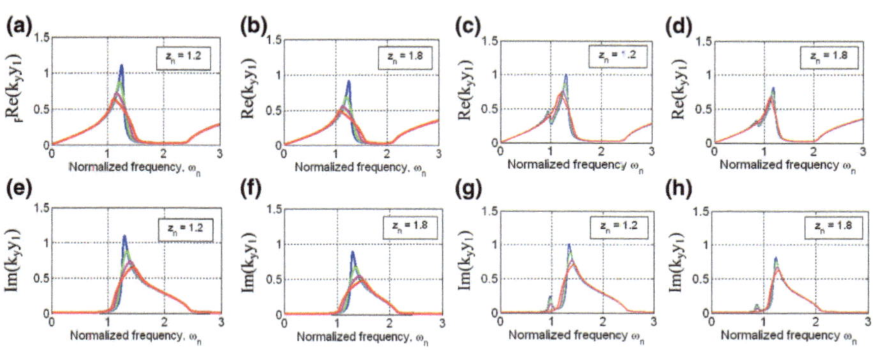

Fig. 6.8 Electromagnetic dispersion curves for the system consisting of the one-dimensional chain of the coupled dipoles and quadrupoles for two spatial periods. **a**, **b**, **e** and **f** depict the dispersion relations for the dipole system, while **c**, **d**, **g**, and **h** depict the dispersion relations for the quadrupole system. The first row depicts the real part of the normalized propagation vector $k_y y_1$, while the bottom row depicts the imaginary part. The values were obtained with the incident angle as parameter (blue—0, green—$\pi/8$, cyan—$\pi/4$, red—$\pi/2$). Note the disappearance of the resonance associated with the quadrupole and magnetic dipole moments at the incident angle of $\pi/2$

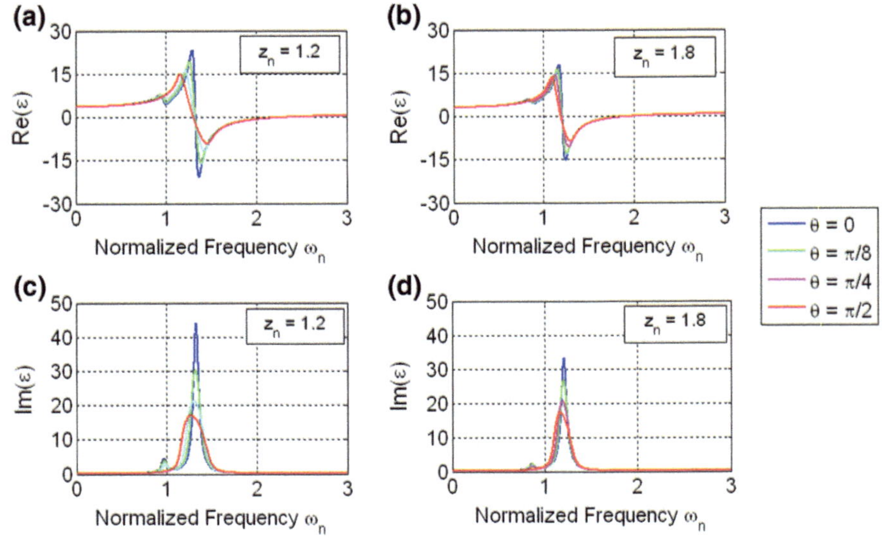

Fig. 6.9 Effective permittivity for the MM with quadrupoles for different angles of incidence—the curves are obtained for the two periodicities: $z_n = 1.2$ and $z_n = 1.8$. The first row depicts curves for the real value of the effective permittivity, while the second row gives the imaginary parts of the effective permittivity. **a** and **c** were obtained for $z_n = 1.2$, while **b** and **d** were obtained for $z_n = 1.8$

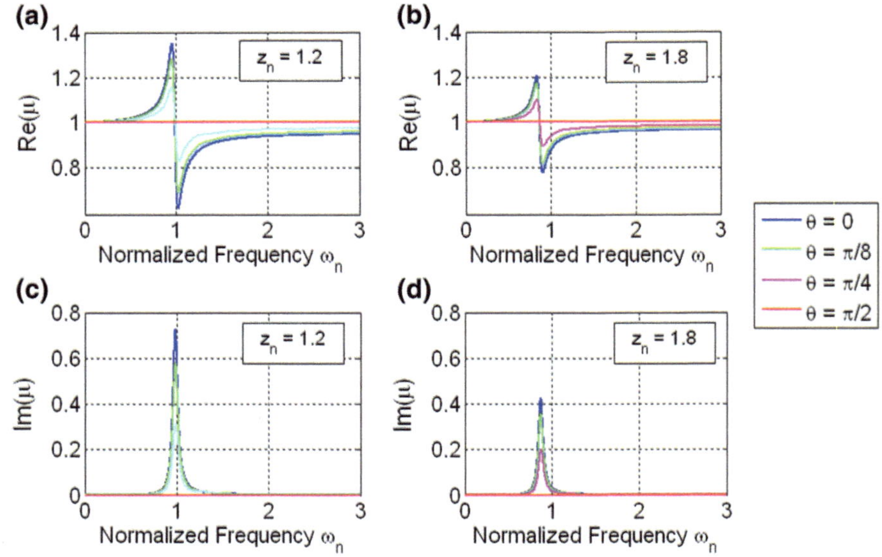

Fig. 6.10 Effective permeability for the MM with quadrupoles for different angles of incidence. The curves are obtained for the two periodicities: $z_n = 1.2$ and $z_n = 1.8$. The first row depicts curves for the real value of the effective permeability, while the second row gives the imaginary parts of the effective permeability; **a** and **c** were obtained for $z_n = 1.2$, while **b** and **d** were obtained for $z_n = 1.8$

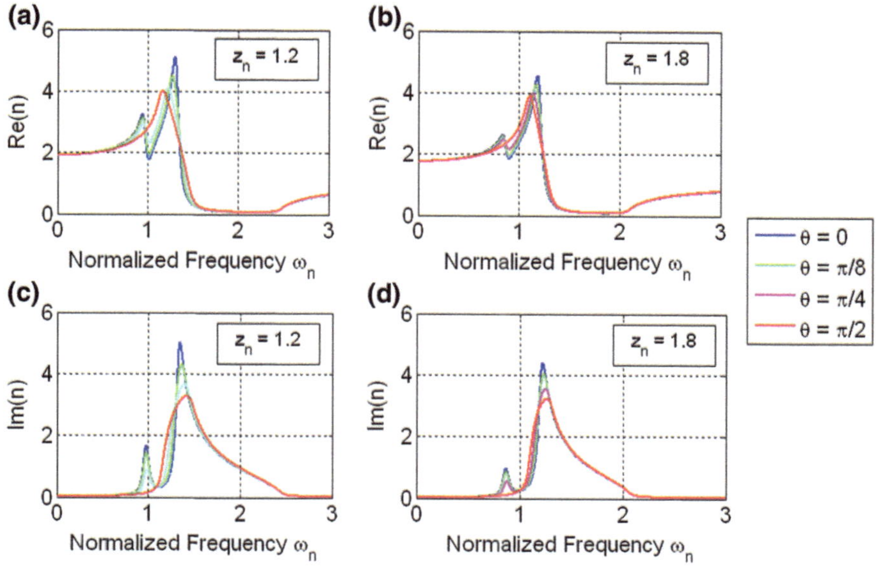

Fig. 6.11 Effective refractive index for the MM with quadrupoles for different angles of incidence —the curves are obtained for the two periodicities: $z_n = 1.2$ and $z_n = 1.8$. The first row depicts curves for the real value of the effective index, while the second row gives the imaginary parts of the effective index; **a** and **c** were obtained for $z_n = 1.2$, while **b** and **d** were obtained for $z_n = 1.8$

period z_n. The angle of incidence is treated as a parameter. Again, the results of this analysis at a periodicity of $z_n = 4.6$, for $\theta = 0$ match perfectly with the results obtained in [11].

In the first row (Fig. 6.12a–c), the resonant frequencies of the effective parameters are plotted as a function of the normalized spatial period z_n. The resonant frequencies for the respective effective parameters asymptotically approach the values obtained in [11]. This asymptotic tendency indicates that as the spatial period becomes larger, the near field interactions between the MA become weaker, and hence the quadrupoles become decoupled as assumed in [11]. It is interesting to note that the position of the anti-symmetric resonance depends on the spatial period. In the second row (Fig. 6.12d–f) the values of the imaginary parts of the respective parameters at resonance are plotted as a function of the spatial period. At $z_n = 5$, the peak values match those as obtained in [11]. Again these correspondences indicate the validity of algorithm and assumptions used in this discussion.

The available spatial frequency spectrum, given by a dispersion relation for particular media, determines the maximum resolution, which can be achieved with this media. Explanation of this statement can be found in any university textbook; qualitatively that the richer the available spatial spectrum the higher resolution could be provided. Hence, roughly speaking, the higher values of the available k vectors are obtained from (6.23) and (6.39), so that media possessing these dispersion relations can in principle achieve a higher resolution than is ordinarily

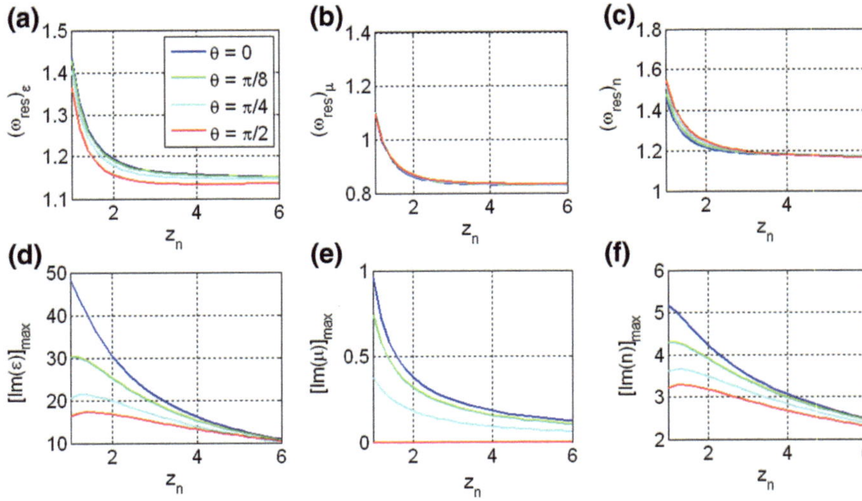

Fig. 6.12 Material parameter curves as a function of period of spacing z_n—the top row contains plots of the resonant frequency as a function of the spatial periodicity, while the bottom row presents plots of the values of the imaginary part of the effective material parameters at the resonant frequencies as a function of z_n : **a**, **d** were obtained for the imaginary part of the effective permittivity, **b**, **e** for the imaginary part of the effective permeability, and **c**, **f** for the imaginary part of the refractive index

allowed. The solutions of these equations are given in Fig. 6.8 for both dipoles and quadrupoles cases. Comparing Fig. 6.8a, b for different coupling strengths reveals that for the higher coupling (Fig. 6.8a) has the higher maximum available value of the real part of the k vector in comparison with the case of lower coupling (Fig. 6.8b). The same relation—higher available k vectors values for the higher coupling—takes place for the quadrupoles as well, see Fig. 6.8c, d. Nevertheless, this enhancement of the real parts of the wave vectors does not allow us to conclude that this effect could lead to the respective resolution enhancement. The problem is that the increase of the maximum available real parts of the wave vectors is accompanied by approximately the same percentage of increase of the losses, i.e. increase of the imaginary parts of the wave vectors, which leads to a suppression of the higher resolution [20].

6.5 Conclusion

The analytical treatment presented here is focused on the extension of the multipole model of MM to the case of MM with the significant coupling between neighboring MA in the lateral direction. The coupling gives rise to a non-local response—that is, the response is mediated by not only the response of a single MA, but also coupling of each MA with its immediate neighborhood. In such a case, spatial modes can be sustained in the MM; excitation the medium at exactly these eigenmodes promises enhanced interaction with the MM. The above analysis confirms the existence of such spatial modes, and throws light onto their general characteristics. Two resonances occur—corresponding to the anti-symmetric and symmetric modes of oscillation of the charges in the MA. The positions of the resonances can be controlled by changing the spatial periodicity of placement of the MA. In the limit of large interspacing periods, the values of resonant frequencies obtained here correspond to the original analysis [11] for non-interacting MA, verifying the validity of the analysis. The changes in the dispersion relation caused by the lateral interaction between MAs are not expected to be used for its resolution enhancement properties due to the fact, that the real part wave vector enhancement is negated by the increase of losses which suppress the effect. Further investigation of the interaction between the MAs in the longitudinal direction must be performed in order to obtain full analysis of the influence of the interaction between the MAs on the dispersion properties of the MMs.

References

1. C.-S. Deng, H. Xu, and Lev Deych, Optical transport and statistics of radiative losses in disordered chains of microspheres. Phys. Rev. A **82**, 041803(R) (2010)
2. W. Weber, G. Ford, Propagation of optical excitations by dipolar interactions in metal nanoparticle chains. Phys. Rev. B **70**, 125429 (2004)
3. M. Quinten, A. Leitner, J. Krenn, F. Aussenegg, Electromagnetic energy transport via linear chains of silver nanoparticles. Opt. Lett. **23**, 1331 (1998)
4. N. Gippius, T. Weiss, S. Tikhodeev, H. Giessen, Resonant mode coupling of optical resonances in stacked nanostructures. Opt. Express **18**, 7569 (2010)
5. N. Feth, M. König, M. Husnik, K. Stannigel, J. Niegemann, K. Busch, M. Wegener, S. Linden, Electromagnetic interaction of split-ring resonators: the role of separation and relative orientation. Opt. Express **18**, 654529 (2010)
6. A. Alù, N. Engheta, Theory of linear chains of metamaterial/plasmonic particles as subdiffraction optical nanotransmission lines. Phys. Rev. B **74**, 205436 (2006)
7. J. Rico-García, J. López-Alonso, A. Aradian, Toy model to describe the effect of positional blocklike disorder in metamaterials composites. JOSA B **29**, 53 (2012)
8. S. Maier, P. Kik, H. Atwater, Optical pulse propagation in metal nanoparticle chain waveguides. Phys. Rev. B **67**, 205402 (2003)
9. A. Alù, N. Engheta, Effect of small random disorders and imperfections on the performance of arrays of plasmonic nanoparticles. New J. Phys. **12**, 013015 (2010)
10. A. Chipouline, J. Petschulat, A. Tuennermann, T. Pertsch, C. Menzel, C. Rockstuhl, F. Lederer, Multipole approach in electrodynamics of Metamaterials. Appl. Phys. A **103**, 899–904 (2011)
11. J. Petschulat, C. Menzel, A. Chipouline, C. Rockstuhl, A. Tünnermann, F. Lederer, T. Pertsch, Multipole approach to metamaterials. Phys. Rev. B **78**, 043811 (2008)
12. C. Simovski, Material parameters of metamaterials (a review). Opt. Spectrosc. **107**, 726 (2009)
13. C. Simovski, On electromagnetic characterization and homogenization of nanostructured metamaterials. J. Opt. **13**, 013001 (2011)

14. E. Tatartschuk, A. Radkovskaya, E. Shamonina, L. Solymar, Generalized Brillouin diagrams for evanescent waves in metamaterials with interelement coupling. Phys. Rev. B **81**, 115110 (2010)
15. A. Radkovskaya, E. Tatartschuk, O. Sydoruk, E. Shamonina, C. Stevens, D. Edwards, L. Solymar, Surface waves at an interface of two metamaterial structures with interelement coupling. Phys. Rev. B **82**, 045430 (2010)
16. A. Radkovskaya, O. Sydoruk, E. Tatartschuk, N. Gneiding, C. Stevens, D. Edwards, E. Shamonina, Dimer and polymer metamaterials with alternating electric and magnetic coupling. Phys. Rev. B **84**, 125121 (2011)
17. E. Shamonina, Magnetoinductive polaritons: hybrid modes of metamaterials with interelement coupling. Phys. Rev. B **85**, 155146 (2012)
18. Z. Jacob, L. Alekseev, E. Narimanov, Optical hyperlens: far-field imaging beyond the diffraction limit. Opt. Express **14**, 8247 (2006)
19. E. Clayton, G.H. Derrick, A numerical solution of wave equations for real or complex Eigenvalues. Aust. J. Phys. **30**, 15 (1977)
20. D. Smith, D. Schurig, M. Rosenbluth, S. Schultz, S. Ramakrishna, J. Pendry, Limitation on subdiffraction imaging with a negative refractive index slab. APL **82**, 1506 (2003)

Chapter 7
Applications of the "Classical" Metamaterial Model—Disordered Metamaterials

7.1 Introduction

In this chapter, the influence of the short-range lateral disorder in the MAs positioning on the effective parameters of the MMs is investigated using the multipole approach. Random variation of the near field quasi-static interaction between MAs in form of double-wires is shown to be the reason for the effective permittivity and permeability changes. The obtained analytical results are compared with known experimental data.

The model for transition from the microscopic to macroscopic system of Maxwell equations, presented here, takes into account all peculiarities of carrier dynamics through the introduction of multipole moments which are represented as the functions of the macroscopic electric and magnetic fields [1]. One of the great advantages of this model is the ability to straightforwardly evaluate the influence of the charge dynamics of the MAs on the effective properties of the MMs. In fact, the multipole moments are calculated through the averaged charge dynamics in the MAs. Any factors influencing the charge dynamics (for example, interaction between the MAs, extra coupling of the MAs with the other objects etc.) causes a change in the multipole expressions, which in turn changes the effective parameters. The interaction between the MAs and hence its influence on the effective permittivity and permeability can be, without difficulty, taken into account in the framework of the model presented here [2].

A review of the optical signal transport in the chain of interacting particles is given in Chap. 6. A natural development of the models of the electromagnetic excitation transport in a chain of particles, but with randomly varying parameters revealed several interesting peculiarities. The problem of wave propagation through disordered systems attracts great attention in both quantum and classical physics [3]. In disordered chains of different dimensions, destructive interference between scattered waves gives rise to an existence of the localized modes, exponentially decaying in space—this effect has been originally found in solid state physics and is

A. Chipouline and F. Küppers, *Optical Metamaterials: Qualitative Models*, Springer Series in Optical Sciences 211, https://doi.org/10.1007/978-3-319-77520-3_7

known as Anderson localization [4]. The existence of delocalized modes that can extend over the sample via multiple resonances and have a transmission close to 1 was found in [5, 6] and experimentally confirmed in [7, 8]. Disorder-induced change of the guiding properties in a chain of plasmonic nano particles under small random uncontrollable disorder was considered in [9] and analogy of the Anderson localization in a chain of such particles was theoretically investigated in [10]. In the analysis presented here the effect of Anderson localization is not considered; nevertheless, it is believed, that the developed here analytical tool turns out to be suitable for the treatment of the similar effect in MMs with different types of disorder.

The influence of various types of disorder on the effective properties of the MMs has been thoroughly investigated as well. Light propagation and Anderson localization in superlattices was theoretically considered in [11, 12] using the model of multilayered system with phenomenological permittivity and permeability (positive and negative) in each of the layers. The effect of the statistical distribution of the sizes of the MAs on the increase of losses in the operation frequency band was considered in [13] using generalized Clausius–Mossotti relation. A significant influence of a small (10%) deviation of the parameters of the microscopic resonances on the propagation wave in a wide frequency range was found in [14] using quasistatic expressions for the effective parameters. Averaging of the Lorenz-type expressions for the effective permittivity and permeability using a phenomenological probability distribution function showed that passband and negative refraction are still present under small positional disorder [15]. Furthermore, the results have been proven experimentally. Interaction in a chain of magnetic particles and its influence on the effective permeability was investigated in [16]. Using the introduced concept of “coherent” and “incoherent” MMs, authors of [17] showed that the influence of disorder on long-range correlated MMs is significantly more pronounced in comparison with the same effect in short-range ordered MMs. Random variation of the interaction between MAs was shown to be the main reason for the disappearance of the long-range correlation and consequently of the “coherent” state [17].

In this chapter, attention is primarily devoted to the extension of the multipole approach to describe in-plane disorder in MMs, which means the randomness in the position of the MAs within the plane of the substrate—see Fig. 7.1. In [9] it was shown that the difference in the electromagnetic properties of the inclusions itself is less important than the disorder in their positions. MMs formed by a self-organization display exactly this kind of disorder [18–21]. Results of experiments with the 2D MMs exhibiting such in-plane disorder [22] are used as a test of the model. The most notable discovery is the fact that although disorder has a deterrent effect on the permittivity, the permeability seems to remain practically unaffected. A theoretical model for such a class of random MMs should reproduce these observations.

The qualitative explanation of the influence of the spatial disorder on the effective parameters follows presently. The positional disorder creates different conditions for the charged dynamics in the MAs due to the interaction between

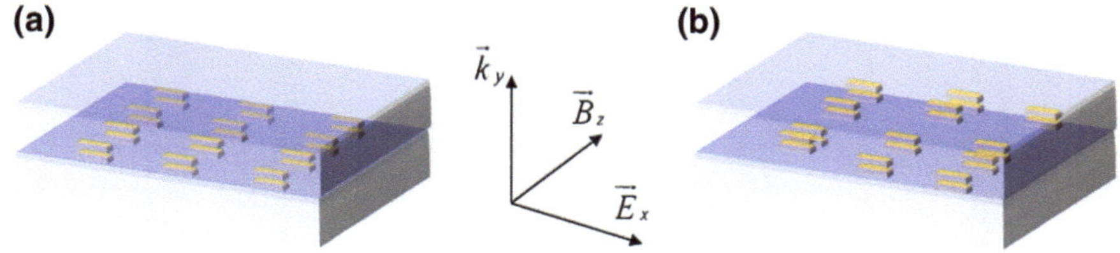

Fig. 7.1 **a** Regular and **b** laterally (along z direction) random positioning of the MAs in a MM. Plane wave propagation is in y direction, electric field polarization coincides with the x axis and with the elongation of the nano-wires. Note that only one layer, which the MM consists of, is shown in the figure [27]

them [23, 24]. This in turn leads to the changes of the averaged dipole, quadrupole, and magnetic dipole moments of the media and results in changes of the effective parameters, which are expressed through these averaged multipole moments [25, 26].

This qualitative hypothesis requires further development of the existing theoretical multipole model, in particular the interaction between the MAs [2] has to be incorporated and adapted to the random character of this interaction. Let us assume that the charge dynamics in the microscopic multipole moments of the MAs depends on the distance δ_k between them (see Fig. 7.2). Following the approach of [25], it is necessary to average the resulting charge dynamics in the multipole moments over all possible representations. In other words, the microscopic multipole moments have to be additionally averaged over all possible distances between the MAs, which mathematically is expressed as an integral over the Probability Distribution Function $\mathrm{PDF}(\delta_k)$, namely:

$$\chi_{\mathrm{macro}}(\omega) = \int \chi_{\mathrm{micro}}(\delta_k, \omega)\mathrm{PDF}(\delta_k)\mathrm{d}\delta_k \qquad (7.1)$$

Here $\chi_{\mathrm{micro}}(\delta_k, \omega)$ is the microscopic multipole moment of the MAs, and $\mathrm{PDF}(\delta_k)$ governs the distribution over all possible separation distances δ_k in a randomly arranged ensemble of the MAs. In case of regular spatial distribution each MA is affected by the same fields, the $\mathrm{PDF}(\delta_k)$ is reduced to a delta function, and averaging (7.1) returns the microscopic multipole moments.

The quest to obtain such $\mathrm{PDF}(\rho_k)$ and the effort to incorporate the effect of disorder into the existing multipole model is discussed in detail now. First, the probability model used to incorporate positional disorder into the multipole theory is described. As a proof of principle and in order to create a systematic model, the approach is then applied to the simple case of randomly arranged dipoles. The treatment is then extended to the case of the randomly arranged quadrupoles. The probabilistic approach is applied to the specific case of randomly positioned MAs, and the obtained results are compared with the experimental observations [22]; the mathematical procedures used to account for the other forms of disorder are highlighted.

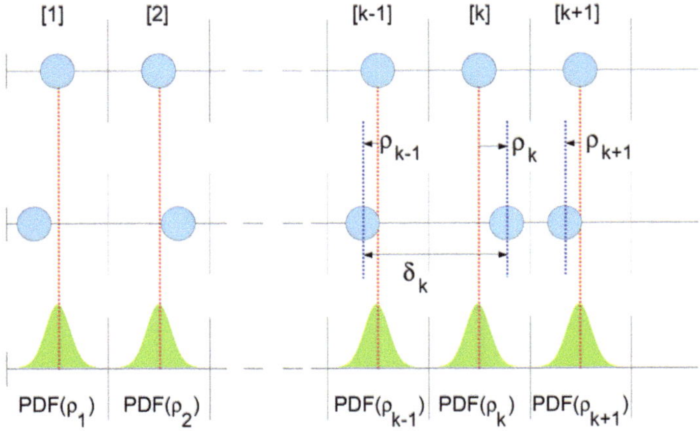

Fig. 7.2 Geometry of the MAs and their respective probability distributions; the spheres show MAs. The first (top) row shows a regular arrangement of the MAs, where each MA occupies the center of a slot with the length equal to the mean period. The second row depicts an arrangement of the MAs exhibiting random uncorrelated positional disorder (denoted by ρ_k), the extent of the disorder being governed by PDF(ρ_k) as shown in the last (down) row. The inter separation δ_k between the two subsequent MAs is a function of the random variables ρ_k and ρ_{k-1}, and the analytic form of PDF(δ_k) can be obtained by the use of the statistical methods if the analytic form of PDF(ρ_k) is given [27]

The main discerning principle of the approach presented here is in the use of the multipole model: the charge dynamics in MAs is primarily considered and calculated taking into account the interaction between MAs, which is expressed as a function of distance between them. Finally, averaging over all possible realization of the MA separation distances gives the expression for the effective parameters. This chapter is primarily devoted to the elaboration of the model and to the effective parameters calculation; further applications of the presented approach (disorder in propagation direction, transition “coherent”-“incoherent” states, influence of the Anderson localization on the effective parameters etc.) will be discussed elsewhere. Interaction between MAs is taken into account most simply by using dipole–dipole near field interaction in the quasi-static limit; extrapolation of the model on the dynamic case is left for the future work. The interaction between quadrupoles is treated the same way, which makes the approach suitable for consideration of the magnetic properties of the MMs. In spite of the excessive simplification of the model with regard to the interaction, our approach treats the effective parameters (especially magnetic response) in a much more correct way then was done before with just the introduction of permeability and/or magnetic susceptibility. Furthermore, it is believed to provide a suitable platform for analytical or semi analytical treatment of the problems, appearing in the case of disordered MMs.

7.2 Modeling of Positional Disorder

The problem of the positional disorder modeling can be tackled in several ways. The most general formulation of the problem requires a Markovian treatment. As an illustration, the one dimensional equivalent of the problem is considered. Supposing that the MAs are introduced one by one on a line of a given length, the probability that a particle will take up a certain position along the line, and hence the probability of a particular inter separation distance, depends not only on the last particle, but also on the history and existing configuration. This is the essence of the Markovian approach. Standard techniques exist for tackling such problems, formulating a rigorous treatment. Nevertheless, due to its complex nature, an alternative simpler mathematical treatment is used and discussed in this chapter.

The math treatment put forward in this work is a consequence of the way in which the MMs were actually produced. In the control experiments masks for random MMs are manufactured by e-beam lithography methods. The writing algorithms can be modified so that instead of a periodic grid, a randomized one is generated by the scanner. The extent of the randomness can be specified and controlled, so that statistically relevant parameters such as the mean period and the variance can be assigned to each mask. Translating the above technological approach to mathematics, one assumes that the MAs are initially placed in well-defined positions of equal separation (see Fig. 7.2) and are subsequently perturbed from their mean positions. The perturbation is described by $\mathrm{PDF}(\rho_k)$. This PDF describes the extent of the perturbation, and also ensures that the displacement is not beyond a certain space slot so that two whole MAs cannot exist in a single unit cell. However, $\mathrm{PDF}(\rho_k)$ is the *positional* disorder distribution function, not a PDF for the MA inter-separation $\mathrm{PDF}(\delta_k)$; the latter has to be found based on the assumed $\mathrm{PDF}(\rho_k)$. Thus, it is required to find the inter-separation $\mathrm{PDF}(\delta_k)$ from a given form of the positional disorder $\mathrm{PDF}(\rho_k)$; $\mathrm{PDF}(\delta_k)$ can then be used in an averaging procedure (7.1) to obtain the required material parameters. In this work, inter-separation $\mathrm{PDF}(\delta_k)$ is obtained by employing a characteristic function approach. The characteristic function $Q_\rho(\omega)$ is by definition a Fourier transform of a given $\mathrm{PDF}(\rho_k)$:

$$Q_{\rho,k}(\omega) = \int_{-\infty}^{\infty} \mathrm{PDF}(\rho_k) \mathrm{e}^{i\omega\rho_k} \mathrm{d}\rho_k \tag{7.2}$$

where $\mathrm{PDF}(\rho_k)$ satisfies the normalization condition:

$$\int_{-\infty}^{\infty} \mathrm{PDF}(\rho_k) \mathrm{d}\rho_k = 1 \tag{7.3}$$

Alternatively, the characteristic function can be considered as an expectation value of the function $\mathrm{e}^{i\omega t}$:

$$Q_{\rho,k}(\omega) = \left\langle e^{i\omega\rho_k} \right\rangle\Big|_{\mathrm{PDF}(\rho_k)} \tag{7.4}$$

The one dimensional equivalent of the problem is formulated as follows. A periodic arrangement of N MAs on a given length L (see Fig. 7.2) with a periodic spacing for the slots as given by $z_0 = L/N$ and the location of the kth MA is $z_k = \frac{z_0}{2}(2k+1)$. Now, the perturbation of the kth MA from its mean position can be given using a random function ρ_k, such that:

$$z_k = \frac{z_0}{2}(2k+1) + \rho_k, \quad -\frac{z_k}{2} < \rho_k < \frac{z_k}{2} \tag{7.5}$$

So, the spacing between the MAs is given by:

$$\Delta_k = z_{k+1} - z_k = z_0 + \left(\rho_{k+1} - \rho_k\right) = z_0 + \delta_k \tag{7.6}$$

Random functions ρ_k and ρ_{k+1} are completely independent from each other. For the present problem, the random function of interest is:

$$\delta_k = \rho_{k+1} - \rho_k \tag{7.7}$$

The mathematical form of $\mathrm{PDF}(\delta_k)$ has to be found. The form of the above probability function can be obtained by using characteristic functions. The characteristic function formed by a sum or difference of two or more PDFs, is nothing but the product of the characteristic functions of the constituent PDFs. That is, if one is interested in a probability distribution function of a variable z, given by:

$$z = y_1 + y_2 + \cdots + y_n \tag{7.8}$$

The characteristic function of z will be given by:

$$Q_Z(\omega) = Q_1(\omega)Q_2(\omega)\ldots Q_N(\omega) \tag{7.9}$$

where $Q_i(\omega)$ are their mutually independent characteristic functions $Q_i(\omega) = \int_{-\infty}^{\infty} \mathrm{PDF}(y_i)e^{i\omega y i}\mathrm{d}y_i$. So the form of z can be simply obtained by performing an inverse Fourier transformation on the product of the characteristic functions.

Defining the characteristic functions for ρ_k and ρ_{k+1} as:

$$Q_{\rho_k}(\omega) = \int_{-\infty}^{\infty} \mathrm{PDF}(\rho_k)\exp(i\rho_k\omega)\mathrm{d}\rho_k \tag{7.10}$$

$$Q_{-\rho,k+1}(\omega) = Q^*_{\rho,k+1}(\omega) = \int_{-\infty}^{\infty} \mathrm{PDF}\left(\rho_{k+1}\right)\exp(-i\rho_{,k+1}\,\omega)\mathrm{d}\rho_{k+1} \tag{7.11}$$

Making use of the above method, the characteristic function $Q_{\delta}(\omega)$ of required $\mathrm{PDF}(\delta_k)$ is:

$$Q_{\delta,k}(\omega) = Q_{\rho,k+1}(\omega) Q^{*}_{\rho,k}(\omega) \tag{7.12}$$

and required $\mathrm{PDF}(\delta_k)$ can be obtained by simply using the convolution theorem:

$$\begin{aligned}\mathrm{PDF}(\delta_k) &= FT^{-1}\left[Q_{\rho_k+1}(\omega) Q^{*}_{\rho_k}(\omega)\right] \\ &= \int_{-\infty}^{\infty} \mathrm{PDF}(\rho_k)\mathrm{PDF}^{*}(\delta_k - \rho_k)\mathrm{d}\rho_k \\ &= \int_{-\infty}^{\infty} \mathrm{PDF}(\rho_k)\mathrm{PDF}(\delta_k - \rho_k)\mathrm{d}\rho_k\end{aligned} \tag{7.13}$$

Therefore, the required $\mathrm{PDF}(\delta_k)$ is the autocorrelation function of the positional disorder $\mathrm{PDF}(\rho_k)$. The integral is taken over the displacement of the MA from its mean position and limited by the finite values of the slot length. The strength of the method is the fact that no explicit assumption is made regarding the form of $\mathrm{PDF}(\rho_k)$ describing the positional disorder.

The mathematical procedure has to ensure that the perturbation does not become so large that the MAs overlap each other. In the analysis, the particles are assumed to be placed on average in the center of the slots of a length equal to the mean spacing period. The particles can randomly move within their own slot, and the extent of the displacement from the center of the slot is given by $\mathrm{PDF}(\rho_k)$. A consequence of such a restraint is that $\mathrm{PDF}(\rho_k)$ is restricted and normalized to a particular slot. Figure 7.5 shows how the autocorrelation function approaches a triangular function from its initial Gaussian form, as the position of the particle within the slot becomes completely random (i.e., $\mathrm{PDF}(\rho_k)$ takes a rectangular form). A simple algebraic form of the probability distribution function cannot be obtained due to this truncation. So the following approach was adopted: the normalized versions of $\mathrm{PDF}(\delta_k)$ for the inter separation were obtained using numerical code, and they were subsequently used for numerical integration [as according to (7.1)] to obtain the relevant effective material parameters.

7.3 Case of Randomly Positioned Dipoles

In this section, using the aforementioned principles, the effect of disorder in a chain of periodically placed dipoles is investigated. The geometry is given in Fig. 7.3. The bold arrow shows the direction of propagation of the electromagnetic wave.

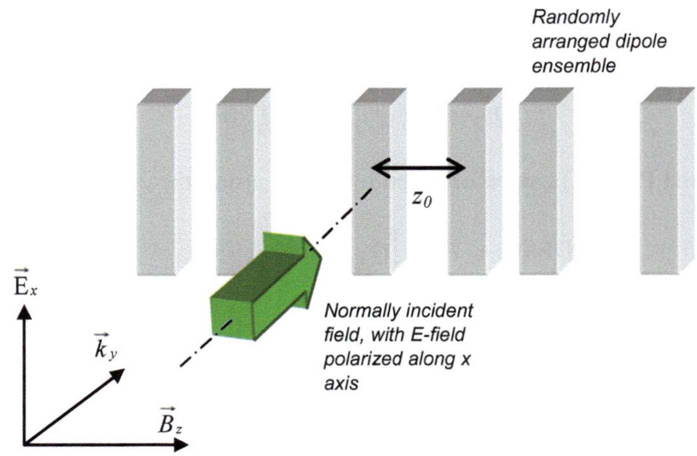

Fig. 7.3 Geometry of propagation for randomly arranged dipole ensemble [27]

The system can be mathematically modeled as follows. Considering the coupling dynamics between two equal adjacent oscillators, one can write the equation describing their dynamics as:

$$\begin{aligned} \frac{\partial^2 x_1}{\partial t^2} + \gamma \frac{\partial x_1}{\partial t} + \omega_0^2 x_1 + \sigma x_2 = \frac{q}{m} E_x \\ \frac{\partial^2 x_2}{\partial t^2} + \gamma \frac{\partial x_2}{\partial t} + \omega_0^2 x_2 + \sigma x_1 = \frac{q}{m} E_x \end{aligned} \tag{7.14}$$

The term on the right side is the same for both oscillators, as the same field impinges on both of them. By substituting the temporal ansatz $x_i(t) = x_i(\omega)\exp(-i\omega t)$, the system can be easily solved for $x_1(\omega)$ and $x_2(\omega)$:

$$\begin{cases} \begin{bmatrix} R & \sigma \\ \sigma & R \end{bmatrix} \begin{bmatrix} x_1(\omega) \\ x_2(\omega) \end{bmatrix} = \frac{q}{m} \begin{bmatrix} E_x(\omega) \\ E_x(\omega) \end{bmatrix} \\ R = \omega_0^2 - \omega - i\gamma\omega \end{cases} \tag{7.15}$$

Thus:

$$\begin{cases} x_1(\omega) = x_2(\omega) = \frac{q/m}{\omega_0^2 - \omega - i\gamma\omega + \sigma} E_x(\omega) \\ \sigma = \sigma(z) = \sigma_0 \frac{z_0^3}{z^3} \end{cases} \tag{7.16}$$

Here σ_0 and z_0 are the coupling constant and the distance between the oscillators, at which the coupling constant is σ_0. It is assumed that the interaction between the oscillators is the near field dipole interaction that stipulates the inverse cubic distance dependence in the second equation in (7.16).

The response of the system can thus be obtained by monitoring the susceptibility of the medium. The polarization of the system can be written as:

$$P_x(z,\omega) = \frac{2q^2\eta}{m}\frac{1}{\omega_0^2 - \omega - i\gamma\omega + \sigma(z)}E_x(\omega) \tag{7.17}$$

so that the effective susceptibility:

$$\chi_x(z,\omega) = \frac{2q^2\eta}{m}\frac{1}{\omega_0^2 - \omega - i\gamma\omega + \sigma(z)} \tag{7.18}$$

To incorporate the effect of disorder, following (7.1), the averaged form of the above susceptibility can be obtained as:

$$\langle\chi(z_0,\omega)\rangle_z = \int_{-\infty}^{\infty} \mathrm{PDF}(\delta,D)\chi(z_0,\delta,\omega)\mathrm{d}\delta \tag{7.19}$$

or:

$$\langle\chi(z_0,\omega)\rangle_z = \frac{2q^2\eta}{m}\int_{-\infty}^{\infty} \mathrm{PDF}(\delta,D)\frac{(z_0+\delta)^3}{\left(\omega_0^2 - \omega - i\gamma\omega\right)(z_0+\delta)^3 + \sigma_0 z_0^3}\mathrm{d}\delta \tag{7.20}$$

where $\mathrm{PDF}(\delta,D)$ is the inter separation PDF and D here quantizes the amount of disorder presented in the system.

7.4 Case of Randomly Positioned Quadrupoles

The extension of the above model to MMs (i.e. taking into account the magnetic response) requires that the interaction between the adjacent MAs is taken into consideration. The system is taken to be similar as the one shown in Fig. 7.3, but the dipoles are now replaced by quadrupoles—Fig. 7.4. The long axis of the cut wires is oriented along x axis. The double-wires forming the quadrupoles are separated along the y direction. The MAs are arranged randomly (in terms of above described random positioning in the respective slots) along z direction. Assuming that a plane electromagnetic wave now propagates through the ensemble along y direction, while its electric vector is polarized along the x direction, the coupled dynamics of two MAs can be modeled via four coupled oscillator equations:

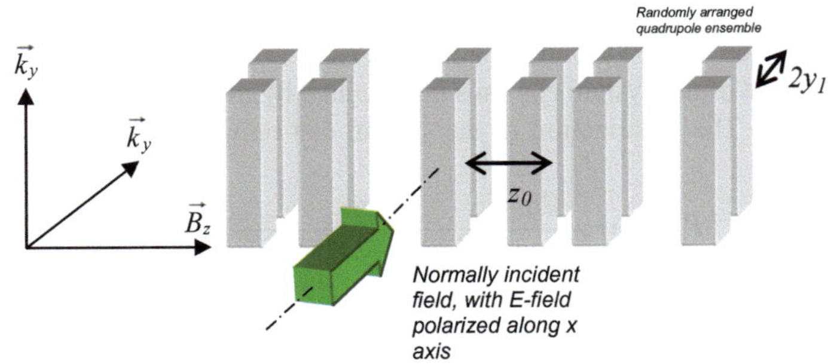

Fig. 7.4 Geometry of propagation for randomly arranged quadrupole ensemble [27]

$$\begin{cases} \frac{\partial^2 x_1}{\partial t^2} + \gamma \frac{\partial x_1}{\partial t} + \omega_0^2 x_1 + a x_2 + b x_3 + c x_4 = \frac{q}{m} E_{x,1} \exp(iky_1) \\ \frac{\partial^2 x_2}{\partial t^2} + \gamma \frac{\partial x_2}{\partial t} + \omega_0^2 x_2 + a x_1 + b x_3 + c x_4 = \frac{q}{m} E_{x,2} \exp(-iky_1) \\ \frac{\partial^2 x_3}{\partial t^2} + \gamma \frac{\partial x_3}{\partial t} + \omega_0^2 x_3 + a x_4 + b x_1 + c x_2 = \frac{q}{m} E_{x,3} \exp(iky_1) \\ \frac{\partial^2 x_4}{\partial t^2} + \gamma \frac{\partial x_4}{\partial t} + \omega_0^2 x_4 + a x_3 + b x_2 + c x_1 = \frac{q}{m} E_{x,4} \exp(-iky_1) \end{cases} \tag{7.21}$$

where:

$$a = \sigma_0, \quad b = \sigma_0 \frac{(2y_1)^3}{z^3}, \quad c = \sigma_0 \frac{(2y_1)^3}{\left((2y_1)^2 + z^2\right)^{3/2}} \tag{7.22}$$

σ_0 is the value of the coupling constant measured for the inter-separation $2y_1$ (this normalization with $2y_1$ instead of z_0 is chosen in order to use the data from [26] where numerical values of σ_0 have been obtained).

The magnitude of the coupling constant varies inversely as the cube of the distance, and so its value can be obtained for other inter-separations—here z and the diagonal distance $((2y_1)^2 + z^2)^{1/2}$. The exponential phase factors in the right side in (7.21) take into account the retardation effect. It is clear that a change in the excitation conditions will affect the form of the right hand side of the above equations, while a change in the configuration of the MAs can be accounted for by a change in the form of the coupling coefficients. The procedure of determining the response of the medium then, remains the same—one seeks to determine effective susceptibilities (corresponding to the symmetric and anti-symmetric modes of oscillation), average them over all possible coupling configurations, and then use the values to ascertain the effective material parameters.

The first step is to find the solution of the above set of equations. They can be transferred to the Fourier domain by using ansatz $x_i(t) = x_i(\omega)\exp(-i\omega t)$, ($i$ = 1, 2, 3, 4), so that the system can be rewritten in a matrix form:

$$\begin{cases} \begin{bmatrix} R & a & b & c \\ a & R & c & b \\ b & c & R & a \\ c & b & a & R \end{bmatrix} \begin{bmatrix} x_1(\omega) \\ x_2(\omega) \\ x_3(\omega) \\ x_4(\omega) \end{bmatrix} = \frac{q}{m} \begin{bmatrix} E_x(\omega)\exp(iky_1) \\ E_x(\omega)\exp(-iky_1) \\ E_x(\omega)\exp(iky_1) \\ E_x(\omega)\exp(-iky_1) \end{bmatrix} \\ R = \omega_0^2 - \omega - i\gamma\omega \end{cases} \tag{7.23}$$

The modes of oscillation of interest are given by $x_1(t) \pm x_2(\omega)$. The system can be solved to obtain the values of $x_1(t)$ and $x_2(\omega)$, and the modes of the system can be written as:

$$\begin{aligned} x_1(t) + x_2(\omega) &= \frac{qE_x}{m} \frac{2\cos(ky_1)}{\omega_0^2 - \omega - i\gamma\omega + (a+b+c)} \\ x_1(t) - x_2(\omega) &= \frac{qE_x}{m} \frac{2\cos(ky_1)}{\omega_0^2 - \omega - i\gamma\omega - (a-b+c)} \end{aligned} \tag{7.24}$$

and hence one can define the effective susceptibility:

$$\chi^{\pm}(z_0, \delta, \omega) = \frac{qE_x}{m} \frac{2\cos(ky_1)}{\omega_0^2 - \omega - i\gamma\omega \pm (a \pm b + c)} \tag{7.25}$$

where the z_0 and δ dependences are due to *a*, *b* and *c*. Due to this form of definition, the functional forms of the polarization, quadrupolar moment and magnetization remain the same:

$$\begin{cases} P = 2q\eta y_1 \begin{pmatrix} 2\chi^+(\omega)\cos(ky_1) \\ 0 \\ 0 \end{pmatrix} E_x(y,\omega) \\ Q = q\eta y_1 \begin{pmatrix} 0 & 2i\chi^-(\omega)\sin(ky_1) & 0 \\ 2i\chi^-(\omega)\sin(ky_1) & 0 & 0 \\ 0 & 0 & 0 \end{pmatrix} E_x(y,\omega) \\ M = q\eta y_1 \begin{pmatrix} 0 \\ 0 \\ 2i\chi^-(\omega)\sin(ky_1) \end{pmatrix} E_x(y,\omega) \end{cases} \tag{7.26}$$

The effect of disorder can then be taken into account by carrying out the extra averaging integration (7.1):

$$\langle \chi^{\pm}(z_0, \omega, D) \rangle = \int_{-\infty}^{\infty} \mathrm{PDF}(\delta, D)\chi^{\pm}(z_0, \delta, \omega)\mathrm{d}\delta \tag{7.27}$$

or, more explicitly:

$$\langle\chi^{\pm}(z_0,\omega,D)\rangle = \int_{-\infty}^{\infty} \mathrm{PDF}(\delta,D)\frac{1}{\left(\omega_0^2-\omega-i\gamma\omega\right)\pm\sigma_0\left(1\pm\frac{(2y_1)^3}{(z_0+\delta)^3}+\frac{(2y_1)^3}{\left((2y_1)^2+(z_0+\delta)^2\right)^{3/2}}\right)}\mathrm{d}\delta \tag{7.28}$$

where z_0 is the mean period. The limits of the integration indicate that the auto-correlation procedure for $\mathrm{PDF}(\delta,D)$ has been already carried out. This integral can be solved numerically for a given value of frequency.

With the effective susceptibility as defined above, one may now consider a planar MM, which is formed by the identical rows of the randomly positioned MAs. The effect of randomness is taken into account by the averaging procedure, and the dispersion relation and the effective material parameters can be written in analogy to [26]. Adhering to the same conditions of geometry and excitation, the following expressions can be utilized:

$$\begin{aligned} k_y^2(\omega) &= \frac{\omega^2}{c^2}\frac{1+A\langle\chi^+(z_0,\omega,D)\rangle}{1+\frac{\omega^2}{c^2}Ay_1^2\left(\frac{1}{2}\langle\chi^+(z_0,\omega,D)\rangle-\langle\chi^-(z_0,\omega,D)\rangle\right)} \\ \varepsilon_{\mathrm{eff}} &= 1+A\langle\chi^+(z_0,\omega,D)\rangle - A\frac{k_y^2y_1^2}{2}\left(\frac{1}{2}\langle\chi^+(z_0,\omega,D)\rangle-\langle\chi^-(z_0,\omega,D)\rangle\right) \\ \mu_{\mathrm{eff}} &= \frac{1}{1-\frac{\omega^2}{c^2}Ay_1^2\langle\chi^-(z_0,\omega,D)\rangle} \end{aligned} \tag{7.29}$$

The above expressions can be easily carried over to a numerical code to obtain the material parameters of interest. The following section presents the results, and compares them with the experimental observations.

7.5 Method of Numerical Implementation

For convenience, the integrations and other expressions have been converted to their normalized versions. The frequencies are normalized with respect to the resonant frequency ω_0 of the independent double-wire, while the distances are normalized with respect to the double-wire spacing $2y_1$. Specifically, the susceptibilities for the case of dipoles and quadrupoles are:

$$\langle\chi^{\pm}(z_0,\omega)\rangle_z = \int_{-\infty}^{\infty} \mathrm{PDF}(\delta_n, D_n)\frac{(1+\delta_n)^3}{\left(1-\omega_n^2 - i\frac{\omega_n}{Q}\right)(1+\delta_n) + \frac{\sigma_0}{\omega_0^2}}\mathrm{d}\delta_n \tag{7.30}$$

$$\langle\chi^{\pm}(z_0,\omega)\rangle_z = \int_{-\infty}^{\infty} \frac{\mathrm{PDF}(\delta_n, D_n)}{\left(1-\omega_n^2 - i\frac{\omega_n}{Q}\right) \pm \sigma_0\left(1 \pm \frac{1}{\left((z_0)_n+\delta_n\right)^3} + \frac{1}{\left(1+\left((z_0)_n+\delta_n\right)^2\right)^{3/2}}\right)}\mathrm{d}\delta_n \tag{7.31}$$

where:

$$\delta_n = \frac{\delta}{2y_1}, \quad (z_0)_n = \frac{z_0}{2y_1}, \quad D_n = \frac{D}{2y_1}, \quad \omega_n = \omega/\omega_0 \tag{7.32}$$

$\mathrm{PDF}(\rho_k, D)$ is assumed to be Gaussian:The results of the analysis for dipoles are

$$\begin{cases} \mathrm{PDF}(\rho_k, D) = \frac{1}{\sqrt{2D}}\exp\left(-\rho_k^2/(2D^2)\right), & -z_0/2 < \rho_k < z_0/2 \\ \mathrm{PDF}(\rho_k, D) = 0, & -z_0/2 < \rho_k, \rho_k > z_0/2 \end{cases} \tag{7.33}$$

The deviation of the dipoles from their mean positions has to be limited within the interval $[-a_0/2, a_0/2]$; recalculation $\mathrm{PDF}(\rho_k, D)$ into $\mathrm{PDF}(\delta_k, D)$ is given by (7.13).

The integrals cannot be performed analytically and is done using the mathematical software MATLAB. Truncation of the positional PDF was achieved by coding. To obtain the autocorrelation of the PDF, a standard subroutine was used. The results of the operations are shown in Fig. 7.5. All the constants used in the

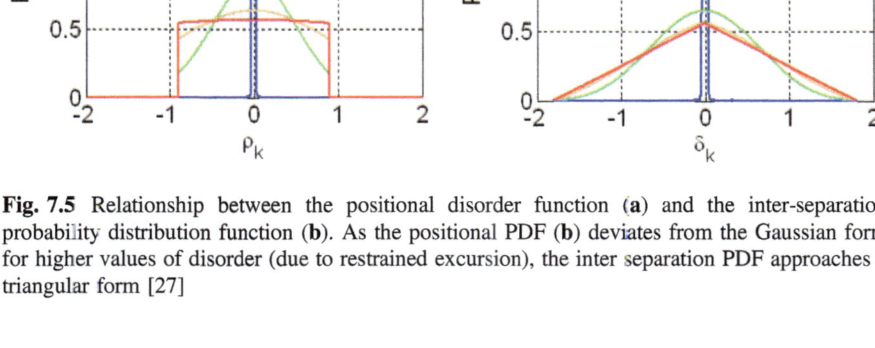

Fig. 7.5 Relationship between the positional disorder function (**a**) and the inter-separation probability distribution function (**b**). As the positional PDF (**b**) deviates from the Gaussian form for higher values of disorder (due to restrained excursion), the inter separation PDF approaches a triangular form [27]

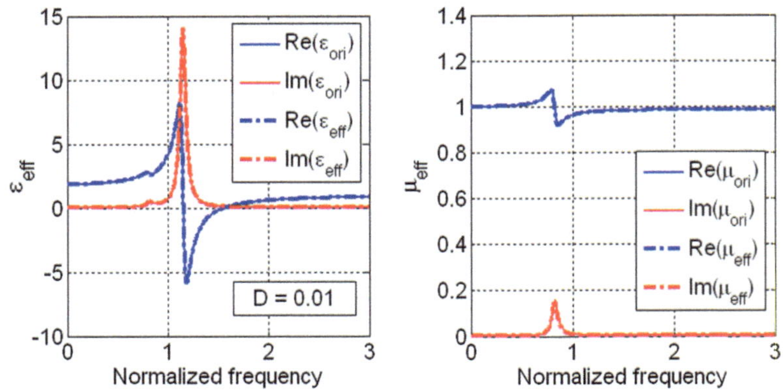

Fig. 7.6 Verification of the code—typical values from [26] were used in the computer code written for the calculation of the effective material parameters for MMs with positional disorder having a very small amount of disorder $D = 0.01$. The results obtained match with those in [26]—this is expected as the nature of coupling considered in the present theory should have negligible influence upon the material parameters for very large spatial periods $z_n > 3$ [27]

analysis were taken from [26]. The spacing between the double-wires, $2y_1$ and the resonant frequency of an isolated cut-wire were taken to be 65 nm and $\omega_0 = 1.39 \times 10^{15}\ \text{rad s}^{-1}$ respectively. The damping coefficient was taken to be $\gamma = 9.42 \times 10^{13}\ \text{rad s}^{-1}$. The mean periodic spacing $(z_0)_n = 1.8$ (the mean spacing between the MAs was taken to be 1.8 times the double-wire spacing $2y_1$). To verify the correct functioning of the code, the results for a very small disorder were compared with the results for a perfectly ordered system (with neighboring MAs interacting with each other)—see Fig. 7.6.

7.6 Results and Discussion

The results of the analysis for dipoles are presented in Fig. 7.7 and the results of the analysis for quadrupoles are presented in Figs. 7.8 and 7.9.

The analysis was carried out for two values of the spacing period z_n, viz. $z_n = 1.2$ (Fig. 7.8) and $z_n = 1.8$ (Fig. 7.9). The positional PDF was taken for the four different values of the standard deviation D, and consequently the inter separation $\text{PDF}(\delta_n, D)$ was obtained using numerical coding in MATLAB. The effective susceptibilities were calculated by numerical implementation of the integration (7.30), (7.31) and then the effective material parameters (7.29) were calculated. The following features are clearly noted:

- For the disordered dipole ensemble, the fall in the permittivity with increasing disorder is clearly visible (Fig. 7.7e–h). The $\text{Im}(\varepsilon)$ curve is symmetric for very small values of the variance (here, $D = 0.01$). However, as the disorder increases the peak shifts towards lower frequencies and the curves broaden and

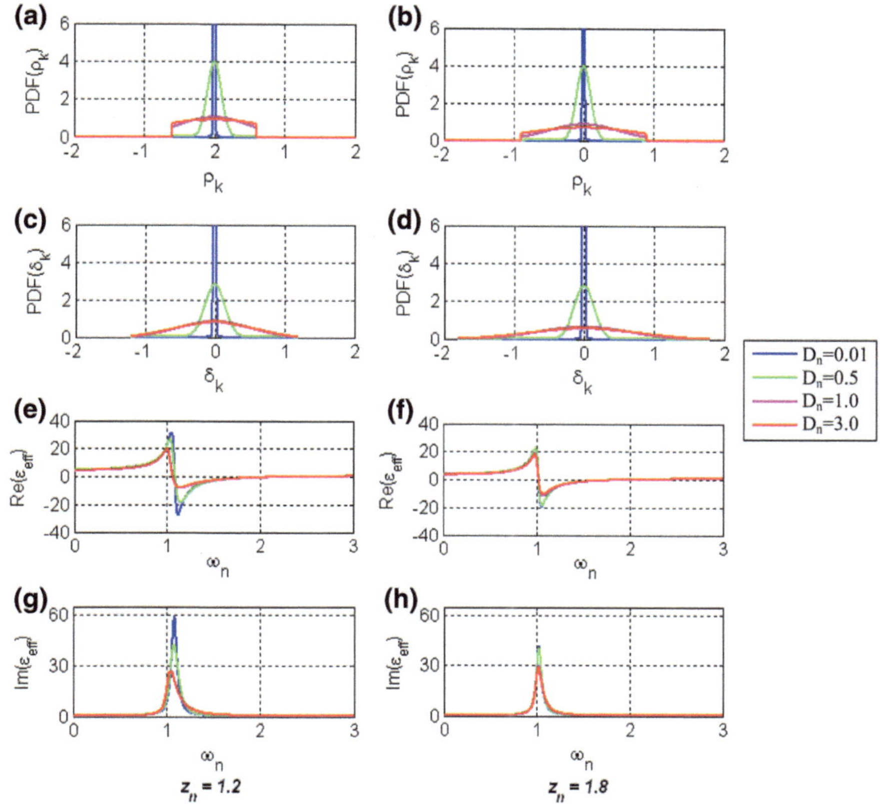

Fig. 7.7 Effective material parameter curves for dipole ensembles exhibiting positional disorder. The effective permittivity and permeability curves for disordered dipole ensembles are presented for different values of disorder. The first column pertains to values obtained for a mean period of z_n = 1.2, while the second column relates to those obtained for a mean period of z_n = 1.8. For the respective periodicities: **a**, **b** the positional disorder function; **c**, **d** the respective inter-separation PDFs; **e**, **f** scaled real part of the permittivity; **g**, **h** scaled imaginary parts of the permittivity. Clearly, increase in disorder brings about a fall in the maximums of the response of the system [27]

become asymmetric. The reason for these observed effects can be explained as follows. A disordered system can be thought of to consist of several different periodic systems. The resonant frequency for each such periodic ensemble depends inversely on its spatial period. If the response of the disordered system is approximated by the sum of the responses of its constituent periodic systems, it becomes evident that the final curve will develop a tail approaching the blue end of the spectrum. The asymmetry can thus be attributed to an inverse power relationship between resonance frequency and inter-separation. The broadening effect is a consequence of particle conservation. On the other hand, the lowest frequency/largest wavelength of the response is not a function of the periodicity, but is actually limited by the eigenfrequency of the independent oscillator. In fact, the resonance frequency approaches the eigenfrequency for a periodic

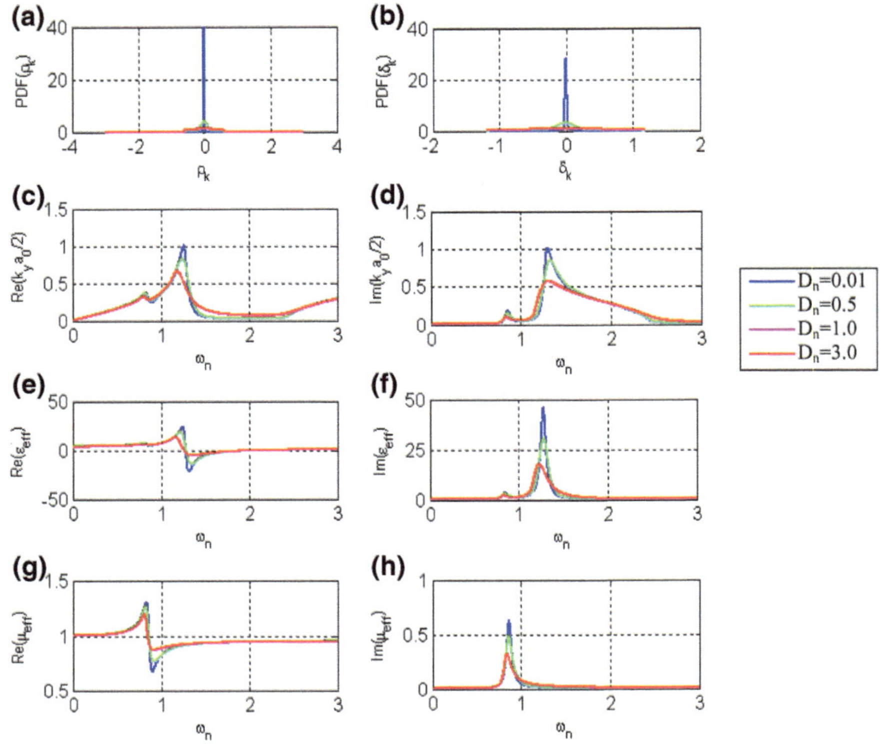

Fig. 7.8 Dispersion and effective material parameter curves for quadrupole ensemble with z_n = 1.2—**a** positional disorder function PDF; **b** inter-separation PDF; **c**, **d** real and imaginary part of k-vector (in normalized units of double-wire separation distance $2y_1$); **e**, **f** real and imaginary parts of effective permittivity; **g**, **h** real and imaginary parts of effective permeability [27]

assembly of dipoles when the spatial period becomes large. Hence, as the disorder in the system increases, the curves are broadened, asymmetric and the peak response shifts towards the eigenfrequency of the independent oscillator.

- In the case of the quadrupole ensemble, a decrease in the value of the electric permittivity is observed as D is increased. This is in agreement with the experimental results. However, there is also a decrease in the value of the magnetic permeability. This decrease is more pronounced for $z_n = 1.2$ in comparison to $z_n = 1.8$. This is an unexpected result, as the magnetic response should remain almost constant. The reason for this discrepancy could lie in the simple form of the probabilistic model chosen to describe the randomness.
- Generally speaking, the final expressions for the permittivity and the permeability were derived under several approximations, associated with (7.1). The observed discrepancy could also be attributed to these approximations. Above all, the fundamental limitations of the multipole theory itself could affect the final results as well. These possibilities should be investigated further.

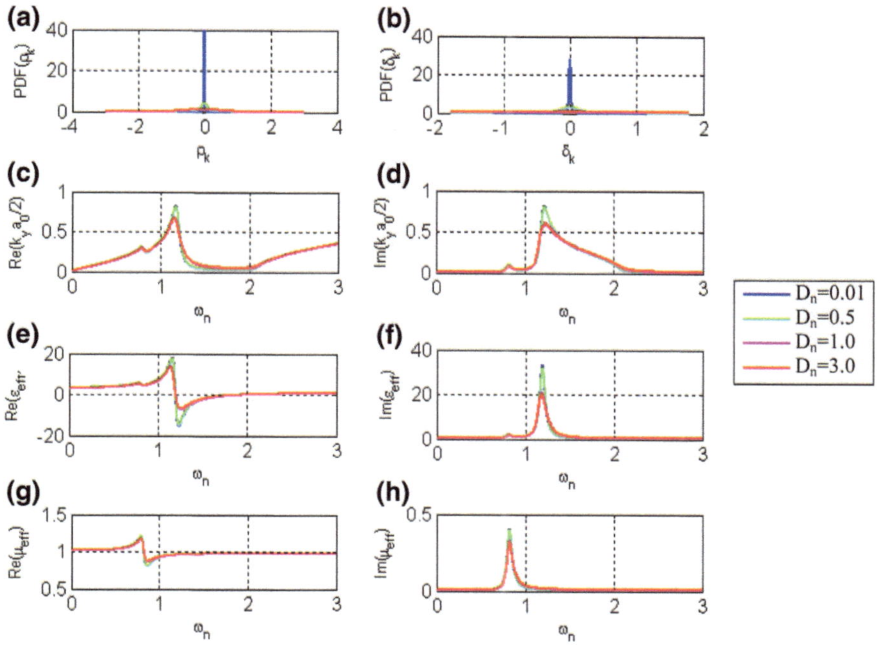

Fig. 7.9 Dispersion and effective material parameter curves for quadrupole ensemble with $z_n = 1.8$—**a** positional disorder function; **b** inter-separation PDF; **c**, **d** real and imaginary part of k-vector (normalized with the double-wire separation distance $2y_1$), **e**, **f** real and imaginary parts of effective permittivity; **g**, **h** real and imaginary parts of effective permeability [27]

In light of the above arguments, it is concluded that as the observed positions of the resonances and the relative magnitudes of the parameters are within the limits of approximation, the analysis is valid, and can be used to roughly predict the properties of MMs with incorporated randomness.

7.7 Other Forms of Disorder

In the preceding analysis, the effect of positional disorder (arising due to aperiodicity) on the averaged material parameters was considered. In a random MM other forms of disorder can also exist. A particular case of interest is positional disorder along the cut-wire axis—see Fig. 7.10. If this form of positional disorder is taken into consideration along with the aperiodicity, the model would then be a step closer to emulate a true self-organized random MM [28]. In the multipole model, the individual cut-wires are replaced by dipoles. In case the quadrupoles are disarrayed, the coupling between them will also be a function of their relative angular positioning.

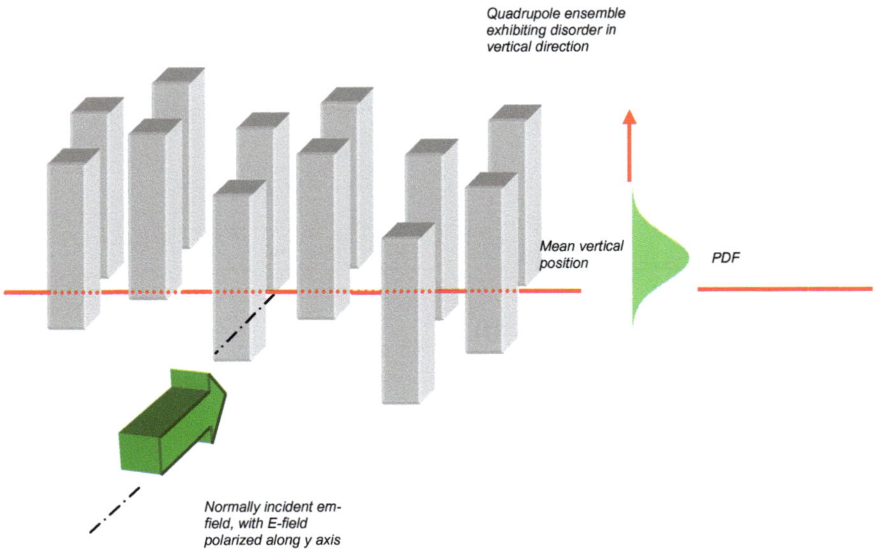

Fig. 7.10 Disorder along the cut-wire axis direction—the figure shows a one dimensional disorder arrangement of MAs. The extent of disorder can be quantified in terms of the angle, the total range of variation being limited to $(-\pi/2, \pi/2)$ [27]

This angular dependence can be introduced into the coupling terms of the dynamic equations. More specifically, the coupling constants b and c in the differential equations will include the angular dependence. All other mathematics remains the same, the averaging procedure can now be carried out between angles $(-\pi/2, \pi/2)$. The curves in Figs. 7.11 and 7.12 summarize the results obtained via the multipole approach.

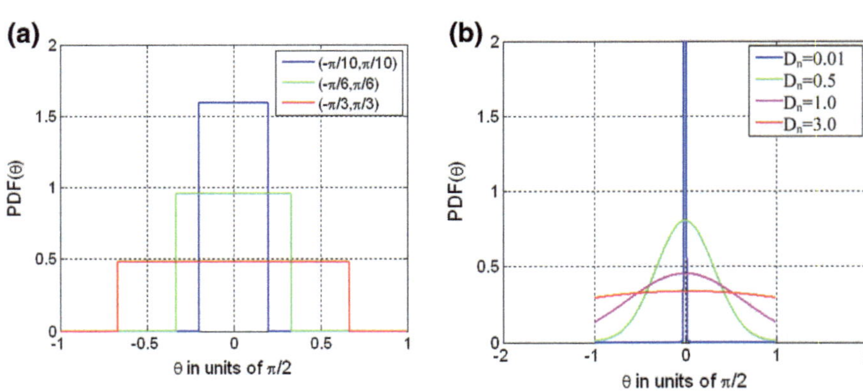

Fig. 7.11 **a** Rectangular and **b** Gaussian forms of distribution function used for governing the positional disorder of the MMs along the lateral direction. The positional disorder is expressed in terms of the relative angle between two neighboring MAs. The effective material parameters of the ensemble are derived and presented in Fig. 7.12 for three different values of disorder [27]

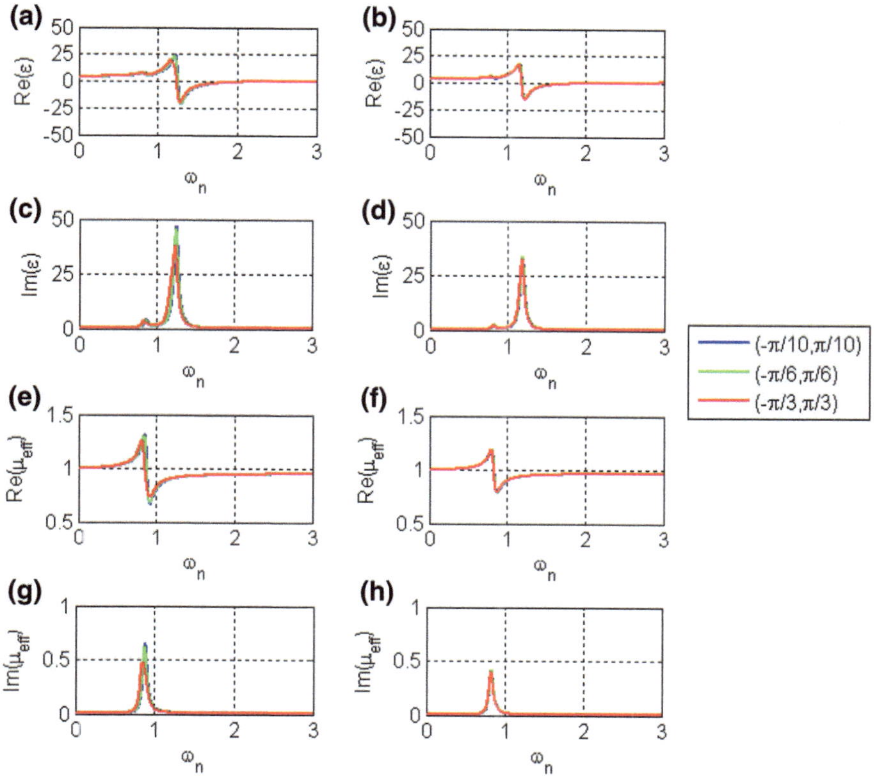

Fig. 7.12 Effective material parameters for MMs exhibiting positional disorder along the lateral direction (y-direction) governed by the rectangular distribution function. The first column gives the material parameters for a MM ensemble having a mean period z_n = 1.2, while the second column is for MMs having a mean period z_n = 1.8 [27]

Two forms of the distribution function were used in the analysis (Fig. 7.11). The extent of disorder is correlated to the relative angular position of the dipoles. The first form of the angular PDF distribution function used was a rectangular function (see Fig. 7.11a). The second form used was a Gaussian distribution (Fig. 7.12b), the random variable being the relative angular position. The function is centered about 0 degrees and the extent of disorder being quantified by the standard deviation D; b and c are multiplied by the term $\cos(\theta)$ to incorporate the angular dependence. Clearly then, when $\theta = \pi/2$, there is no interaction between the cut-wires.

The Reimannian integration is limited between the values $(-\pi/2, \pi/2)$. The constants were again taken from the original reference [24], and the mean periodicity was set to $z_n = 1.8$. The results (Figs. 7.11 and 7.13) show that both the effective permittivity and permeability are clearly affected by the angular disorder. In a similar fashion, in-plane and out-of-plane skew disorders of MAs can also be accounted for by the model with appropriate changes to the coupling terms.

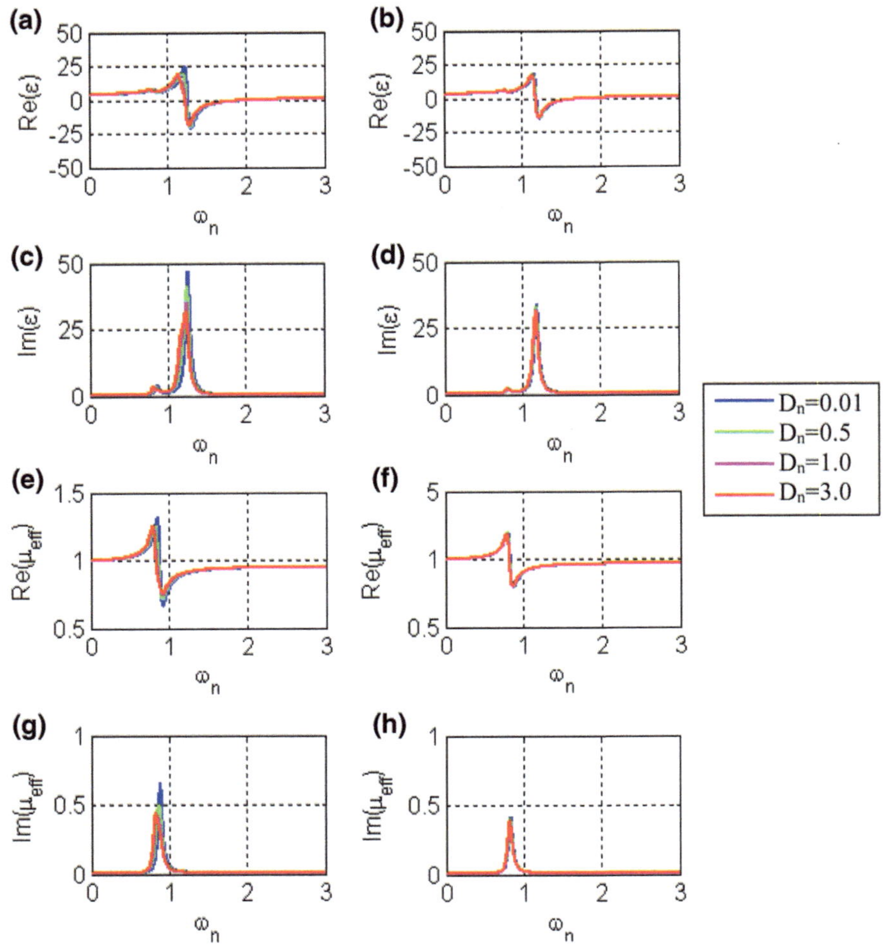

Fig. 7.13 Effective material parameters for MMs exhibiting positional disorder along the lateral direction (y-direction) governed by the Gaussian distribution function (Fig. 7.11). The first column gives the material parameters for a MM ensemble having a mean period $z_n = 1.2$, while the second column is for MMs having a mean period $z_n = 1.8$ [27]

7.8 Conclusion

In extending the multipole approach [1] to the case of random MMs, the effect of spatial distribution of the MAs was taken into account by considering the near field coupling between neighboring MAs. The effective susceptibility was expressed as a function of the inter-separation between MAs—the ensemble averaged susceptibility was then obtained as an expectation value, weighted by the probability distribution function of all possible inter-separations. In the present work, the disorder was considered only along one direction (in-plane, perpendicular to cut-wire long axis), assuming that adjacent rows of MAs do not interact with each other. Results obtained by the numerical implementation of the equations confirm the experimental findings that increasing disorder has a more pronounced effect on the effective electrical permittivity than on the effective magnetic permeability. For smaller periodicities, however, the electrical permittivity and magnetic permeability are affected equally. This conflicting result may be caused by a coupling of the quadrupole moments of neighboring MAs. This has not been explicitly considered in the present version of the model. Also, other factors such as the effect of incident polarization, or the coupling between adjacent rows have not been considered in the present theory. The understanding gained from the study of this simple case can now be used to account for the above specific and more involved cases.

References

1. A. Chipouline, J. Petschulat, A. Tuennermann, T. Pertsch, C. Menzel, C. Rockstuhl, F. Lederer, Multipole approach in electrodynamics of Metamaterials. Appl. Phys. A **103**, 899–904 (2011)
2. A. Chipouline, S. Sugavanam, J. Petschulat, T. Pertsch, Metamaterials with interacting metaatoms (2012). http://arxiv.org/abs/1205.6839
3. J. Pendry, Light finds a way through maze. Physics **1**, 20 (2008)
4. P. Anderson, Absence of diffusion in certain random lattices. Phys. Rev. **109**(5), 1492 (1958)
5. J. Pendry, Quasi-extended electron states in strongly disordered systems. J. Phys. C Solid State Phys. **20**(5), 733 (1987)
6. A. Tartakovskii, M. Fistul, M. Raikh, I. Ruzin, Hopping conductivity of metal-semiconductormetal contacts. Sov. Phys. Semicond. **21**, 370 (1987)
7. J. Bertolotti, S. Gottardo, D.S. Wiersma, M. Ghulinyan, L. Pavesi, Optical necklace states in Anderson localized 1D systems. PRL **94**(11), 113903 (2005)
8. K.Y. Bliokh, Y.P. Bliokh, V. Freilikher, A.Z. Genack, B. Hu, P. Sebbah, Localized modes in open one dimensional dissipative random systems. PRL **97**(24), 243904 (2006)
9. A. Alù, N. Engheta, Effect of small random disorders and imperfections on the performance of arrays of plasmonic nanoparticles. New J. Phys. **12**, 013015 (2010)
10. F. Rüting, Plasmons in disordered nanoparticle chains: localization and transport (2011). arXiv:1102.2705v1
11. D. Mogilevtsev, F. Pinheiro, R. dos Santos, S. Cavalcanti, L. Oliveira, Light propagation and Anderson localization in disordered superlattices containing dispersive metamaterials: effects of correlated disorder. Phys. Rev. B **84**, 094204 (2011)
12. W. Tan, Y. Sun, Z.-G. Wang, H. Chen, H.-Q. Lin, Transparency induced by coupled resonances in disordered metamaterials. Opt. Express **17**, 24371 (2009)
13. L. Jylhä, I. Kolmakov, S. Maslovski, S. Tretyakov, Modeling of isotropic backward-wave materials composed of resonant spheres. J. Appl. Phys. **99**, 043102 (2006)
14. M. Gorkunov, S. Gredeskul, I. Shadrivov, Y. Kivshar, Effect of microscopic disorder on magnetic properties of metamaterials. Phys. Rev. E **73**, 056605 (2006)
15. X. Zhou, X. Zhao, Y. Liu, Disorder effects of left-handed metamaterials with unitary dendritic structure cell. Opt. Express **16**, 7674 (2008)
16. J. Rico-García, J. López-Alonso, A. Aradian, Toy model to describe the effect of positional blocklike disorder in metamaterials composites. JOSA B **29**, 53 (2012)
17. N. Papasimakis, V.A. Fedotov, Y.H. Fu, D.P. Tsai, N.I. Zheludev, Coherent and incoherent metamaterials and order-disorder transitions. Phys. Rev. B **80**, 041102(R) (2009)

18. A. Boltasseva, V.M. Shalaev, Fabrication of optical negative-index metamaterials: recent advanced and outlook. Metamaterials **2** (2008)
19. J. Wright, O. Worsfold, C. Whitehouse, M. Himmelhaus, Ultra at ternary nanopatterns fabricated using colloidal lithography. Adv. Mater. **18**, 421 (2006)
20. P. Hanarp, D. Sutherland, J. Gold, B. Kasemo, Nanostructured model biomaterial surfaces prepared by colloidal lithography. Nanostruct. Mater. **12**, 429 (1999)
21. R. Glass, M. Moeller, J.P. Spatz, Block copolymer micelle nanolithography. Nanotechnology **14**, 1153 (2003)
22. C. Helgert, C. Rockstuhl, C. Etrich, E.-B. Kley, A. Tuennermann, F. Lederer, T. Pertsch, Effective properties of amorphous metamaterials. Phys. Rev. B **79**, 233107 (2009)
23. N. Gippius, T. Weiss, S. Tikhodeev, H. Giessen, Resonant mode coupling of optical resonances in stacked nanostructures. Opt. Express **18**, 7569 (2010)
24. N. Feth, M. König, M. Husnik, K. Stannigel, J. Niegemann, K. Busch, M. Wegener, S. Linden, Electromagnetic interaction of split-ring resonators: the role of separation and relative orientation. Opt. Express **18**, 654529 (2010)
25. P. Mazur, B. Nijboer, On the statistical mechanics of matter in an electromagnetic field. I. Physica **XIX**, 971 (1953)
26. J. Petschulat, C. Menzel, A. Chipouline, C. Rockstuhl, A. Tünnermann, F. Lederer, T. Pertsch, Multipole approach to metamaterials. Phys. Rev. B **78**, 043811 (2008)
27. A. Chipouline, S. Sugavanam, J. Petschulat, T. Pertsch, Extension of the multipole approach to random metamaterials. Adv. Optoelectron. **2012**, Article ID 161402 (2012)
28. D. Pawlak, S. Turczynski, M. Gajc, K. Kolodziejak, R. Diduszko, J. Smalc, I. Vendik, How far are we from making metamaterials by self organization: the microstructure of highly anisotropic particles with an srr-like geometry. Adv. Func. Mater. **20**, 1116 (2010)

Chapter 8
Applications of the "Classical" Metamaterial Model—Nonlinear Metamaterials: Multipole (Second Order) and Third Order Nonlinearities

8.1 Introduction

In this chapter, the nonlinear optical response of MMs evoked by second-order multipoles is analyzed for the split-ring resonator, although the introduced formalism can be applied to arbitrary structures. The equations that describe nonlinear light propagation are derived where special emphasis is put on second-harmonic generation. This contribution basically aims at stretching versatile and existing concepts to describe light propagation in nonlinear media toward the realm of MMs.

The third harmonic generated by a double-layer fishnet MM was investigated experimentally and theoretically as well. An analytical model based on the nonlinear dynamics of the electrons inside the gold shows excellent agreement with experimental and numerical results.

The simultaneous consideration of the magnetic dipole and the electric quadrupole is required by nature since both occur in the same order of the multipole expansion. Though this has been extensively discussed in the literature [1–6], the quadrupole moment is frequently dropped. Besides the linear properties that can be covered by this expansion the extension of the multipole description lads to the quadratic nonlinear optical regime. Since the multipole expansion is truncated beyond second-order terms, the study is focused on quadratic nonlinear effects associated with these second-order multipoles. We mention that this procedure of introducing nonlinearity is known from the early works in nonlinear optics [7], Pershan (1963) and is supported by several papers that observed multipole induced nonlinear optical effects in various plasmonic nanostructures [8–16].

The formalism developed here is general and can be straightforwardly applied to any geometry. The SRR was chosen because first experiments on the second-harmonic (SH) generation (SHG) were already reported in this structure; although in a configuration amenable for nanofabrication where the induced magnetic dipole is non-radiating [17, 18].

A. Chipouline and F. Küppers, *Optical Metamaterials: Qualitative Models*, Springer Series in Optical Sciences 211, https://doi.org/10.1007/978-3-319-77520-3_8

In order to predict the second-harmonic generation by purely analytical means, the undepleted pump approximation (UDPA) was applied and the associated equations were derived within the multipole model. An excellent agreement with the numerically derived solutions was observed which justifies the application of this approximation for further predictions.

8.2 Nonlinear Wave Equations

Our investigation starts with the wave equation incorporating multipoles up to second order [5]:

$$\frac{\partial^2 E_x(y,\omega)}{\partial y^2} + \frac{\omega^2}{c^2}\left(E_x(y,\omega) + 4\pi P_x(y,\omega)\right) + \frac{i4\pi\omega}{c}\frac{\partial M_z(y,\omega)}{\partial y} = 0 \tag{8.1}$$

Now in terms of the plasmonic eigenmodes of interest the respective meta-atom has to be mapped onto the point multipoles: electric dipoles and quadrupole $P_x(y,\omega)$ and magnetic dipoles $M_z(y,\omega)$. In order to observe both, an electric and a magnetic response, the SRR is uprightly oriented [19] (see Fig. 8.1). To cover the fundamental electric and magnetic modes [20], sketched by the black solid and dashed lines in Fig. 8.1, respectively, four auxiliary positive and negative charges with predefined spatial degrees of freedom are required as indicated in Fig. 8.1. In passing we note that the carrier configuration shown in Fig. 8.1b represents a suitable solution that allows us to reproduce the fundamental modes as sketched in Fig. 8.1a but might not provide the only possible arrangement accounting for these dynamics. Similar to the numerical discrete dipole [21] or multiple multipole model [22] the number of applied supplementary carriers increases the resolution in terms of the consideration of higher order modes.

We selected the configuration shown in Fig. 8.1b since it constitutes a manageable carrier number, describing the fundamental plasmonic properties properly. With the knowledge about these directional constraints for the carrier dynamics the microscopic definitions of the multipole moments is given by (3.52):

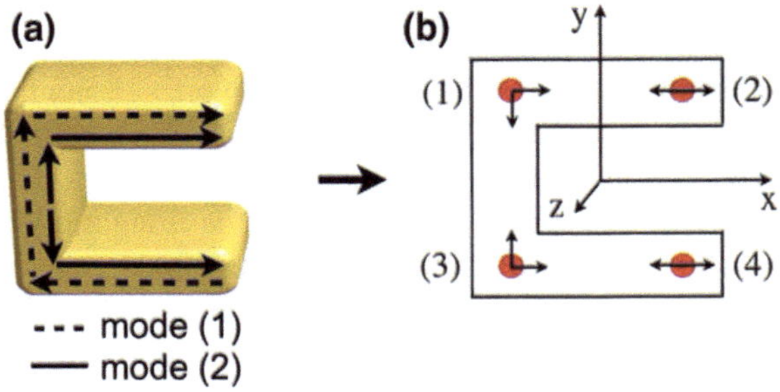

Fig. 8.1 **a** SRR meta-atom and the intrinsic currents for the fundamental electric (black solid line) and magnetic (black dashed line) mode. **b** The associated auxiliary charge distribution (red points) with predefined degrees of freedom (black arrows) [23]

$$\begin{cases} \vec{P}(\vec{R},t) = \eta\left\langle \sum_{s}^{\text{all charges}} q_s \vec{r}_s \right\rangle - \nabla \bullet Q(\vec{R}, t) \\ Q_{ij}(\vec{R}, t) = \frac{\eta}{2}\left\langle \sum_{s}^{\text{all charges}} q_s r_{i,s} r_{j,s} \right\rangle \\ \vec{M}(\vec{R}, t) = \frac{\eta}{2c}\left\langle \sum_{s}^{\text{all charges}} q_s \left[\vec{r}_s, \frac{\partial \vec{r}_s}{\partial t}\right] \right\rangle \end{cases} \tag{8.2}$$

The equations of motion $r_{i,s}(t)$ for each charge q_s contain all information about the plasmonic eigenmodes. Even at this early stage it can be seen that if $r_{i,s}(t) \sim E(r, t)$ second-order multipoles will immediately evoke nonlinear contributions since they involve terms $r_{i,s}^2(t)$. For the carrier configuration proposed here the associated terms $r_{i,s}(t)$ are:

$$\begin{array}{ll} r_1^+(t) = (-x_0, y_0, 0), & r_1^-(t) = (-x_0 - x_1, y_0 - x_1, 0) \\ r_2^+(t) = (x_0, y_0, 0), & r_2^-(t) = (x_0 - x_1, y_0, 0) \\ r_3^+(t) = (-x_0, -y_0, 0), & r_3^-(t) = (-x_0 - x_2, -y_0 + x_2, 0) \\ r_4^+(t) = (x_0, -y_0, 0), & r_4^-(t) = (x_0 - x_2, -y_0, 0) \end{array} \tag{8.3}$$

In (8.3) the superscripts $\pm$ denote whether the position vector is associated with a positive or a negative charge. All positive carriers are fixed at the positions $\pm x_0$, $\pm y_0$, while negative carriers are allowed to oscillate around these sites described by $x_{1,2}(t)$. The distinction between the carrier oscillations in both SRR wires is vital for realizing the two plasmonic eigenmodes (Fig. 8.1a). These carrier oscillations evoked by an external electromagnetic field and intrinsic Coulomb interaction may be described by a set of coupled oscillator equations as (see (4.14)):

$$\begin{cases} \frac{\partial^2 x_1(t)}{\partial t^2} + \gamma \frac{\partial x_1(t)}{\partial t} + \omega_0^2 x_1(t) + \sigma x_2(t) = \frac{q}{m} E_{x1,\text{loc}} \\ \frac{\partial^2 x_2(t)}{\partial t^2} + \gamma \frac{\partial x_2(t)}{\partial t} + \omega_0^2 x_2(t) + \sigma x_1(t) = \frac{q}{m} E_{x2,\text{loc}} \end{cases} \tag{8.4}$$

In (8.4) γ represents the damping, ω_0 is the eigenfrequency while α describes the coupling strength between the carriers in the SRR arms. The physical origin of this coupling is the Coulomb interaction of carriers in the horizontal SRR arms excited by an electric field parallel to the arms and the carriers in the vertical arm that are excited by the local fields of the horizontally oscillating charges, producing a current inside the entire SRR. Moreover this coupling between two identical oscillators results in a splitting into symmetric and antisymmetric oscillation modes. Substituting (8.3) into (8.2) the remaining multipole moments are obtained as:

$$\begin{cases} P_x(y,t) = 2\eta q(x_1 + x_2) - \frac{\partial Q_{xy}}{\partial y} \\ Q_{xy}(y,t) = \frac{\eta q}{2}(x_1 - x_2)[(2y_0 - x_0) - (x_1 + x_2)] \\ M_z(y,t) = -\frac{\eta q}{2c}(2y_0 + x_0)\frac{\partial}{\partial t}(x_1 - x_2) \end{cases} \tag{8.5}$$

From (8.5) it can be deduced that all multipoles depend either on the sum or the difference of $x_1(t)$ and $x_2(t)$. Especially for the symmetric carrier oscillation in the SRR arms $x_1(t) = x_2(t)$ all second-order moments vanish and only two identical electric dipoles parallel to the SRR arms remain (symmetric mode). In turn, an antisymmetric oscillation $x_1(t) = -x_2(t)$ excites both second order multipoles and a longitudinal electric dipole only since the electric dipoles in the top and bottom SRR arms are canceling each other (antisymmetric mode). Thus, the charge alignment chosen meets all requirements to describe the desired plasmonic eigenmodes and their dynamics are determined. Decomposing the electric field into plane waves at the fundamental (FF) and the SH frequency:

$$E_x(y,t) = E_\omega \exp i[k(\omega)y - \omega t] + E_{2\omega} \exp i[k(2\omega)y - 2\,\omega t] + c.c. \tag{8.6}$$

the solutions to the oscillator equations (8.4) read as:

$$x_1(t) \pm x_2(t) = x_\omega^\pm E_\omega \exp i[k(\omega)y - \omega t] + x_{2\omega}^\pm E_{2\omega} \exp i[k(2\omega)y - 2\,\omega t] + c.c. \tag{8.7}$$

where the amplitudes are given by:

$$\begin{aligned} x_\omega^+ &= 2\chi_\omega^+ \cos[k(\omega)y_0]E_\omega \\ x_\omega^- &= 2i\chi_\omega^- \sin[k(\omega)y_0]E_\omega \end{aligned} \tag{8.8}$$

with the introduced quasi susceptibility:

$$\chi_\omega^\pm = \frac{q}{m} \frac{1}{\omega_0^2 - \omega^2 - i\gamma\omega \pm \sigma} \tag{8.9}$$

The respective equations for the SH field follow by substituting ω by 2ω in (8.8). In contrast to ordinary electric dipole interaction we observe a frequency splitting in the quasi susceptibility $\chi^\pm$ evoked by the two frequency degenerated eigenmodes. Now, (8.5)–(8.9) can be inserted into (8.1) yielding a set of nonlinear eigenvalue equations with second-order nonlinear source terms:

$$\begin{aligned} &\left[\frac{\partial^2}{\partial y^2} + \frac{\omega^2}{c^2}(1 + 4\pi(p_\omega - u_\omega)) + \frac{i4\pi\omega m_\omega}{c}\frac{\partial}{\partial y}\right] E_\omega \exp i[k(\omega)y] \\ &= -\frac{4\pi\omega^2}{c^2} u_{\omega;2\omega,-\omega} \frac{\partial}{\partial y}\left\{E_\omega^* E_{2\omega} \exp(i[k(2\omega) - k^*(\omega)]y)\right\} \\ &\left[\frac{\partial^2}{\partial y^2} + \frac{4\omega^2}{c^2}(1 + 4\pi(p_{2\omega} - u_{2\omega})) + \frac{i4\pi\omega m_{2\omega}}{c}\frac{\partial}{\partial y}\right] E_{2\omega} \exp i[k(2\omega)y] \\ &= -\frac{16\pi\omega^2}{c^2} u_{\omega;2\omega,\omega} \frac{\partial}{\partial y}\left\{E_\omega^2 \exp i[2k(\omega)y]\right\} \end{aligned} \tag{8.10}$$

where the following abbreviations have been used for the linear:

$$\begin{aligned} p_\omega &= \frac{2\eta\, q x_\omega^+}{2} \\ m_\omega &= i\omega \frac{\eta\, q}{2}(2y_0 + x_0)x_\omega^- \\ u_\omega &= \frac{\eta\, q}{2}(2y_0 - x_0)x_\omega^- \end{aligned} \tag{8.11}$$

and the nonlinear multipole source terms:

$$\begin{aligned} u_{\omega;2\omega,-\omega} &= \frac{\eta\, q\left(x_\omega^- x_{2\omega}^{+*} - x_{2\omega}^- x_\omega^{+*}\right)}{2} \\ u_{\omega;2\omega,\omega} &= \frac{\eta\, q x_\omega^- x_\omega^+}{2} \end{aligned} \tag{8.12}$$

The exact solution to this eigenvalue equation would result in a *nonlinear dispersion relation* with $k(2\omega, E_{\omega,2\omega})$, $k(\omega, E_{\omega,2\omega})$ and a fixed ratio $E_\omega / E_{2\omega} E_{2\omega}$. The left hand side of (8.10) contains a part well known from dipole interaction p_ω but in addition contributions which stem from the second-order multipole response u_ω, m_ω. Additionally the quadrupole moment causes a nonlinear term on the right-hand side. Interestingly, in this model the nonlinear response of the magnetic dipole produces no nonlinear contributions, which is supported by rigorous simulations for a corresponding SRR configuration [24]. There the magnetic nonlinear contributions have been shown to be much smaller in comparison to a convective electric current [25] which is equivalent to the quadrupole contribution in our approach. Furthermore, it is mentioned that in contrast to usual second order nonlinear optics [26] here the nonlinear source term incorporates the first spatial derivative induced by the quadrupole moment.

8.3 Linear Optical Properties: Effective Material Parameters

In order to validate the predictions of the model we start with the investigation of the linear properties. To this end the nonlinear source terms (8.12) have been dropped which yields two decoupled linear eigenvalue equations for the FF and the SH wave. The obtained linear wave equations describe the field propagation in an effective medium determined by its multipolar contributions. Thus the correlated wave vector $k(\omega)$, i.e., the dispersion relation, represents a self-consistent solution and contains all physical information to describe light propagation in the respective MM. This can be understood in complete analogy to the exact parameter retrieval which assumes also a homogeneous wave propagation inside a MM slab.

By neglecting the nonlinear source terms $u_{\omega;2\omega,-\omega}$ and $u_{\omega;2\omega,\omega}$ in (8.10) we get the linear wave equation:

$$\left[\frac{\partial^2}{\partial y^2} + \frac{\omega^2}{c^2}(1 + 4\pi(p_\omega - u_\omega)) + \frac{i4\pi\omega m_\omega}{c}\frac{\partial}{\partial y}\right] \exp i[k(\omega)y] = 0 \tag{8.13}$$

Details of the dispersion relation and effective parameters calculation can be found in Petschulat et al. [23], here only final results are presented in Fig. 8.2.

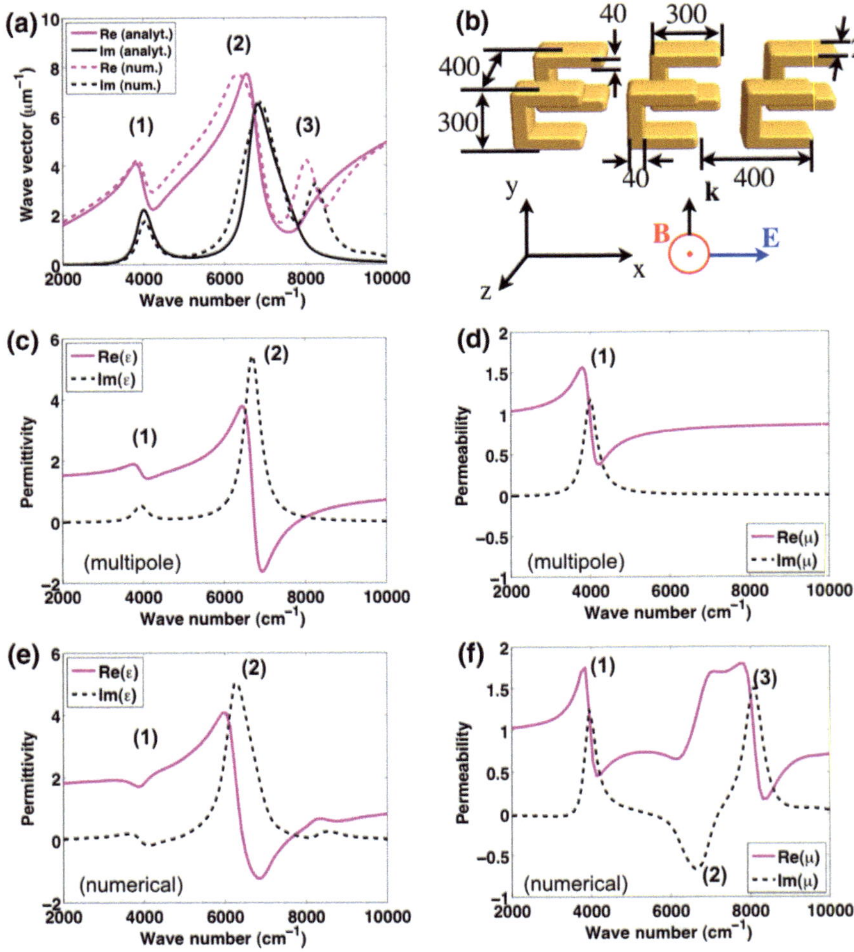

Fig. 8.2 **a** Comparison of the dispersion relation—numerical model (dashed), analytical model (solid) **b** single MM layer (gold SRR) used for the numerical simulation (dimensions in nm); spectral dependence of **c** effective permittivity and **d** effective permeability from the multipole model; spectral dependence of **e** effective permittivity and **f** effective permeability from the numerical simulations [23]

8.4 Nonlinear Optical Properties: Second Harmonic Generation

To study the nonlinear behavior induced by the fundamental modes of the present meta-atom, we resort to the linear dispersion relation and treat the nonlinearity as perturbation rather than solving (8.10) exactly. As usual we rely on the slowly varying envelope approximation (SVEA) [26]. Within this approximation the fast spatial oscillation $\exp i[k(\omega)y]$ is separated from a slowly varying amplitude $A(y)$, which contains all information about the generation and depletion of the fundamental $E_\omega = A_\omega(y)\exp i[k(\omega)y]$ and the second harmonic $E_{2\omega} = A_{2\omega}(y)\exp i[k(2\omega)y]$.

8.4.1 Exact Numerical Solution

At first the solution to the wave equations incorporating the SVEA ansatz has been performed numerically. Therefore we simplify this system by introducing the following substitutions:

$$\begin{aligned}\delta_\omega &\equiv \frac{\omega^2}{c^2}(1+4\pi p_\omega)\\ \beta_\omega &\equiv \frac{4\pi}{c^2}\left(\omega^2 u_\omega - i\omega m_\omega\right)\\ \psi_{\omega;2\omega,-\omega} &\equiv \frac{4\pi\,\omega^2}{c^2}u_{\omega;2\omega,-\omega}\\ \psi_{2\omega;\omega,\omega} &\equiv \frac{4\pi\,\omega^2}{c^2}u_{2\omega;\omega,\omega}\end{aligned} \tag{8.14}$$

Now the eigenvalue equations take the following form:

$$\begin{aligned}&\left[\frac{\partial^2}{\partial y^2} + \delta_\omega - \beta_\omega \frac{\partial}{\partial y}\right] E_\omega \exp i[k(\omega)y]\\ &= -\psi_{\omega;2\omega,-\omega}\frac{\partial}{\partial y}\left\{E_\omega^* E_{2\omega}\exp(i[k(2\omega)-k^*(\omega)]y)\right\}\\ &\left[\frac{\partial^2}{\partial y^2} + \delta_{2\omega} - \beta_{2\omega} \frac{\partial}{\partial y}\right] E_{2\omega} \exp i[k(2\omega)y]\\ &= -\psi_{\omega;2\omega,\omega}\frac{\partial}{\partial y}\left\{E_\omega^2 \exp i[2k(\omega)y]\right\}\end{aligned} \tag{8.15}$$

The solution to these equations are two coupled nonlinear dispersion relations, one for the fundamental and one for the second-harmonic wave, which depend on both fields. In order to avoid the solution of this involved system we treat the nonlinearity as a perturbation and resort to the linear dispersion relation. To study the effect of the nonlinear source terms we apply the slowly varying envelope approximation where the linear fields are weighted by a slowly varying amplitude functions $A(y)$. Replacing the constant amplitudes $E_{\omega,2\omega}$ by $A_{\omega,2\omega}(y)$ we obtain upon substitution and upon neglecting of the second-order derivatives (SVEA):

$$\begin{aligned}\frac{\partial}{\partial y}A_{\omega}(y) &= -\frac{\psi_{\omega;2\omega,-\omega}}{2ik(\omega)-\beta_{\omega}}\left\{A_{2\omega}(y)\frac{\partial}{\partial y}A^{*}_{\omega}(y)+A^{*}_{\omega}(y)\frac{\partial}{\partial y}A_{2\omega}(y)\right.\\ &\qquad\left.+\,i[k(2\omega)-k^{*}(\omega)]A^{*}_{\omega}(y)A_{2\omega}(y)\right\}\\ &\quad\times\exp i[k(2\omega)-k^{*}(\omega)-k(\omega)]y\\ \frac{\partial}{\partial y}A_{2\omega}(y) &= -\frac{\psi_{\omega;2\omega,\omega}}{2ik(2\omega)-\beta_{2\omega}}\left\{2A_{\omega}(y)\frac{\partial}{\partial y}A_{\omega}(y)+2ik(\omega)A^{2}_{2\omega}(y)\right\}\\ &\quad\times\exp i[2k(\omega)-k(2\omega)]y\end{aligned} \tag{8.16}$$

This final system has been solved numerically (Fig. 8.3).

The FF wave evolution E_{ω} is determined by the two eigenmode resonances (both indicated by the resonances in the red lined dispersion relation), where a

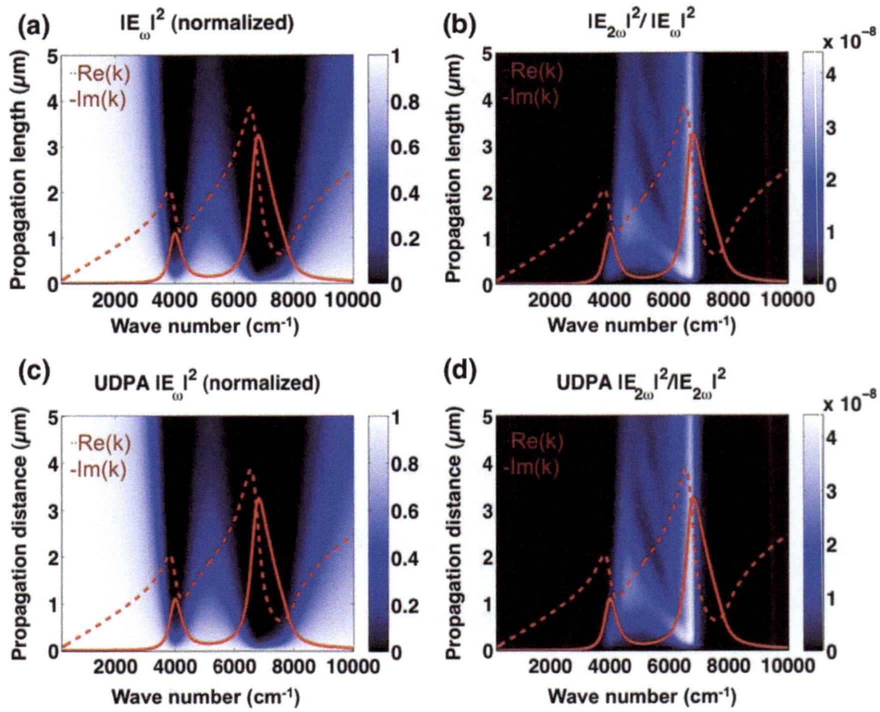

Fig. 8.3 Evolution of normalized electric field intensity for **a** The FF and **b** the SH as a function of the wave number of the fundamental. The red lines indicate the real (dashed) and the imaginary part (solid) of the linear dispersion relation; **c** and **d** the corresponding results for the undepleted pump approximation (UDPA) [23]

strong damping is observed. For frequencies out of the spectral domain of these resonances the FF wave propagates without excessive losses, as expected. For the SH wave a strong contribution at the fundamental magnetic and electric resonance can be observed. In these calculations the SHG signal originating from the electric resonance around 6400 cm^{-1} seems to be much stronger than in the spectral vicinity of the magnetic resonance 4000 cm^{-1}. This originates from the strong damping of the SHG wave for the magnetic resonance (which propagates at 8000 cm^{-1}) because in this spectral domain an enhanced damping occurs due to the presence of the electric resonance. This changes dramatically for the SH wave induced by the electric mode, since at the SH frequency the imaginary part of $k(2\omega)$ is close to zero. Thus the second harmonic originating from the electric resonance propagates almost without damping.

8.4.2 Undepleted Pump Approximation

In order to double-check the results and to take the weak conversion efficiency into account the UDPA has been applied [26]. Within this approximation the fundamental wave (the pump) remains unaffected by the generated second harmonic wave. This can be expressed by setting $\psi_{\omega;2\omega,-\omega}$ in the first equation as well as the first derivative of $A_\omega(y)$ in the second equation to zero. This results in:

$$
\begin{aligned}
\frac{\partial}{\partial y}A_{2\omega}(y) &= -\frac{\psi_{2\omega;\omega,\omega}}{2ik(2\omega)-\beta_{2\omega}}2ik(\omega)A_{2\omega}^2(y)\times\exp i[2k(\omega)-k(2\omega)]y \\
A_{2\omega}(y) &= -\frac{\psi_{2\omega;\omega,\omega}}{2ik(2\omega)-\beta_{2\omega}}\frac{2k(\omega)}{2k(\omega)-k(2\omega)}A_{2\omega}^2(y) \\
&\quad\times\{1-\exp i[2k(\omega)-k(2\omega)]y\}
\end{aligned}
\tag{8.17}
$$

Comparing the numerically determined electric field for the fundamental and the second harmonic wave to those of the undepleted pump approximation (see Fig. 8.3) one can clearly deduce that the UDPA describes the propagation for both waves almost exactly. Furthermore (8.24) permits to calculate analytically the conversion efficiency from fundamental to second harmonic intensity, e.g., for a slab consisting of a single layer of SRRs. Our calculations for a single slab of upright oriented SRRs predict intensity conversion efficiency in the order of 10^{-8}–10^{-9}. Compared with the values reported in Feth et al. [24] our estimated conversion efficiency is two orders of magnitude larger. This can be explained first by the different orientation since we investigated uprightly oriented SRRs for which second-order multipoles are radiating. These multipoles were considered as the nonlinear sources in this work and consequently an enhanced nonlinear interaction is expected. Second, any measured SHG intensity is supposed to be smaller due to structural imperfections of the individual meta-atoms that can even be observed within the comparison of numerically determined linear properties and its experimental counterparts. Thus it is different to predict the exact conversion efficiency for fabricated MMs by theory. The calculations presented here provide a physical motivation, an estimated order of magnitude as well as an expected dispersive dependence for second-order nonlinear effects occurring for such special types of nanostructures. It is important to note that for predicting the nonlinear response only parameters are required that are fixed by comparison with the linear effective material interaction. In passing we comment that such a procedure; the determination of the nonlinear material properties based on the linear material parameters is known as Miller's delta as well established rule in nonlinear optics [27, 28].

8.5 Third Harmonic Generation from Fishnet Structure

8.5.1 Measurement of Third Harmonic Generation

For the nonlinear measurements a setup based on an optical parametric amplifier (OPA) was used operating at wavelengths of 1.49, 1.54, 1.56 and 1.60 μm and having an average output power of 3 mW focused to a 300 μm-spot from the air side of the sample.

The OPA was pumped by a Nd:YAG laser with pulse duration of 5 ps and a repetition rate of 5 kHz. The resulting fluence took values up to 700 μJ/cm^2 in the plane of the sample. The sample was placed on a 6-axis positioning stage such that during the angular spectroscopy the beam is always focused into the same spot. The forward propagating THG signal pulses were detected by a photomultiplier tube and gate-integrated by an oscilloscope. We used the *p-p* polarization configuration —illuminating with *p*-polarized light and selecting only the *p*-polarized part of forward propagating light before the detector. For all measurements spectral filtering (Schott RG610 and BG40) before the detector was used for picking up the desired wavelength. The extinction curves of the used filters provided that the third harmonic response was orders of magnitude larger than signals at other wavelength, i.e. at the pump wavelength. The averaged THG signal from the pure SiO_2 substrate measured outside the MM area was approximately one order of magnitude lower than that from the MM area. Contributions from the substrate were therefore safely neglected. The principle setup is shown in Fig. 8.4. For numerical simulation an extension of the FMM which includes the nonlinear interaction was used [30]. The method relies on the undepleted pump approximation that ignores the feedback of the nonlinearity induced field to the pump field [31]. The approach allows solving the problem completely rigorously and permits a reliable prediction of the diffracted amplitudes of the third harmonic fields. The third harmonic intensity was measured and simulated in the forward zeroth diffraction order with the fundamental wavelength exciting the magnetic resonance. The angular spectra of THG are provided in Fig. 8.5a–d for the fundamental wavelengths of 1.49, 1.54, 1.56 and 1.60 μm, respectively.

Fig. 8.4 The setup for angular spectroscopy of the third harmonic generation (THG) intensity. The pump polarization is set to p and p-polarized third harmonic radiation is detected with a photomultiplier tube (PMT). The diffraction in the x direction is not shown [29]

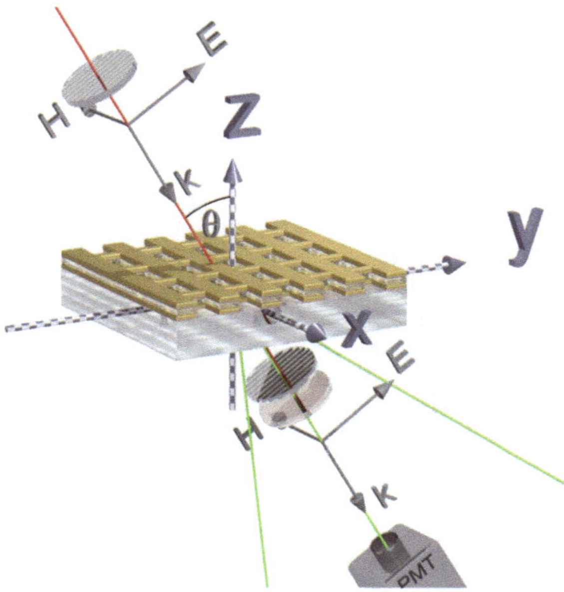

The magnetic resonance position for normal incidence is 1.54 μm. The maximum of the THG signal is seen at angles of incidence around 20°. The appearance of this maximum is detailed in the discussion section and is believed to be caused by the interference of THG from the individual layers forming the fishnet MM. The simulation shows an agreement with the experimental values. The THG signal is expressed in a pump power-independent fashion as derived from the numerical calculations; the absolute values of the THG signal are valid only for the simulation results while for the experimental data they are of the same order of magnitude. The estimation of the experimental value of the effective nonlinear susceptibility is $\chi^{(3)}_{1111} = 10^{-18}\,\mathrm{m}^2\,\mathrm{V}^2$, which is the same order of magnitude as the reference value of bulk gold [32].

8.5.2 Discussion and Modeling of Third Harmonic Generation

Plasmon-enhanced THG at the magnetic resonance of fishnet MMs was reported previously [12]. It was shown that the THG spectra obey the principles of the local-field enhanced nonlinear response. It was proposed that the wavelength dispersion of the THG efficiency is defined by the spectral line of the magnetic resonance cubed. The maximum of THG at the angles of about 20° can neither be explained by means of dispersion of the local field factor at the fundamental frequency, see Fig. 8.5e–h, nor with the linear transmission characteristics at the third

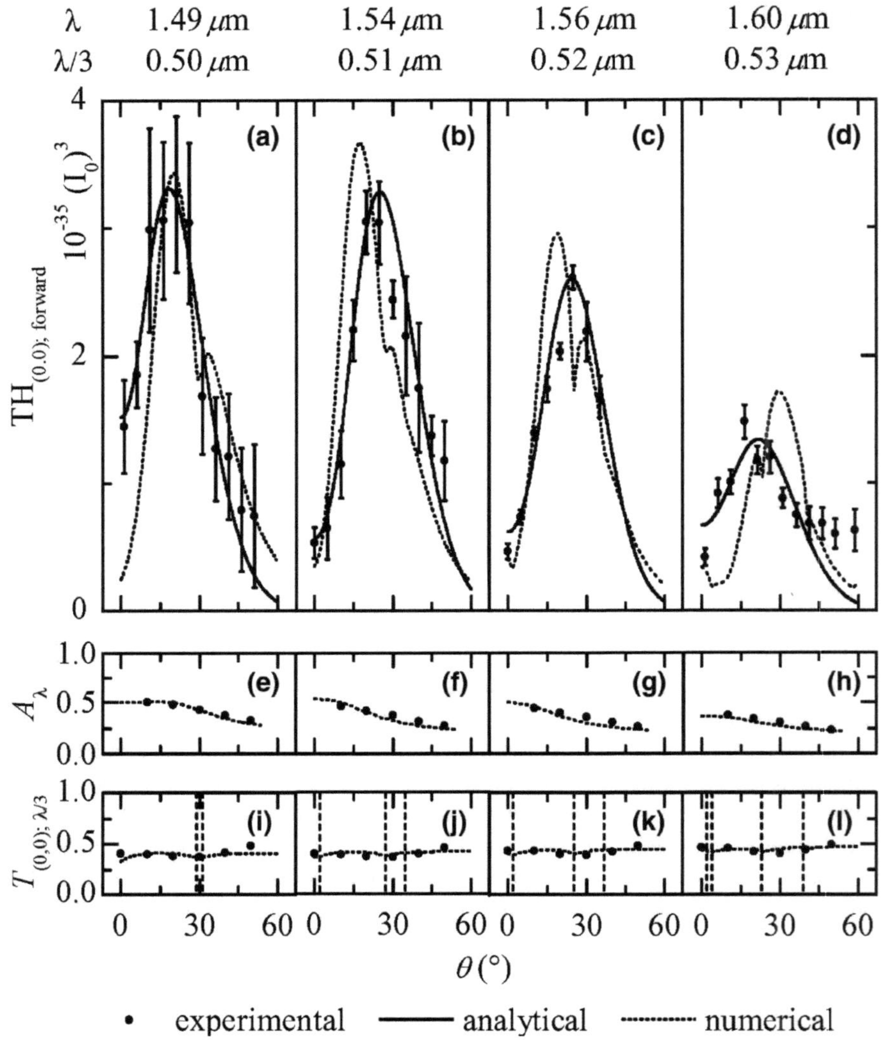

Fig. 8.5 Subplots (**a–d**) show the third harmonic signal as a function of the angle of incidence for different wavelengths in the spectral vicinity to the magnetic resonance. For comparison (**e–h**) show the linear absorption A_λ at the same fundamental wavelengths and (**i–l**) shows the linear transmission T at the corresponding third harmonic wavelengths. The vertical dashed lines indicate the angular positions of the appearance and the disappearance of diffraction orders. The black dots represent the experimental data and the dotted lines represent the simulation results. The solid lines are curves calculated with (8.39). This equation represents an analytical model which describes the nonlinear response of coupled oscillators [29]

harmonic wavelength, see Fig. 8.5i–l. Finally, the position of the maximum does not coincide with the angular position of the propagating diffraction order appearance as illustrated with the vertical dashed lines in Fig. 8.5i–l.

In this section we show that, firstly, this feature is caused by retardation effects, and secondly, it is specific to the anti-symmetric electric current structure of the magnetic resonance. The observed third harmonic radiation is considered to be caused by the nonlinear polarization of gold due to anharmonic electron movement. Nonlinearities of other substances of the MM are neglected since their $\chi^{(3)}$-tensor components are several orders of magnitude smaller than that of bulk gold: $\chi^{(3)}_{1111}(\mathrm{SiO_2}) = 4.6 \times 10^{-23}\,\mathrm{m^2/V^2}$, $\chi^{(3)}_{1111}(\mathrm{MgO}) = 1 \times 10^{-22}\,\mathrm{m^2/V^2}$, and $\chi^{(3)}_{1111}$ $(\mathrm{SiO_2}) = 7.5 \times 10^{-19}\,\mathrm{m^2/V^2}$ [32–34]. Without further discussion of the specific source of that third-order nonlinearity, we describe the motion of electrons of gold at the third harmonic wavelength within the conducting layers of the MM using a model of weakly coupled oscillators. Within the chosen model the phase difference between the oscillators in the two layers dictates whether the resonance is anti-symmetric—currents in the two layers are antiparallel to each other, Fig. 8.6a —or symmetric—currents are parallel, Fig. 8.6b.

Here we discuss the phase difference between the sources of third harmonic radiation. The sources of the radiation are oscillations in the gold layers at the third harmonic frequency. We use a model of coupled oscillators with a nonlinear extension. Uncompensated charges are induced at the edges of the thick wires of the MM by the external electromagnetic field with a polarization along the thin wires as shown in Fig. 8.6 [35]. Charge conservation implies $q_1(t) = -q_3(t)$ and $q_2(t) = -q_4(t)$. Harmonic oscillations of the charge densities in two coupled layers can be described as a superposition of two eigen modes of the system—the first one corresponds to co-directional currents in the layers and the second one corresponds to counter-directional ones [5]. Consider $x_1(t) = q_1(t) - q_3(t) = 2q_1(t)$ for the uncompensated charge at the upper fishnet layer and $x_2(t) = q_2(t) - q_4(t) = 2q_2(t)$ for the lower fishnet layer. The linear dynamics of these values is described by the model of coupled harmonic oscillators:

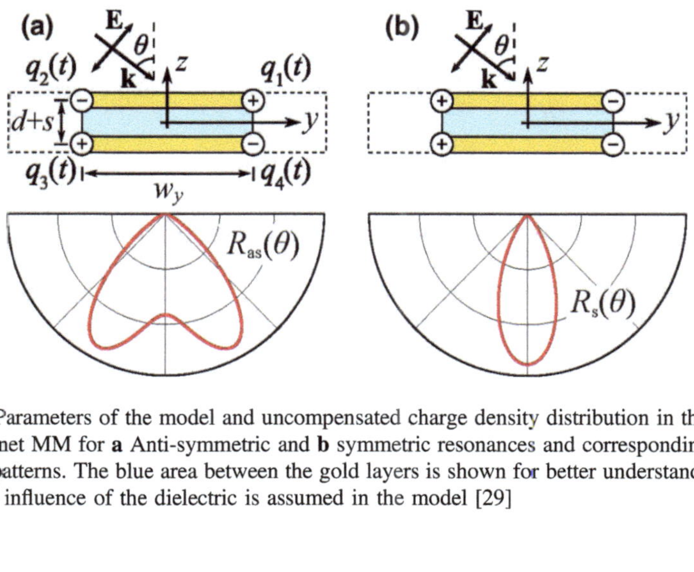

Fig. 8.6 Parameters of the model and uncompensated charge density distribution in the unit cell of the fishnet MM for **a** Anti-symmetric and **b** symmetric resonances and corresponding far-field radiation patterns. The blue area between the gold layers is shown for better understanding of the layout, no influence of the dielectric is assumed in the model [29]

$$\begin{cases} \frac{\partial^2 x_1(t)}{\partial t^2} + \gamma \frac{\partial x_1(t)}{\partial t} + \omega_0^2 x_1(t) + \sigma x_2(t) = f \exp(i\omega t) \\ \frac{\partial^2 x_2(t)}{\partial t^2} + \gamma \frac{\partial x_2(t)}{\partial t} + \omega_0^2 x_2(t) + \sigma x_1(t) = f \exp(i\omega t + \varphi_0) \end{cases} \quad (8.18)$$

Here γ is the damping constant, ω_0 is the central frequency of the resonance for an isolated layer, α is the coupling constant, f is the oscillator strength and φ_0 is the difference of phases of the exciting fields caused by the retardation. The dynamics of the antisymmetric mode $X(t) = x_1(t) - x_2(t)$ is described by:

$$\frac{\partial^2 X(t)}{\partial t^2} + \gamma \frac{\partial X(t)}{\partial t} + \omega_0^2 X(t) + \sigma X(t) = f(1 - \exp(i\varphi_0)) \exp(i\omega t) \quad (8.19)$$

The solution of the equation in the frequency domain is expressed as:

$$X(\omega) = \frac{f(1 - \exp(i\varphi_0))}{\omega_0^2 - \omega^2 + 2i\gamma\omega - \sigma} \quad (8.20)$$

In the case when Q-factor of the modes is high enough for the condition $\sqrt{\sigma} \gg \gamma$ to be held the asymmetric mode implies $x_1(\omega) + x_2(\omega) \approx 0$ and $\arg(x_1(\omega)) - \arg(x_2(\omega)) \approx \pi$ as a consequence. Now we consider a nonlinear addition to the electron movements:

$$\begin{cases} \frac{\partial^2 x_1(t)}{\partial t^2} + \gamma \frac{\partial x_1(t)}{\partial t} + \omega_0^2 x_1(t) + \sigma x_2(t) + \alpha x_1^3(t) = f \exp(i\omega t) \\ \frac{\partial^2 x_2(t)}{\partial t^2} + \gamma \frac{\partial x_2(t)}{\partial t} + \omega_0^2 x_2(t) + \sigma x_1(t) + \alpha x_2^3(t) = f \exp(i\omega t + \varphi_0) \end{cases} \quad (8.21)$$

where $\alpha \ll \gamma^2\omega_0^4/f^2$. This restriction corresponds to the experimentally observed low conversion ($\sim 10^{-11}$) from the fundamental field to the third harmonic field and allows one to use the perturbation theory approach. At the magnetic resonance apply $x_1(t) = -x_2(t)$ and only one equation have to be considered:

$$\frac{\partial^2 x_1(t)}{\partial t^2} + \gamma \frac{\partial x_1(t)}{\partial t} + \omega_0^2 x_1(t) + \sigma x_2(t) + \alpha x_1^3(t) = f \exp(i\omega t) \quad (8.22)$$

The approximate solution is reduced to two terms:

$$x_1(t) = x_1^0(\omega) \exp(i\omega t) + x_1'(\omega) \exp(i3\omega t) \quad (8.23)$$

After substituting the solution into (8.18) and calculating the multipliers of $\exp(i\omega t)$ and $\exp(i3\omega t)$, one gets:

$$x_1^0(\omega) = \frac{f}{\omega_0^2 - \omega^2 + 2i\gamma\omega - \sigma} \quad (8.24)$$

and:

$$x_1'(\omega) = \frac{\alpha}{\omega_0^2 - (3\omega)^2 + 6i\gamma\omega - \sigma}\left(x_1^0(\omega)\right)^3 \tag{8.25}$$

Analogously one gets:

$$x_2^0(\omega) = \frac{f}{\omega_0^2 - \omega^2 + 2i\gamma\omega - \sigma} \tag{8.26}$$

and:

$$x_2'(\omega) = \frac{\alpha}{\omega_0^2 - (3\omega)^2 + 6i\gamma\omega - \sigma}\left(x_2^0(\omega)\right)^3 \tag{8.27}$$

Since the first multipliers in (8.22) and (8.24) are not resonant and have the same phase, the phase difference $\arg\left(x_1'(\omega)\right) - \arg\left(x_2'(\omega)\right)$ is defined by the second multipliers. These multipliers are equal to $\left(x_1^0(\omega)\right)^3$ and $\left(x_2^0(\omega)\right)^3$ for the upper and lower layers, respectively. As a consequence $\arg\left(x_1'(\omega)\right) - \arg\left(x_2'(\omega)\right) = 3(\arg(x_1(\omega)) - \arg(x_2(\omega))) = 3\pi$, which means that at the THG frequency the electrons move inside two gold layers out of phase.

With this knowledge we write down the dynamical equations for the charge density at the third harmonic frequency:

$$\begin{aligned} \rho^{\mathrm{as}}(r,t) = q_0\cos(3\omega t) * \left[\delta\left(y - \frac{w_y}{2}\right) - \delta\left(y + \frac{w_y}{2}\right)\right] \\ * \left[\delta\left(z - \frac{d+s}{2}\right) - \delta\left(z + \frac{d+s}{2}\right)\right] \end{aligned} \tag{8.28}$$

and for the current density:

$$\begin{aligned} j_y^{\mathrm{as}}(r,t) = 3\omega\, q_0\sin(3\omega t) * \left[\Theta\left(y - \frac{w_y}{2}\right) - \Theta\left(y + \frac{w_y}{2}\right)\right] \\ * \left[\delta\left(z - \frac{d+s}{2}\right) - \delta\left(z + \frac{d+s}{2}\right)\right] \end{aligned} \tag{8.29}$$

for the antisymmetric resonance and the dynamical equations for the charge density:

$$\begin{aligned} \rho^{s}(r,t) = q_0\cos(3\omega t) * \left[\delta\left(y - \frac{w_y}{2}\right) - \delta\left(y + \frac{w_y}{2}\right)\right] \\ * \left[\delta\left(z - \frac{d+s}{2}\right) + \delta\left(z + \frac{d+s}{2}\right)\right] \end{aligned} \tag{8.30}$$

and for the current density:

$$\begin{aligned} j_y^s(r,t) &= 3\omega\, q_0 \sin(3\omega\, t) * \left[\Theta\left(y - \frac{w_y}{2}\right) - \Theta\left(y + \frac{w_y}{2}\right)\right] \\ &* \left[\delta\left(z - \frac{d+s}{2}\right) + \delta\left(z + \frac{d+s}{2}\right)\right] \end{aligned} \tag{8.31}$$

for the symmetric resonance. Here $\delta(y)$ is the Dirac delta function, $\Theta(y)$ is the Heaviside step function and q_0 is the amplitude of the uncompensated charge oscillations at third harmonic frequency. The latter depends on the magnitude of the nonlinear polarization and is proportional to the $\chi^{(3)}$ components and the local field factors at the third harmonic frequency $L_{3\omega}(\theta)$ and fundamental frequency $L_{\omega}(\theta)$ cubed. The solution of the potential equation:

$$\left(\Delta - \frac{1}{c^2}\frac{\partial^2}{\partial t^2}\right)\vec{A}(r,t) = -\frac{4\pi}{c^2}\vec{j}(r,t) \tag{8.32}$$

is sought. The problem is considered two-dimensional, i.e. x-independent. Firstly, we consider the antisymmetric resonance. The solution of (8.32) could be expressed with the retarded potential:

$$\vec{A}(r,t) = \frac{1}{c^2}\int \frac{\vec{j}(r', t - |r' - r|/c)}{|r' - r|}\,\mathrm{d}V \tag{8.33}$$

Since $\mathbf{H} = \mathrm{curl}\mathbf{A}$, the magnetic field distribution in the far field ($r \gg r'$) is expressed in the cylindrical coordinates by substitution of (8.29) into (8.33) as follows:

$$\begin{aligned} H_x(r,t) &= \frac{3\omega\, q_0 \sin\beta}{\pi\, r \cos\beta} * \sin\left(\frac{kw_y \cos\beta}{2}\right) * \sin\left(\frac{k(d+s)\sin\beta}{2}\right) * \sin(3\omega\, t - kr) \\ H_y &= H_z = 0 \end{aligned} \tag{8.34}$$

where $k = \frac{3\omega}{c}$ and $\beta = \theta + \pi/2$. The angular radiation pattern $R(\theta)$ is defined by the averaged electromagnetic intensity which the unit cell of the MM emits per unit solid angle as a function of radiation angle. It is expressed as follows:

$$R(\beta) = \frac{\mathrm{d}P}{\mathrm{d}\beta} = r[r[E \times H]] \tag{8.35}$$

For a plane wave applies $r[r[E \times H]] = rH^2$. By substitution of (8.34) into (8.35) and time averaging we get the angular radiation pattern for the anti-symmetric resonance:

$$R_{\mathrm{as}}(\beta) \propto \left[q_0 \tan\beta * \sin\left(\frac{kw_y \cos\beta}{2}\right) * \sin\left(\frac{k(d+s)\sin\beta}{2}\right)\right]^2 \tag{8.36}$$

The radiation pattern can be evaluated for the symmetric resonance in the same way by use of (8.31) and (8.33):

$$R_s(\beta) \propto \left[q_0 \tan\beta * \sin\left(\frac{kw_y \cos\beta}{2}\right) * \cos\left(\frac{k(d+s)\sin\beta}{2}\right)\right]^2 \tag{8.37}$$

The polar plots in Fig. 8.6 show the normalized angular dependences of THG calculated using (8.36) and (8.37) for the antisymmetric and symmetric resonances, respectively, for the same parameters. The dependence of $q_0 \sim |L_\omega(\theta)|^3$ can be expressed for the magnetic resonance with a Lorentz line in the angular domain which arises from the angular dispersion of the resonance position:

$$L_\omega(\theta) \propto \left[\left(\omega_0^0 + \frac{\partial\omega_0}{\partial\theta}\theta\right)^2 - \omega^2 + 2i\gamma\omega\right]^{-1} \tag{8.38}$$

The central frequency of the resonance $\omega_0(\theta)$ is substituted by the truncated Taylor expansion in the form of $\omega_0(\theta) = \left(\omega_0^0 + \frac{\partial\omega_0}{\partial\theta}\theta\right)$. The angular radiation pattern of the third harmonics is straightforwardly connected to the angular dependence of THG. The third harmonic radiation is emitted from each unit cell of the MM with the relative phase which depends on the angle of incidence of the pump. Radiation from each cell interferes to compose the diffraction pattern. The intensity of each diffraction lobe depends on the angle of diffraction via the radiation pattern dependence. If only the zeroth diffraction order is detected then the diffraction angle equals the angle of incidence and thus the radiation pattern is probed by measuring the angular dependence of THG. Now we use (8.36) and (8.38) to calculate the data on angular-dependent THG from the fishnet MM. The function used is expressed as follows:

$$I(\theta) = B\left[|L(\theta)|^3 \cot(\theta) \sin\left(\frac{kw_y \sin\theta}{2}\right) \sin\left(\frac{k(d+s)\cos\theta}{2}\right)\right]^2 \tag{8.39}$$

The parameters in (8.39) are determined from the linear measurements. The angular dispersion of the resonance central frequency is $\frac{\partial\omega_0}{\partial\theta} \approx 3 \times 10^{12}\,\mathrm{rad}/(\mathrm{deg} \times \mathrm{s})$ and $\gamma = 0.15 \pm 0.01\,\mathrm{ps}^{-1}$ (corresponds to $\Delta\lambda_{FWHM} = 220\,\mathrm{nm}$). The parameter B stands for a calibration coefficient that was not measured precisely. For w_y the SEM-measured value was taken and $(d + s)$ was set to 250 nm. The angular dependent third harmonic intensity function (8.39) is plotted in Fig. 8.5a–d with solid lines. A good quantitative correspondence is observed between the experimental data, the numerically calculated data and the modeled dependence. From all the parameters only $(d + s)$ differs from the experimentally measured one. The main reason is general oversimplification of the model, i.e. not taking the real phase velocity of the third harmonic radiation inside the MM into account, considering pure symmetric/antisymmetric modes, assuming infinitely dense charge and current distributions, etc. Nevertheless, the model gives an explicit way how one can distinguish between symmetric and anti-symmetric resonances of the MM by means of its nonlinear optical response. For the symmetric resonance no local extremum is observed at oblique incidence whereas the maximum is present in the case of the anti-symmetric resonance. In terms of effective $\chi^{(3)}$ tensor components of the MM this means that the $\chi^{(3)}_{yyyy}$ component of the medium at the magnetic resonance is less pronounced than that at the electric resonance. In correspondence with the general concept of MMs it makes possible to tailor the relation between different tensor components by the proper choice of the MM resonance and its parameters.

8.6 Sources of Nonlinearity in Maxwell Equations: General Consideration

The question about the origin of the nonlinear response of media on microscopic level is of fundamental interest. Maxwell equations (ME) are essentially linear in term of mutual relations between the fields. From the other side, ME consist of not only four known equations for the field, but also contains fifth one for the charge dynamics:

$$\begin{cases} \mathrm{rot}\,\vec{e} = \frac{i\omega}{c}\vec{h} \\ \mathrm{div}\,\vec{h} = 0 \\ \mathrm{div}\,\vec{e} = 4\pi\rho = 4\pi\sum\limits_k q_k\delta(\vec{r}-\vec{r}_k) \\ \mathrm{rot}\,\vec{h} = -\frac{i\omega}{c}\vec{e} + \frac{4\pi}{c}\vec{j} = -\frac{i\omega}{c}\vec{e} + \frac{4\pi}{c}\sum\limits_k q_k\langle\vec{v}_k\delta(\vec{r}-\vec{r}_k)\rangle \\ \frac{\mathrm{d}\vec{v}_k}{\mathrm{d}t} = \frac{q_k}{m_k}\vec{e} + \frac{q_k}{m_k c}\left[\vec{v}_k * \vec{h}\right] \end{cases} \tag{8.40}$$

Here $\vec{e}$ and $\vec{h}$ are the microscopic electric and magnetic fields, respectively, $\rho = \sum_i q_i\delta(\vec{r}-\vec{r}_i)$ is the charge density, $\vec{q}_i$, $\vec{p}_i = m_i\vec{v}_i$, m_i, $\vec{r}_i$ and $\vec{v}_i$ are the charges, momenta, masses, coordinates and velocities of charges, $\vec{j} = \sum_i \vec{v}_i q_i\delta(\vec{r}-\vec{r}_i)$ is the microscopic current density, ω and c are the frequency and the velocity of light in vacuum. It is assumed that system (8.40) is strictly valid without any approximations.

First, the fifth equation is nonlinear i.e. the momentum depends on the field nonlinearly: the velocities are proportional (to the first approximation) to the fields, and the momentum consequently is proportional to the field in square. This nonlinearity appears due to the interaction with the magnetic field. The last, as it is known, is the relativistic effect; e.g. one can see that the magnetic field is proportional to the velocities divided by the velocity of light and disappears as the velocity of light tends to infinity. Hence, this source of nonlinearity (which could be named "relativistic nonlinearity") becomes significant only in the case of very high velocities of the charges and usually can be safely neglected and (8.40) becomes:

$$\begin{cases} \operatorname{rot}\vec{e} = \frac{i\omega}{c}\vec{h} \\ \operatorname{div}\vec{h} = 0 \\ \operatorname{div}\vec{e} = 4\pi\rho \\ \operatorname{rot}\vec{h} = -\frac{i\omega}{c}\vec{e} + \frac{4\pi}{c}\vec{j} \\ \frac{\mathrm{d}\vec{v}_k}{\mathrm{d}t} = \frac{q_k}{m_k}\vec{e}_{\mathrm{ext}} + \frac{q_k}{m_k}\vec{e} \end{cases} \tag{8.41}$$

Here the electric field is subdivided by an external one and the field generated by the considered charges. System (8.41) is obviously linear, e.g. it is not expected that the charge dynamics (and consequently scattered fields produced by this dynamics) excited by an external field (irrespective to its intensity) would depend on this external field nonlinearly. The fifth equation $\frac{\mathrm{d}\vec{v}_k}{\mathrm{d}t} = \frac{q_k}{m_k}\vec{e}_{\mathrm{ext}} + \frac{q_k}{m_k}\vec{e}$ is the only form fully compatible with ME in a sense of ME elaborated in Landau and Lifshitz [36]. It has to be emphasized, that ME are compatible with classical relativistic mechanics, but, rigorously speaking, are NOT compatible with quantum mechanics. It means in turn, that ME can describe a system of free charges (interacting with each other or not), but can NOT describe a system of bounded (in a sense of stable system as atom or molecule i.e. in a sense of quantum mechanics) charges. Hence the mentioned above "relativistic nonlinearity" is the only one which can be successfully considered in the frame of ME.

In order to describe a bound states (charges in atoms/molecules), usually instead of fifth equation in (8.40) a harmonic oscillator equation is used:

$$\frac{\partial^2\vec{r}_k}{\partial t^2} + \gamma_k\frac{\partial\vec{r}_k}{\partial t} + \omega_k^2\vec{r}_k = \frac{q_k}{m_k}\vec{e} \tag{8.42}$$

Eigen frequency ω_k describes the internal atomic (quantum) structure; the presence of the stable states described by (8.42) can NOT be introduced in the frame of classical approach. This equation is in turn linear and should not cause any nonlinear response. System (8.41) in this case becomes:

$$\begin{cases} \operatorname{rot}\vec{e} = \frac{i\omega}{c}\vec{h} \\ \operatorname{div}\vec{h} = 0 \\ \operatorname{div}\vec{e} = 4\pi\rho \\ \operatorname{rot}\vec{h} = -\frac{i\omega}{c}\vec{e} + \frac{4\pi}{c}\vec{j} \\ \frac{\partial^2\vec{r}_k}{\partial t^2} + \gamma_k\frac{\partial\vec{r}_k}{\partial t} + \omega_k^2\vec{r}_k = \frac{q_k}{m_k}\vec{e}_{\mathrm{ext}} + \frac{q_k}{m_k}\vec{e} \end{cases} \tag{8.43}$$

i.e. remains linear. Note, that the presence of the limited dynamics (8.42) contradicts to the postulate of stability of the atom/molecule—the oscillating electron emits light, loses energy and consequently tends to collapse. This known contradiction actually caused originally the necessity of the quantum mechanical approach. Here this contradiction will not be further considered and (8.43) is expected to be valid.

In order to introduce a nonlinearity, usually a nonlinear term is added to the dynamic equations, namely:

$$\frac{\partial^2 \vec{r}_k}{\partial t^2} + \gamma_k \frac{\partial \vec{r}_k}{\partial t} + \omega_k^2 \vec{r}_k + \sum_p \beta_{k,p} \vec{r}_k^p = \frac{q_k}{m_k} \vec{e}_{\text{ext}} \tag{8.44}$$

Order of nonlinearity p determines the order of the final nonlinear response of matter.

The presented above consideration can be found in any textbook. The conclusion about the linear response of (8.41) is commonly accepted, and the way of introducing of nonlinearity (8.44) is supposed to be undoubtedly right.

From the other side, this seems to be wrong. System of ME (8.41) appears to be more sophisticated. The first two equations in (8.41) obviously linear, but the third and fourth ones $\text{div}\vec{e} = 4\pi\rho$ and $\text{rot}\vec{h} = -\frac{i\omega}{c}\vec{e} + \frac{4\pi}{c}\vec{j}$ are linear only in a sense of the linearity between fields and ρ and $\vec{j}$. If the dynamics depends linearly on an external field, it DOES NOT mean that the solution for the scattered field is also linear. The usually observed linearity of the response is stipulated by the fact that the scattered field is caused in most cases by a dipole emission, which is internally linear. Higher multipoles cause a natural nonlinearity, but the higher than dipole moments are extremely rare in optical domain.

In the nearfield, potential is governed by a simple equation, which is in turn a general solution of Maxwell equations (see Fig. 8.7):

$$\nabla\varphi = -\sum_k q_k \delta(r_k) \tag{8.45}$$

Solution of (8.46) is known:

$$\varphi(R_0) = \sum_k \frac{q_k}{|R_0 - r_k|} \tag{8.46}$$

This is the equation which gives rise of the multipole expansion after the expansion of function $\frac{q_k}{|R_0 - r_k|}$. In (8.46) r_k are proportional to an external electric field and in CW case (harmonic oscillation around equilibrium point) can be written as:

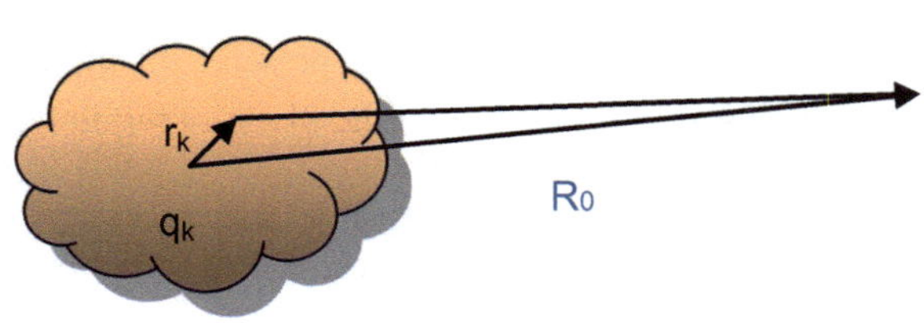

Fig. 8.7 Solution at the point of observation R_0 from the charge dynamics in shadowed space region

$$
\begin{aligned}
r_k &= r_{k0} + \Delta r_{k0} = r_{k0} + \frac{q_k E_0 \exp(i\omega t)}{m\left(\left(\omega_0^2 - \omega^2\right) + 2i\gamma\omega\right)} \\
\Delta r_{k0} &= \frac{q_k E_0 \exp(i\omega t)}{m\left(\left(\omega_0^2 - \omega^2\right) + 2i\gamma\omega\right)}
\end{aligned}
\tag{8.47}
$$

Here q_k is the charge, E_0 is the amplitude of the harmonic electric field acting on the charge, m is the mass of the charge, and the respective is for harmonic oscillator model (8.42). In turn (8.46) becomes:

$$
\varphi(R_0) = \sum_k \frac{q_k}{\left|(R_0 - r_{k0}) - \frac{q_k E_0 \exp(i\omega t)}{m\left(\left(\omega_0^2 - \omega^2\right) + 2i\gamma\omega\right)}\right|}
\tag{8.48}
$$

Till $R_0 \neq r_{k0}$ expression (8.48) provides higher harmonics. If we start to make a standard multipole expansion, the first two terms (zero order—total charge, is assumed to be zero and first order dipole) do not produce the higher harmonics, all business starts from the quadrupole & magnetic dipole, and even magnetic dipole does not contribute into the nonlinearity—only quadrupole one. The same for higher orders. The next order, octupole, can be expressed as a sum of the non reducable tensors of all orders, namely first (toroidal, no nonlinearity), then second (magnetic quadrupole), and actually octupole moments.

This question about “multipole nonlinearity” requires further investigation.

8.7 Conclusion

In summary, a self-consistent physical model that permits to describe the linear response of MA geometries by their intrinsic plasmonic eigenmodes has been presented. The occurring specific carrier dynamics have been mimicked by an auxiliary carrier alignment interacting with the incident radiation. The knowledge of these charge oscillations allows the application of the multipole expansion which provides the eigenvalue equation for electromagnetic waves propagating in such a composite MM. Moreover we have shown that the specific convective carrier oscillations together with the quadrupole moment inherently introduce nonlinear material interactions. Considering the SHG process, our calculations show the expected enhanced signal both for the electric and the magnetic resonance and provide a microscopic and physical understanding of them. For further investigations it is important that the nonlinear response can be determined only from knowing the linear response, which is accessible by comparing the dispersion relation or any other effective material property to the introduced multipole model.

A magnetic resonance contribution to third-order optical nonlinearities of the fishnet MM was shown. It was achieved by means of measurements of the third harmonic signal in forward direction from a fishnet sample and numerical simulations with a nonlinear FMM. Interference of radiation from separated third harmonic sources is shown to emerge as a local maximum in the angular spectra of the third harmonic signal found at oblique incidence. Antisymmetric oscillations of currents, which are the intrinsic property of magnetic resonances, are found to be responsible for the particular radiation pattern. Based on this an analytical model was build. The angular characteristic of the third harmonic response from the experiment, the FMM and the analytical model were compared. A quantitative correspondence between these data sets is observed. The results contribute to a better understanding of the possibilities of the nonlinear properties of optical MMs with plasmonic resonances of different symmetries.

It was shown that the "multipole nonlinearity" is a natural consequence of the standard solution of ME for the expansion orders higher than the dipole one.

References

1. N. Sharma, PRL **98**, 217402 (2007)
2. D.J. Cho, F. Wang, X. Zhang, Y.R. Shen, Phys. Rev. B **78**, 121101(R) (2008)
3. Y. Shin, A. Chavez-Pirson, Y. Lee, Opt. Lett. **25**, 171 (2000)
4. D. Guzatov, V. Klimov, M. Pikhota, Laser Phys. (2009). https://doi.org/10.1134/PS1054660X09170083
5. J. Petschulat, C. Menzel, A. Chipouline, C. Rockstuhl, A. Tünnermann, F. Lederer, T. Pertsch, Multipole approach to metamaterials. Phys. Rev. B **78**, 043811 (2008)
6. V. Pustovit, J. Sotelo, G. Niklasson, JOSA A **19**, 513 (2002)
7. N. Bloembergen, *Nonlinear Optics* (Benjamin Press, New York, 1965)
8. S. Kujala, B.K. Canfield, M. Kauranen, Y. Svirko, J. Turunen, PRL **98**, 167403 (2007)
9. D. Bethune, Opt. Lett. **6**, 287 (1981)
10. G. Bachelier, I. Russier-Antoine, E. Benichou, C. Jonin, P.-F. Brevet, JOSA B **25**, 955 (2008)
11. J. Jayabalan, P. Manoranjan, A. Banerjee, K.C. Rustagi, Phys. Rev. B **77**, 045421 (2008)
12. E. Kim, F. Wang, W. Wu, Z. Yu, Y.R. Shen, Phys. Rev. B **78**, 113102 (2008)
13. A. Maluckov, L. Hadzievski, N. Lazarides, G. Tsironis, Phys. Rev. E **77**, 046607 (2008)
14. I. Shadrivov, A. Kozyrev, D. Weide, Y. Kivshar, Appl. Phys. Lett. **93**, 161903 (2008)
15. I. Shadrivov, A. Kozyrev, D. Weide, Y. Kivshar, Opt. Express **16**, 20266 (2008)
16. A. Zharov, I. Shadrivov, Y. Kivshar, PRL **91**, 037401 (2003)
17. M.W. Klein, C. Enkrich, M. Wegener, S. Linden, Science **313**, 502 (2006)
18. M.W. Klein, M. Wegener, N. Feth, S. Linden, Opt. Express **15**, 5238 (2007)
19. C. Rockstuhl, T. Zentgraf, E. Pshenay-Severin, J. Petschulat, A. Chipouline, J. Kuhl, T. Pertsch, H. Giessen, F. Lederer, Opt. Express **15**, 8871 (2007)
20. C. Rockstuhl, T. Paul, F. Lederer, T. Pertsch, T. Zentgraf, T. Meyrath, H. Giessen, Phys. Rev. B **77**, 035126 (2008)
21. B.T. Draine, P.J. Flatau, JOSA A **11**, 1491 (1994)
22. C. Hafner, *The Generalized Multipole Technique for Computational, Electromagnetics* (Artech House, Boston, 1990)
23. J. Petschulat, A. Chipouline, A. Tüunnermann, T. Pertsch, C. Menzel, C. Rockstuhl, F. Lederer, Phys. Rev. A **80**, 063828 (2009)
24. N. Feth et al., Opt. Lett. **33**, 1975 (2008)
25. Y. Zeng, W. Hoyer, J. Liu, S. Koch, J. Moloney, Phys. Rev. B **79**, 235109 (2009)
26. P. Butcher, D. Cotter, *The Elements of Nonlinear Optics* (Cambridge University Press, Cambridge, 1990)

27. R. Miller, APL **5**, 17 (1964)
28. M. Choy, R. Byer, Phys. Rev. B **14**, 1693 (1976)
29. J. Reinhold, M. Shcherbakov, A. Chipouline, V. Panov, C. Helgert, T. Paul, C. Rockstuhl, F. Lederer, E.-B. Kley, A. Tünnermann, A. Fedyanin, T. Pertsch, The contribution of the magnetic resonance to the third harmonic generation from a fishnet metamaterial. Phys. Rev. B **86**, 115401 (2012)
30. T. Paul, C. Rockstuhl, F. Lederer, JOSA B **27**, 1118 (2010)
31. R.W. Boyd, *Nonlinear Optics*, 2nd edn. (Academic Press, San Diego, CA, 2003)
32. N. Bloembergen, W.K. Burns, M. Matsuoka, Opt. Commun. **1**, 195 (1969)
33. R. Adair, L.L. Chase, S.A. Payne, J. Opt. Soc. Am. B **4**, 875 (1987)
34. R. Adair, L.L. Chase, S.A. Payne, Phys. Rev. B **39**, 3337 (1989)
35. A. Mary, S. Rodrigo, F. Garcia-Vidal, L. Martin-Moreno, PRL **101**, 103902 (2008)
36. L.D. Landau, E.L. Lifshitz, *Electrodynamics of Continuous Media*, 2nd edn. (Pergamon Press, New York, 1960). Chap. IX

Chapter 9
Multipole Approach for Homogenization of Metamaterials: "Quantum" Metamaterials

9.1 Introduction: Quantum Dynamics Versus Classical One

Before discussing the problems of coupled dynamics of the classical and quantum objects, it is worth to make several remarks about compatibility of quantum mechanics and ME. Surprisingly, they are not compatible. Reminding [1], ME have been elaborated in the frame of the classical dynamics for the charges interacting with the fields. In system (2.4) the dynamical equation is written based on the classical mechanics and, rigorously speaking, remains valid only for free (but may be interacting) charges. The dynamics of charges in atoms/molecules have to be described by the quantum mechanical tools. Nevertheless, classical harmonic oscillator equation is widely used to model charges in atoms/molecules, in this case system (2.4) becomes:

$$\begin{cases} \operatorname{rot}\vec{e} = \frac{i\omega}{c}\vec{h} \\ \operatorname{div}\vec{h} = 0 \\ \operatorname{div}\vec{e} = 4\pi\rho \\ \operatorname{rot}\vec{h} = -\frac{i\omega}{c}\vec{e} + \frac{4\pi}{c}\vec{j} \\ \frac{\partial^2\vec{r}_k}{\partial t^2} + \gamma_k\frac{\partial\vec{r}_k}{\partial t} + \omega_k^2\vec{r}_k = \frac{q_k}{m_k}\vec{e}_{\text{ext}} + \frac{q_k}{m_k}\vec{e} \end{cases} \tag{2.4a}$$

here $\vec{e}_{\text{ext}}e_{\text{ext}}$ is the external electric field (actually, it could be any other forces other than interaction with the field $\vec{e}$). It is easy to see, that in this case the atom's stability requirement is not satisfied due to the fact, that in the frame of (2.4a) the electrons will lose energy and finally stops (annihilates with protons) as soon as an external forces are zero (i.e. $\vec{e}_{\text{ext}} = 0$). It is worth noting that the electron loses energy even with no internal damping described by coefficient γ_k, but also due to the emission of light due to acceleration.

The accepted here approach is in the substitution of the classical dynamics by quantum one. Instead of the last equation in (2.4) the density matrix formalism is

A. Chipouline and F. Küppers, *Optical Metamaterials: Qualitative Models*, Springer Series in Optical Sciences 211, https://doi.org/10.1007/978-3-319-77520-3_9

used. In this case the stability of atoms is naturally guarantied. From the other side it is necessary to emphasize that the quantum mechanical equations are not elaborated together with ME, as it takes place in the case of classic mechanics in (2.4). It should not cause some fundamental problems excepting the case of radiation losses—in quantum mechanics it is described by two phenomenological constants, which already include the radiation losses. Detailed discussion will be done in Chap. 14.

9.2 Coupled Dynamics of Plasmonic Resonator and Quantum Elements: General Approach

9.2.1 Model Formulation

In this chapter, an analytical model for describing complex dynamics of a hybrid system consisting of resonantly coupled classical resonator and quantum structures is presented. Classical resonators in this model correspond to plasmonic nanoresonators of various geometries, as well as other types of nano- and microstructures, optical response of which can be described classically. Quantum structures are represented by atoms or molecules, or their aggregates (for example, quantum dots, carbon nanotubes, dye molecules, polymer or bio molecules etc.), which can be accurately modeled only with the use of the quantum-mechanical approach. Our model is based on the set of equations that combines well-established density matrix formalism appropriate for quantum systems, coupled with harmonic-oscillator equations ideal for modeling sub-wavelength plasmonic resonators. As a particular example of application of our model, it is shown that the saturation nonlinearity of carbon nanotubes increases multifold in the resonantly enhanced near field of a MM and compare the results with the experimental data (Chap. 10). Using the developed approach, regular and stochastic dynamics of the nanolaser (spaser) is considered, and generalization of Schawlow-Towns expression is elaborated (Chap. 11). The dynamics of the plane wave propagating in the MMs (where the MAs are the coupled plasmonic and quantum systems) is considered (Chap. 12).

Accurate description of the dynamics of interacting classical systems is a fundamental problem. The current approach is to use a set of coupled equations for two (or more) harmonic oscillators, which can normally be solved under appropriate approximations. It combines mathematical simplicity with adequate physical insight and has been adopted in various branches of science ranging from optics to nuclear physics. If the interacting systems are quantum their dynamics can be satisfactory described in the framework of quantum mechanics based on the Schrödinger equation or density matrix approach, for instance. However, for describing classical and quantum systems coupled together a special approach is required. It was originally developed to model the dynamics of lasers where the classical system is normally represented by an optical (mirror) resonator, while the quantum system—by amplifying medium [2]. The basic idea was that the quantum formalism allowed accurate calculation of the medium's polarizability, while the latter could be used in the classical Maxwell equations describing electromagnetic fields in the optical resonator.

With the rapid development of nanotechnology it has become possible to engineer and study hybrid quantum-classical systems at the nanometer scale such as metallic nanoresonators and their arrays (i.e. MMs) combined with quantum dots, carbon nanotubes or dye molecules [3–6]. While optical response of a metallic nanoresonator is affected by plasmonic excitations and shape effects, its rather complicated dynamics can still be satisfactory modeled by the harmonic oscillator equations with appropriately chosen parameters [7]. As it will be shown below, this allows us to extend the quantum-classical treatment to modeling analytically a wide range of optical and plasmonic effects in the hybrid quantum MMs, such as loss compensation, enhancement of nonlinear response and luminescence, etc. Furthermore, the model can be used to describe the dynamics of superconducting Josephson-junction-based MMs, as well as SQUIDs coupled to an RF strip resonator [4].

Although a wide range of numerical approaches describing rigorously the internal quantum dynamics of molecules have been developed, including time dependent density function theory, multi-configurational self-consistent field method, polarizable quantum mechanical/molecular mechanical method, capacitance-polarizability interaction model and discrete interaction model/quantum mechanics (see [8–12] and references therein), our model takes advantage of the phenomenological approach, which allows relatively simple analytical treatment and provides deeper insight into the physical behavior of coupled quantum-classical systems. The parameters of such a model can always be found from fitting experimental data and/or using rigorous numerical approaches mentioned above.

Another advantage of the proposed model is that it takes into account additional important nonradiative relaxation channels due to stochastic interaction with the environment, which is naturally included in the adopted density matrix approach through two phenomenological relaxation times for polarization and population.

Here a Quantum System (QS) placed in the near-field zone of a Classical electromagnetic System (CS) is considered. The field produced by the CS, E_{CS}, affects QS that in turn acts on CS with its field In addition, there is an external field E_{ext} of the incident light, which interacts with both CS and QS; see Fig. 9.1.

The actual number of the harmonic-oscillator equations required to adequately describe CS depends on its particular structure [7]. For the illustration purpose the analysis will be first restricted to just one dipole like MA (Fig. 9.2a), which is described by a single harmonic-oscillator equation), which should not limit the generality of our approach. The dynamics of QS is modelled using the density matrix formalism.

In general case the quantum dynamics of QS that is assumed to be in contact with a thermostat environment, is described by the following set of ordinary differential equations [13]:

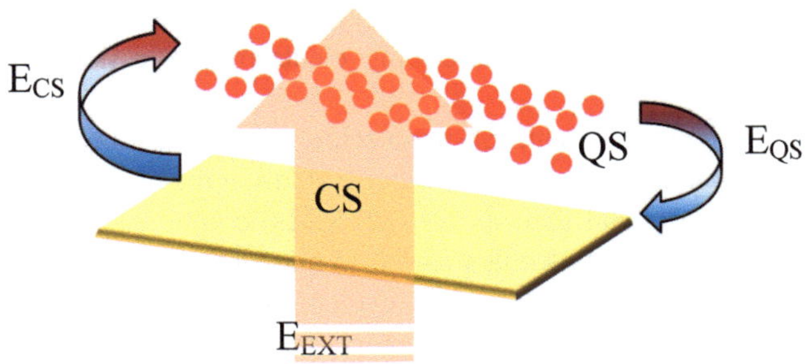

Fig. 9.1 Schematic representation of the interaction between plasmonic nanoresonator (Classic System—CS, yellow block) covered with a layer of quantum systems (Quantum System—QS, red circles). E_{CS} is the field produced by CS and acting on QS, E_{QS} is the field produced by QS and acting on CS, E_{EXT} is the external filed field acting on both CS and QS

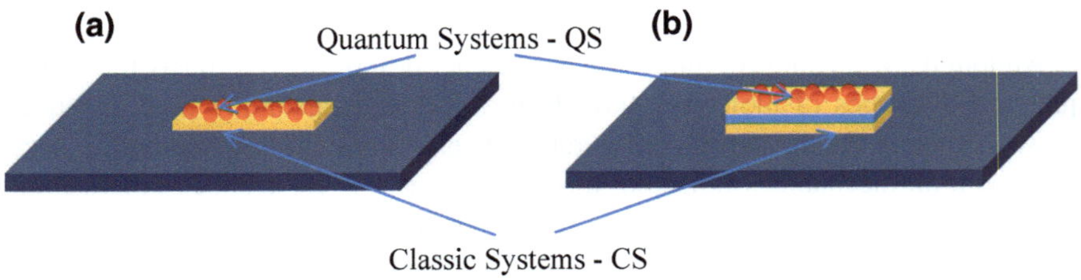

Fig. 9.2 Schematic of the modelled active hybrid MAs with quantum ingredients: a plasmonic nanoresonator (Classic System—CS, yellow blocks) covered with a layer of quantum ingredients (Quantum System—QS, red circles). **a** Dipole-like MA (one nanoresonator, **b** quadrupole-like MA (two coupled nanoresonators, separated by dielectric layer)

$$\begin{cases} \frac{d\rho_{nn}}{dt} + \sum\limits_{m} (k_{nm}\rho_{nn} - k_{mn}\rho_{mm}) = -\frac{i}{\hbar}\sum\limits_{m} (H_{nm}\rho_{mn} - H_{mn}\rho_{nm}) \\ \frac{d\rho_{kl}}{dt} + i\omega_{kl}\rho_{kl} + \frac{\rho_{kl}}{\tau_{kl}} = -\frac{i}{\hbar}\sum\limits_{m} (H_{km}\rho_{ml} - H_{ml}\rho_{km}) \end{cases} \tag{9.1}$$

Here k_{nm} and τ_{kl} are energy and phase relaxation constants respectively, ω_{kl} is frequency of the transition from k to l, ρ_{nn} and ρ_{kl} ($\rho_{kl} = \overset{*}{\rho_{kl}}$) are diagonal and non-diagonal elements of the density matrix, H_{kl} is a Hamiltonian matrix element responsible for interaction of the quantum system with the external field.

In framework of this formalism the averaged polarization density is expressed through non-diagonal density matrix elements:

$$P_{kl} = N\mu_{kl}(\rho_{kl} + \rho_{lk}) \tag{9.2}$$

μ_{kl} is the dipole moment of a quantum system, which is proportional to overlap integral between psi-functions of both levels, N is the quantum systems concentration.

In the case of resonant interaction the internal QS dynamics can be modelled to a first approximation by a two-level system subjected to a pump:

$$\begin{cases} \frac{d\rho_{12}}{dt} - i\omega_{21}\rho_{12} + \frac{\rho_{12}}{\tau_2} = -\frac{iH_{12}(\rho_{22}-\rho_{11})}{\hbar} \\ \frac{d\rho_{22}}{dt} + \frac{\rho_{22}}{\tilde{\tau}_1} = -\frac{iH_{12}(\rho_{12}-\rho_{21})}{\hbar} + W\rho_{11} \\ \rho_{22} + \rho_{11} = 1 \end{cases} \tag{9.3}$$

Here ρ_{22}, ρ_{11} and ρ_{12}, $\overset{*}{\rho}_{12}$ are the diagonal and non-diagonal matrix density elements, respectively; τ_2 and $\tilde{\tau}_1$ are the constants describing phase and energy relaxation processes due to the interaction with a thermostat; $\omega_{21} = (E_2 - E_1)/\hbar$ is the transition frequency between levels 2 and 1; H_{12} is the Hamiltonian matrix element responsible for interaction of QS with the external fields; W is the phenomenological pump rate—this could model pumping QS. It is also convenient to introduce new variables $N = \rho_{22} - \rho_{11}$ and $N_0 = \frac{(W\tilde{\tau}_1 - 1)}{(W\tilde{\tau}_1 + 1)}$ so that:

$$\begin{cases} \frac{d\rho_{12}}{dt} + i\omega_{12}\rho_{12} + \frac{\rho_{12}}{\tau_2} = -\frac{iH_{12}N}{\hbar} \\ \frac{dN}{dt} + \frac{N-N_0}{\tau_1} = -\frac{2iH_{12}\left(\rho_{12}-\rho_{12}^*\right)}{\hbar} \\ \tau_1 = \frac{\tilde{\tau}_1}{W\tilde{\tau}_1 + 1} \end{cases} \tag{9.4}$$

In order to describe dynamics of the plasmonic nanoresonator the following harmonic-oscillator equation is used:

$$\frac{d^2x}{dt^2} + 2\gamma\frac{dx}{dt} + \omega_0^2 x = \chi(E_{ext} + E_{QS}) \tag{9.5}$$

Here γ and ω_0 are the loss coefficient and resonance eigenfrequency, E_{ext} and E_{QS} are the external electric field and field generated by QS respectively, and χ is the effective kinetic inductance of the nanoresonator. The dimensionless variable x corresponds here to one the dynamic characteristics of the oscillator, which will be identified later.

From (9.4) and (9.5) one can obtain:

$$\begin{cases} \frac{d\rho_{12}}{dt} + i\omega_{12}\rho_{12} + \frac{\rho_{12}}{\tau_2} = -\frac{iH_{12}N}{\hbar} \\ \frac{dN}{dt} + \frac{N-N_0}{\tau_1} = -\frac{2iH_{12}\left(\rho_{12}-\rho_{12}^*\right)}{\hbar} \\ \frac{d^2x}{dt^2} + 2\gamma\frac{dx}{dt} + \omega_0^2 x = \chi(E_{ext} + E_{QS}) \end{cases} \tag{9.6}$$

In order to make the next step it is necessary to determine the nature of the interaction between CS and QS and write down expressions for H_{12} and We assume that the fields in the near-field zone of both systems are predominantly electric and produced by the effective electric dipole moments d. Electric field of an oscillating dipole is proportional to the d:

$$E \sim d \tag{9.7}$$

Correspondingly, electric field generated by the dipole moment of QS d_{QS} at the location of CS can be written as:

$$E_{QS} \sim d_{QS} \sim \mu_{QS}(\rho_{12} + \rho_{21}) \tag{9.8}$$

where μ_{QS} is the dipole moment of QS.

According to the same relation the local electric field of CS is:

$$E_{CS} \sim d_{CS} \sim \mu_{CS} x \tag{9.9}$$

where μ_{CS} is the effective dipole moment of CS. From (9.9) it follows that the dimensionless variable χ has basically the same meaning as the non-diagonal element of the density matrix, namely the dimensionless polarization. It is worth noting that (9.8) and (9.9) assume both QS and CS as point-like dipoles. Despite the evident importance of addressing the overlap between the spatially inhomogeneous field of the plasmonic nanoresonator and localization of the quantum system, it is believed that this corresponds to the next level of complication that is not essential for adequate modeling of the response dynamics. The Hamiltonian of interaction H_{12} is defined by the following expressions:

$$\begin{cases} H_{12} = -\mu_{QS}(E_{ext} + E_{CS}) = -\left(\mu_{QS}E_{ext} + \alpha_x x\right) \\ \alpha_x \sim \mu_{QS}\mu_{CS} \end{cases} \tag{9.10}$$

Substituting (9.8) and (9.10) into (9.6) one obtains:

$$\begin{cases} \frac{d\rho_{12}}{dt} - i\omega_{21}\rho_{12} + \frac{\rho_{12}}{\tau_2} = \frac{i\left(\mu_{QS}E_{ext} + \alpha_x x\right)N}{\hbar} \\ \frac{dN}{dt} + \frac{N - N_0}{\tau_1} = \frac{2i\left(\mu_{QS}E_{ext} + \alpha_x x\right)(\rho_{12} - \rho_{21})}{\hbar} \\ \frac{d^2x}{dt^2} + 2\gamma\frac{dx}{dt} + \omega_0^2 x - \alpha_\rho(\rho_{12} + \rho_{21}) - \chi E_{ext} = 0 \\ N_0 = \frac{(W\tilde{\tau}_1 - 1)}{(W\tilde{\tau}_1 + 1)}, \quad \tau_1 = \frac{\tilde{\tau}_1}{W\tilde{\tau}_1 + 1} \\ \alpha_\rho \sim \mu_{QS}\chi \\ \alpha_x \sim \mu_{QS}\mu_{CS} \end{cases} \tag{9.11}$$

Here N_0 is the population inversion due to pump (in the absence of pump $N_0 = -1$); $N_0 > 0$ corresponds to the regime of amplification, $N_0 < 0$—to losses. Both eigenfrequencies ω_{21} and ω_0 are the resonance frequencies of QS and CS respectively and can vary independently. Rotating wave approximation for the system (9.11) is introduced through the following notations:

$$\begin{cases} \rho_{12} = \frac{\tilde{\rho}_{12}}{2}\exp(i\omega t) \\ x = \frac{1}{2}\left(\tilde{x}(t)\exp(-i\omega t) + \tilde{x}(t)^*\exp(i\omega t)\right) \\ E_{ext} = \frac{1}{2}\left(A(t)\exp(-i\omega t) + A(t)^*\exp(i\omega t)\right) \end{cases} \tag{9.12}$$

resulting in:

$$\begin{cases} \frac{d\tilde{\rho}_{12}}{dt} + \tilde{\rho}_{12}\left(\frac{1}{\tau_2} + i(\omega - \omega_{21})\right) = \frac{i\alpha_x \tilde{x}^* N}{\hbar} + \frac{i\mu_{QS} A^* N}{\hbar} \\ \frac{dN}{dt} + \frac{(N - N_0)}{\tau_1} = \frac{i\alpha_x\left(\tilde{x}\tilde{\rho}_{12} - \tilde{x}^*\tilde{\rho}_{12}^*\right) + i\mu_{QS}\left(A\tilde{\rho}_{12} - A^*\tilde{\rho}_{12}^*\right)}{2\hbar} \\ 2(\gamma - i\omega)\frac{d\tilde{x}}{dt} + \left(\omega_0^2 - \omega^2 - 2i\omega\gamma\right)\tilde{x} = \alpha_\rho \tilde{\rho}_{12}^* + \chi A \end{cases} \quad (9.13)$$

(9.13) are the master set of equations describing regular dynamics of interacting QS and CS. Taking into account stochastic noise sources, the set becomes:

$$\begin{cases} \frac{d\tilde{\rho}_{12}}{dt} + \tilde{\rho}_{12}\left(\frac{1}{\tau_2} + i(\omega - \omega_{21})\right) = \frac{i\alpha_x \tilde{x}^* N}{\hbar} + \frac{i\mu_{QS} A^* N}{\hbar} + \xi_\rho \\ \frac{dN}{dt} + \frac{(N - N_0)}{\tau_1} = \frac{i\alpha_x\left(\tilde{x}\tilde{\rho}_{12} - \tilde{x}^*\tilde{\rho}_{12}^*\right) + i\mu_{QS}\left(A\tilde{\rho}_{12} - A^*\tilde{\rho}_{12}^*\right)}{2\hbar} \\ 2(\gamma - i\omega)\frac{d\tilde{x}}{dt} + \left(\omega_0^2 - \omega^2 - 2i\omega\gamma\right)\tilde{x} = \alpha_\rho \tilde{\rho}_{12}^* + \chi A + \xi_x \end{cases} \quad (9.14)$$

Here ξ_ρ and ξx are the stochastic Langevin terms, which take into account spontaneous emission and thermal fluctuations respectively (the stochastic term influence description can be found in [14]).

Set of equations (9.14) can describe the following experimental situations.

9.2.2 Nano-Laser (Spaser) [3, 8, 14, 15]

In this case $N_0 = \frac{W\tau_1 - 1}{W\tau_1 + 1} > 0$ and $A = 0$, (9.14) gives transition and stationary dynamics of a nanolaser (spaser):

$$\begin{cases} \frac{d\tilde{\rho}_{12}}{dt} + \tilde{\rho}_{12}\left(\frac{1}{\tau_2} + i(\omega - \omega_{21})\right) = \frac{i\alpha_x \tilde{x}^* N}{\hbar} + \xi_\rho \\ \frac{dN}{dt} + \frac{(N - N_0)}{\tau_1} = \frac{i\alpha_x\left(\tilde{x}\tilde{\rho}_{12} - \tilde{x}^*\tilde{\rho}_{12}^*\right)}{2\hbar} \\ 2(\gamma - i\omega)\frac{d\tilde{x}}{dt} + \left(\omega_0^2 - \omega^2 - 2i\omega\gamma\right)\tilde{x} = \alpha_\rho \tilde{\rho}_{12}^* + \xi_x \end{cases} \quad (9.15)$$

With the stochastic Langevin terms one can calculate laser bandwidth in analog with well-known Schawlow-Towns approach [9].

9.2.3 Luminescence Enhancement [5, 10]

The problem of the luminescent enhancement can be described in the frame of the presented here quantum approach. Note that in this approach ME have not been used. Alternatively, the problem can be considered using the classical theory, where system (2.4a) gives solution for the radiation losses. It is worth reminding again that for the problem of luminescent enhancement these two approaches cannot be mixed i.e. ME cannot be combined with the density matrix formalism.

The both cases, classic and quantum ones, will be considered in details in Chap. 13. Here the main differences between both approaches have to be mentioned.

First, the density matrix dynamics does not require ME—the radiative relaxation is caused by an interaction with the virtual photons, which are not described by ME. In this case, the luminescent enhancement are stipulated by increasing of the density of states in the vicinity of the nanoobjects.

In the case of relaxation of a single quantum system, interaction with the electrons in the nanoobject has to be considered as a transient interaction with free electron gas. In fact, the interaction time (photon emission time) is much shorter than the eigen mode formation time, and all electrons interacting with the emitted photon are incoherent.

In the case of CW operation (CW pump, steady state), the eigen modes are formed and interaction with the electrons are no more stochastic—it is a regular coupled dynamics of eigen modes causing stimulated emission. It affects the level populations and consequently number of spontaneously emitted photons. Purcell coefficients P (see below) describing relaxation time modification (due to the increased density of states) remain the same for the both (transient and CW) cases.

Developed here approach actually work for turning off CW only. In fact, it is assumed, that the quantum system is fully inverted $N(t = 0) = 1$, but (in contrast to the case of spaser) there is no pump $N_0 = -1$. From the other side, it is assumed that the eigen modes in the nanobject are formed and the dynamics is described by (9.16). Realistic situation, corresponding to (9.16) is the abrupt turning of the CW operation mode (abrupt turning of the pump). It is worth noting again, that (9.16) does not describe the situation of a short pumping pulse (e.g. one photon pump) followed by the relaxation process.

$$\begin{cases} \frac{d\tilde{\rho}_{12}}{dt} + \tilde{\rho}_{12}\left(\frac{1}{\tau_2} + i(\omega - \omega_{21})\right) = \frac{i\alpha_x \tilde{x}^* N}{\hbar} \\ \frac{dN}{dt} + \frac{(N - N_0)}{\tau_1} = \frac{i\alpha_x\left(\tilde{x}\tilde{\rho}_{12} - \tilde{x}^*\tilde{\rho}_{12}^*\right)}{2\hbar} \\ 2(\gamma - i\omega)\frac{d\tilde{x}}{dt} + \left(\omega_0^2 - \omega^2 - 2i\omega\gamma\right)\tilde{x} = \alpha_\rho \tilde{\rho}_{12}^* \end{cases} \tag{9.16}$$

It has to be emphasized that the Purcell effect affects both relaxation times τ_2 and τ_1, which appear from the interaction with thermostat $\frac{1}{\tau_1} = \frac{1}{\tau_{1,\mathrm{nr}}} + \frac{P}{\tau_{1,r}}$, $\frac{1}{\tau_2} = \frac{1}{\tau_1} + \Delta = \frac{1}{\tau_{1,\mathrm{nr}}} + \frac{P}{\tau_{1,r}} + \Delta$($\tau_{1,\mathrm{nr}}$—is the nonradiative relaxation time, $\tau_{1,r}$—is the radiative relaxation time, $P > 1$—is the Purcell factor, and Δ—is the extra term, giving difference between τ_1 and τ_2).

9.2.4 Nonlinear Response Enhancement [6, 16]

The nonlinearity of QS appears due to the saturation effect and basically does not require either positive N_0 or nanoresonator. Enhancement of the saturation is caused by an addition channel: external field transfers energy to QS through the nanoresonator in addition to the direct pumping. Taking into account the field enhancement effect near the plasmonic nanoresonator, the model adequately describes the increased strength of the nonlinear response experimentally observed in [6]:

$$\begin{cases} \frac{d\tilde{\rho}_{12}}{dt} + \tilde{\rho}_{12}\left(\frac{1}{\tau_2} + i(\omega - \omega_{21})\right) = \frac{i\alpha_x \tilde{x}^* N}{\hbar} + \frac{i\mu_{QS} A^* N}{\hbar} \\ \frac{dN}{dt} + \frac{(N - N_0)}{\tau_1} = \frac{i\alpha_x\left(\tilde{x}\tilde{\rho}_{12} - \tilde{x}^*\tilde{\rho}^*_{12}\right) + i\mu_{QS}\left(A\tilde{\rho}_{12} - A^*\tilde{\rho}^*_{12}\right)}{2\hbar} \\ 2(\gamma - i\omega)\frac{d\tilde{x}}{dt} + \left(\omega_0^2 - \omega^2 - 2i\omega\gamma\right)\tilde{x} = \alpha_\rho \tilde{\rho}^*_{12} + \chi A \end{cases} \quad (9.17)$$

9.2.5 Enhancement of Magnetic Dipolar Response [17]

Marginal modification of system (9.13) allows us to model the enhancement of high-order multipole response in the hybrid MM. In particular, complex nanoresonators (like double-wire or split-ring resonators) support anti-symmetric mode of excitation, which is responsible for magnetic dipolar response [7]. It can be adequately described by two (instead of one) coupled harmonic oscillator equations. In the case of sufficiently strong pumping $N_0 = \frac{W\tilde{\tau}_1 - 1}{W\tilde{\tau}_1 + 1} > 0$ the energy transferred from the appropriately positioned QS will support excitation of the anti-symmetric mode—see (9.19).

9.2.6 Quantum Magnetic Metamaterials [17]

Combining active QS (such as quantum dots) with the specially designed plasmonic nanoresonators can lead to magnetization at optical frequencies (see also Sect. 9.2.5 above) produced not only by the plasmonic modes, but also modes of coherently coupled QS. Such hybrid structures could serve as building blocks for the lossless MMs with strong magnetic response at optical frequencies.

9.2.7 Linear and Nonlinear Response of SQUIDs [18, 19]

The behavior of SQUID coupled with an RF resonator [19] also falls in the range of phenomena described by the model. In this particular case the dynamics of SQUID is governed by the direct interaction with the resonator without influence from the external field:

$$\begin{cases} \frac{d\tilde{\rho}_{12}}{dt} + \tilde{\rho}_{12}\left(\frac{1}{\tau_2} + i(\omega - \omega_{21})\right) = \frac{i\alpha_x \tilde{x}^* N}{\hbar} \\ \frac{dN}{dt} + \frac{(N - N_0)}{\tau_1} = \frac{i\alpha_x\left(\tilde{x}\tilde{\rho}_{12} - \tilde{x}^*\tilde{\rho}_{12}^*\right) + i\mu_{QS}\left(A\tilde{\rho}_{12} - A^*\tilde{\rho}_{12}^*\right)}{2\hbar} \\ 2(\gamma - i\omega)\frac{d\tilde{x}}{dt} + \left(\omega_0^2 - \omega^2 - 2i\omega\gamma\right)\tilde{x} = \alpha_\rho\tilde{\rho}_{12}^* + \chi A \end{cases} \tag{9.18}$$

9.3 Extension on the Case of Double Wires Based Metaatoms (Metaatoms with Magnetic Response)

Among possible applications of the nanolaser it was proposed to achieve generation using non emitting (dark) modes of the plasmonic resonators. It was claimed that the lasing with the dark modes should have lower threshold, and consequently has to be achieved at lower pump levels [20]. Here a combination of cut wires (see Fig. 9.3b) and interaction of this structure with the QS is considered. The system of two coupled oscillators possesses symmetric (dipole like) and asymmetric (quardupole-like) modes with different respective eigenfrequencies ω_{sym} and ω_{asym}. The transition frequency of the QS ω_{21} can be adjusted in order to match the respective eigenfrequency and consequently provide maximum interaction efficiency. In order to elaborate the respective system of equations in analog with (9.14), it is necessary to substitute the single harmonic oscillator by two coupled harmonic oscillators, as it has been done for the double wires MAs in [7]. It is assumed also that only one nanoresonator is coupled with the QS, and the dynamics of the second nanoresonator is driven by the coupling with the first one; the slowly varying approximation remains the same (9.12), and the both coupled nanoresonators are assumed to be equivalent, i.e. eigenfrequencies ω_0 and loss coefficients γ for the both nanoresonators are the same. The resulting system of equations is:

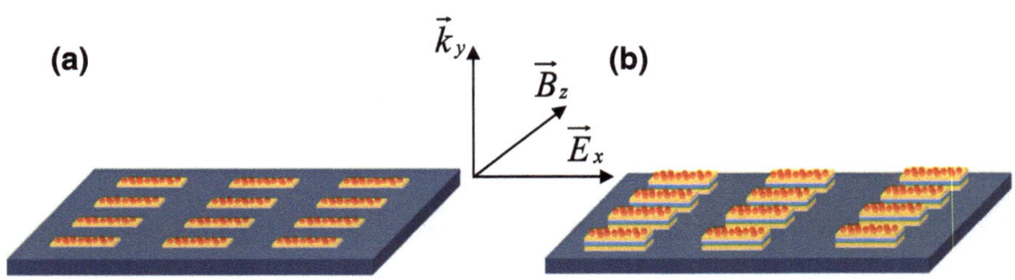

Fig. 9.3 Schematic of the modelled active hybrid MM with QS: **a** dipole-like MA (one nanoresonator), **b** quadrupole-like MA (two coupled nano-resonators, separated by dielectric layer). The electric field is polarized along the long side of the wires. QSs are shown by the red circles on the top of the upper nanoresonators

$$\begin{cases} \frac{d\tilde{\rho}_{12}}{dt} + \tilde{\rho}_{12}\left(\frac{1}{\tau_2} + i(\omega - \omega_{21})\right) = \frac{i\alpha_x \tilde{x}_1^* N}{\hbar} + \frac{i\mu_{QS} A^* N}{\hbar} + \xi_\rho \\ \frac{dN}{dt} + \frac{(N - N_0)}{\tau_1} = \frac{i\alpha_x\left(\tilde{x}_1\tilde{\rho}_{12} - \tilde{x}_1^*\tilde{\rho}_{12}^*\right) + i\mu_{QS}\left(A\tilde{\rho}_{12} - A^*\tilde{\rho}_{12}^*\right)}{2\hbar} \\ 2(\gamma - i\omega)\frac{d\tilde{x}_1}{dt} + \left(\omega_0^2 - \omega^2 - 2i\omega\gamma\right)\tilde{x}_1 + \sigma x_2 = \alpha_\rho \tilde{\rho}_{12}^* + \chi A_1 + \xi_{x1} \\ 2(\gamma - i\omega)\frac{d\tilde{x}_2}{dt} + \left(\omega_0^2 - \omega^2 - 2i\omega\gamma\right)\tilde{x}_2 + \sigma x_1 = \chi A_2 + \xi_{x2} \end{cases} \tag{9.19}$$

The phenomenological constant σ describes coupling between the nanoresonators through the near field, and $A_{1,2}$ are the fields acting on the upper and lower nanoresonator. The symmetric and asymmetric oscillation modes (keeping in mind symmetric and asymmetric eigenfrequencies $\omega_{sym} = \sqrt{\omega_0^2 + \sigma}$ and $\omega_{asym} = \sqrt{\omega_0^2 - \sigma}$ respectively) can be straightforwardly introduced according to:

$$\begin{cases} m_s = \tilde{x}_1 + \tilde{x}_2 \\ m_a = \tilde{x}_1 - \tilde{x}_2 \end{cases} \tag{9.20}$$

In these new variables system (9.19) becomes:

$$\begin{cases} \frac{d\tilde{\rho}_{12}}{dt} + \tilde{\rho}_{12}\left(\frac{1}{\tau_2} + i(\omega - \omega_{21})\right) = \frac{i\alpha_x\left(m_s^* + m_a^*\right)N}{2\hbar} + \frac{i\mu_{QS} A_1^* N}{\hbar} + \xi_\rho \\ \frac{dN}{dt} + \frac{(N - N_0)}{\tau_1} = \frac{i\alpha_x\left((m_s + m_a)\tilde{\rho}_{12} - \left(m_s^* + m_a^*\right)\tilde{\rho}_{12}^*\right) + i\mu_{QS}\left(A_1\tilde{\rho}_{12} - A_1^*\tilde{\rho}_{12}^*\right)}{4\hbar} \\ 2(\gamma - i\omega)\frac{dm_s}{dt} + \left(\omega_0^2 - \omega^2 - 2i\omega\gamma + \sigma\right)m_s = \alpha_\rho \tilde{\rho}_{12}^* + \chi(A_1 + A_2) + \xi_{ms} \\ 2(\gamma - i\omega)\frac{dm_a}{dt} + \left(\omega_0^2 - \omega^2 - 2i\omega\gamma - \sigma\right)m_a = \alpha_\rho \tilde{\rho}_{12}^* + \chi(A_1 - A_2) + \xi_{ma} \end{cases} \tag{9.21}$$

In case of the absence of the external field A, system (9.21) describes dynamics of the "multipole spaser", where one of the mode m_s is coupled with the external space (bright mode) while the other one m_a is to the first approximation not coupled with the far field zone (dark mode) and generates magnetic moment and magnetic field between the nanoresonators:

$$\begin{cases} \frac{d\tilde{\rho}_{12}}{dt} + \tilde{\rho}_{12}\left(\frac{1}{\tau_2} + i(\omega - \omega_{21})\right) = \frac{i\alpha_x\left(m_s^* + m_a^*\right)N}{2\hbar} + \xi_\rho \\ \frac{dN}{dt} + \frac{(N - N_0)}{\tau_1} = \frac{i\alpha_x\left((m_s + m_a)\tilde{\rho}_{12} - \left(m_s^* + m_a^*\right)\tilde{\rho}_{12}^*\right)}{4\hbar} \\ 2(\gamma_s - i\omega)\frac{dm_s}{dt} + \left(\omega_0^2 - \omega^2 - 2i\omega\gamma_s + \sigma\right)m_s = \alpha_\rho \tilde{\rho}_{12}^* + \xi_{ms} \\ 2(\gamma_a - i\omega)\frac{dm_a}{dt} + \left(\omega_0^2 - \omega^2 - 2i\omega\gamma_a - \sigma\right)m_a = \alpha_\rho \tilde{\rho}_{12}^* + \xi_{ma} \end{cases} \tag{9.22}$$

The dark mode m_a has less radiative losses $\gamma_a < \gamma_s$ and thus has lower generation threshold (see Chap. 12).

9.4 Modeling of Metamaterials Made of Plasmonic Metaatoms Coupled with Quantum Elements

In this part a propagation equation for the EM wave in a MMs with MAs, consisting of coupled plasmonic resonators and QS will be elaborated. The approach is the natural extension of the multipole one, developed in this work and allows us to investigate bulk properties of the MMs with gain and saturation type of nonlinearity and in the presence of a magnetic response. This approach creates a solid basis which can be used for qualitative consideration of all most important problems appearing in case of consideration of the plane wave propagation in MMs with gain or, more generally, in case of MAs consisting of plasmonic nanoresonators, coupled with QS. The MMs with MAs depicted in Fig. 9.2 are considered.

Elaboration of the propagation equation is the same as in case of the passive MAs [7]. The charge dynamics of the MAs becomes in this case more complicated and is described by system (9.14) for dipole-like MAs (Fig. 9.2a) or (9.22) for quadrupole-like MAs (Fig. 9.3b). From the other side, the propagation equation for the field and the calculation algorithm for the multipoles remain the same (see also 4.3), and the full system of equations for the case of quadrupole-like particles (Fig. 9.3b) and CW operation is:

$$\begin{cases} \frac{\partial^2 E_x}{\partial y^2} + \frac{\omega^2}{c^2}\left(E_x + 4\pi P_x(y,\rho_{12},\omega)\right) + \frac{i4\pi\omega}{c}\frac{\partial M_z(y,\rho_{12},\omega)}{\partial y} = 0 \\ \mathrm{P}_x(y,\rho_{12},\omega) = \eta q m_{\mathrm{s}} - \frac{\partial Q_{xy}}{\partial y} \\ Q_{xy}(y,\rho_{12},\omega) = \eta q y_1 m_{\mathrm{a}} \\ \mathrm{M}_z(y,\rho_{12},\omega) = \frac{i\omega\eta q y_1}{c} m_{\mathrm{a}} \\ \frac{\mathrm{d}\tilde{\rho}_{12}}{\mathrm{d}t} + \tilde{\rho}_{12}\left(\frac{1}{\tau_2} + i(\omega-\omega_{21})\right) = \frac{i\alpha_x\left(m_{\mathrm{s}}^* + m_{\mathrm{a}}^*\right)N}{2\hbar} + \frac{i\mu_{\mathrm{QS}}E_{x,1}^*N}{\hbar} + \xi_\rho \\ \frac{\mathrm{d}N}{\mathrm{d}t} + \frac{(N-N_0)}{\tau_1} = \frac{i\alpha_x\left((m_{\mathrm{s}}+m_{\mathrm{a}})\tilde{\rho}_{12} - \left(m_{\mathrm{s}}^*+m_{\mathrm{a}}^*\right)\tilde{\rho}_{12}^*\right) + 2i\mu_{\mathrm{QS}}\left(E_{x,1}\tilde{\rho}_{12} - E_{x,1}^*\tilde{\rho}_{12}^*\right)}{4\hbar} \\ 2(\gamma - i\omega)\frac{dm_{\mathrm{s}}}{dt} + \left(\omega_0^2 - \omega^2 - 2i\omega\gamma + \sigma\right)m_{\mathrm{s}} = \alpha_\rho\tilde{\rho}_{12}^* + \chi\left(E_{x,1}+E_{x,2}\right) + \xi_{\mathrm{ms}} \\ 2(\gamma - i\omega)\frac{\mathrm{d}m_{\mathrm{a}}}{\mathrm{d}t} + \left(\omega_0^2 - \omega^2 - 2i\omega\gamma - \sigma\right)m_{\mathrm{a}} = \alpha_\rho\tilde{\rho}_{12}^* + \chi\left(E_{x,1}+E_{x,2}\right) + \xi_{\mathrm{ma}} \end{cases} \tag{9.23}$$

here A is the electric field in the propagating wave, A_1 and A_2 are the electric field at upper and lower nanowires respectively, P_x is the medium polarizability caused by a dipole and quadrupole contributions, while Q and M describe impact of higher order multipoles, giving rise the MMs effects. In this equation all multipole terms are functions of the non-diagonal element ρ_{12}, which comes to the multipole moments through the interaction term in the mode dynamics equations of system (9.23). For the case of dipole-like MAs (Fig. 9.2a) system (9.23) is reduced to:

$$\begin{cases} \frac{\partial^2 E_x}{\partial y^2} + \frac{\omega^2}{c^2}(E_x + 4\pi P_x(y, \rho_{12}, \omega)) = 0 \\ \mathrm{P}_x(y, \rho_{12}, \omega) = \eta q m_s \\ 2(\gamma - i\omega)\frac{\mathrm{d}m_s}{\mathrm{d}t} + \left(\omega_0^2 - \omega^2 - 2i\omega\gamma\right)m_s = \alpha_\rho \tilde{\rho}_{12}^* + 2\chi E_x + \xi_{ms} \\ \frac{\mathrm{d}\tilde{\rho}_{12}}{\mathrm{d}t} + \tilde{\rho}_{12}\left(\frac{1}{\tau_2} + i(\omega - \omega_{21})\right) = \frac{i\alpha_x m_s^* N}{2\hbar} + \xi_\rho \frac{i\mu_{QS} E_x^* N}{\hbar} \\ \frac{\mathrm{d}N}{\mathrm{d}t} + \frac{(N - N_0)}{\tau_1} = \frac{i\alpha_x\left(m_s\tilde{\rho}_{12} - m_s^*\tilde{\rho}_{12}^*\right) + 2i\mu_{QS}\left(E_x\tilde{\rho}_{12} - E_x^*\tilde{\rho}_{12}^*\right)}{4\hbar} \end{cases} \quad (9.24)$$

The results of solution of (9.23) will be given in Chap. 12.

9.5 Conclusion

The multipole approach, initially suggested and developed for the MM with the classic MA (MA with charge dynamics successfully described in the frame of the classical physics) has been extended on the case of the quantum MM, where quantum physics appeared to be necessary for adequate modelling of the dynamics of MAs. The MAs, consisting of coupled plasmonic nanoresonator and QS, described by the density matrix formalism, is shown to be an appropriate object, which serves as a basis for modelling various physically realizable systems.

References

1. L.D. Landau, E.L. Lifshitz, *Field Theory*, 2nd edn. (Pergamon Press, New York, 1960)
2. A. Yariv, *Quantum Electronics*, 2nd edn. (Wiley, London, 1975)
3. M. Noginov, G. Zhu, A. Belgrave, R. Bakker, V. Shalaev, E. Narimanov, S. Stout, E. Herz, T. Suteewong, U. Wiesner, Demonstration of a spaser-based nanolaser. Nature **460**, 1110 (2009)
4. R. Oulton, V. Sorger, T. Zentgraf, R.-M. Ma, C. Gladden, L. Dai, G. Bartal, X. Zhang, Plasmon lasers at deep subwavelength scale. Nature **461**, 629 (2009)
5. K. Tanaka, E. Plum, J.Y. Ou, T. Uchino, N. Zheludev, Multi-fold enhancement of quantum dot luminescence in a plasmonic metamaterial. PRL **105**, 227403 (2010)
6. A. Nikolaenko, F. Angelis, S. Boden, N. Papasimakis, P. Ashburn, E. Fabrizio, N. Zheludev, Carbon nanotubes in a photonic metamaterials. PRL **104**, 153902 (2010)
7. J. Petschulat, C. Menzel, A. Chipouline, C. Rockstuhl, A. Tünnermann, F. Lederer, T. Pertsch, Multipole approach to metamaterials. Phys. Rev. B **78**, 043811 (2008)
8. J. Gersten, A. Nitzan, Spectroscopic properties of molecules interacting with small dielectric particles. J. Chem. Phys. **75**(3), 1139 (1981)
9. N. Liver, A. Nitzan, K. Freed, Radiative and nonradiative decay rates of molecules absorbed on clusters of small dielectric particles. J. Chem. Phys. **82**(8), 3831 (1985)
10. S. Vukovic, S. Corni, B. Mennucci, Fluorescence enhancement of chromophores close to metal nanoparticles. Optimal setup revealed by polarizable continuum model. J. Phys. Chem. C **113**, 121 (2009)

11. S. Morton, L. Jensen, A discrete interaction model/quantum mechanical method for describing response properties of molecules absorbed on metal nanoparticles. J. Chem. Phys. **133**, 074103 (2010)
12. A. Gonzales, S. Corni, B. Mennucci, Surface enhanced fluorescence within a metal nanoparticle array: the role of solvent and Plasmon couplings. J. Phys. Chem. C **115**, 5450 (2011)
13. V.M. Fain, *Quantum Radio Physics, Vol. 1: Photons and Nonlinear Media*. Sovetskoe Radio (1972) (in Russian)
14. A.S. Chirkin, A.V. Chipouline, Generalized expression for the natural width of the radiation spectrum of quantum oscillators. JETP Lett. **93**, 114 (2011)
15. A.F. Koenderink, On the use of Purcell factors for plasmon antennas. Opt. Lett. **35**, 4208 (2010)
16. A. Chipouline, C. Simovski, S. Tretyakov, Basics of averaging of the Maxwell equations for bulk materials. Metamaterials **6**, 77 (2012)
17. A. Chipouline, V. Fedotov, Towards quantum magnetic metmaterials, in *Proceedings Nanometa, 2011*, THU4s.3 (96), 2011
18. G. Oelsner, S.H.W. van der Ploeg, P. Macha, U. Hübner, D. Born, S. Anders, E. Il'ichev, H.-G. Meyer, M. Grajcar, S. Wünsch, M. Siegel, A. Omelyanchouk, O. Astafiev, Weak continuous monitoring of a flux qubit using coplanar waveguide resonator. Phys. Rev. B **81**, 172505 (2010)
19. Y. Greenberg, A. Izmalkov, M. Grajcar, E. Il'ichev, H.-G. Meyer, M.H.S. Amin, A.-M. van den Brink, Low frequency characterization of quantum tunneling in flux qubits. Phys. Rev. B **66**, 214525 (2002)
20. N. Zheludev, S. Prosvirin, N. Papasimakis, V. Fedotov, Lasing spaser. Nat. Photonics **2**, 351 (2008)

Chapter 10
Application of the Model of "Quantum" Metamaterials: Metamaterial Caused Enhancement of Nonlinear Response

10.1 Modeling of Metamaterials Caused Enhancement of Nonlinear Response

10.1.1 Model Adaptation

In this chapter, the first demonstration of exceptional light-with-light optical switching performance of the carbon nanotube MM—hybrid nanostructure of plasmonic MM hybridized with semiconducting single-walled carbon nanotubes (CNT) is provided. Modulation depth of 10% in the near-IR with sub-500 fs response time is achieved with the pump fluency of just 10 μJ/cm^2, which is order of magnitude lower than in previously reported artificial nanostructures. Since spectral position of the excitonic response and MM plasmonic resonance can be adjusted by using CTNs of different diameter and scaling MM design, the giant nonlinear response of the hybrid MM—in principle—can be engineered to cover the entire second and third telecom windows, from O to U-band.

Application of our approach is illustrated for the modelling of enhanced nonlinear optical response demonstrated recently in a plasmonic MMs combined with carbon nanotubes (CNT) [1], see Fig. 10.1. In such a hybrid quantum-classical system the MM structure works as a light concentrator enhancing optical fields locally, which are coherent with the incident field and also affect the response dynamics of CNT. In the resonance case the intensity of the local fields can become significantly higher than that of the incident wave and therefore substantially affect the dynamics of CNT response. The nonlinearity of CNT appears due to the saturation induced by the direct pumping of such a two-level-like quantum system and basically requires neither positive N_0 nor presence of a nanoresonator. The enhancement of the nonlinearity is caused by the addition pumping channel, where the external field transfers energy to CNT through the nanoresonator. This is described by the term $\frac{i\alpha_x x^* N}{\hbar}$ in the first equation of system (9.17).

A. Chipouline and F. Küppers, *Optical Metamaterials: Qualitative Models*, Springer Series in Optical Sciences 211, https://doi.org/10.1007/978-3-319-77520-3_10

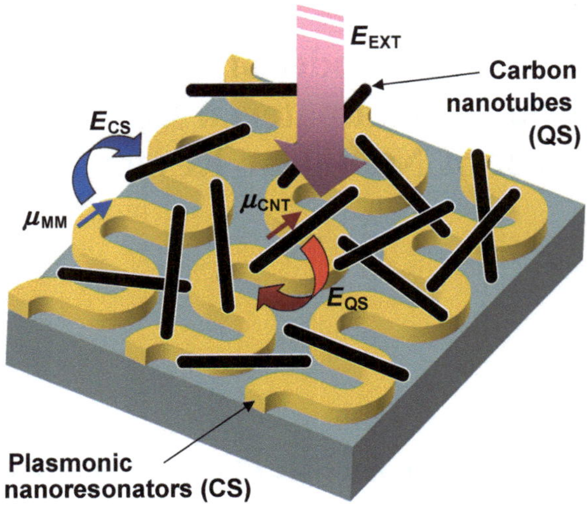

Fig. 10.1 Schematic of the modelled active hybrid MM with quantum ingredients: an array of plasmonic nanoresonators (Classic System—CS) covered with a layer of carbon nanotubes (Quantum System—QS). E_{CS} is the local field acting on the carbon nanotubes, which is produced by the dipole moments induced in the MM nanoresonators; E_{QS} is the local field acting on the nanoresonators, which is produced by the dipole moments induced in the carbon nanotubes [2]

In order to solve (9.17) analytically we assume that the plasmonic resonator is driven mainly by the external electric field:

$$\begin{cases} \frac{d\rho_{12}}{dt} + \rho_{12}\left(\frac{1}{\tau_2} + i(\omega - \omega_{21})\right) = \frac{i\alpha_x x^* N}{\hbar} + \frac{i\mu_{QS} A^* N}{\hbar} \\ \frac{dN}{dt} + \frac{(N - N_0)}{\tau_1} = \frac{i\alpha_x\left(x\rho_{12} - x^*\rho_{12}^*\right) + i\mu_{QS}\left(A\rho_{12} - A^*\rho_{12}^*\right)}{2\hbar} \\ 2(\gamma - i\omega)\frac{dx}{dt} + \left(\omega_0^2 - \omega^2 - 2i\omega\gamma\right)x = \chi A \end{cases} \quad (10.1)$$

Both eigenfrequencies ω_{21} and ω_0 are the resonance frequencies of CNT and the MM respectively and can vary independently; $\alpha_x \sim \mu_{CNT}\mu_{MM}$, $\alpha_\rho \sim \mu_{CNT}\chi$, μ_{CNT} is the CNT dipole moment for the band gap transition, μ_{MM} is the effective MM dipole moment.

Below we consider the case of CW excitation as an approximation of the pumping regime used in [1]:

$$\begin{cases} \rho_{12}\left(\frac{1}{\tau_2} + i(\omega - \omega_{21})\right) = \frac{i\alpha_x x^* N}{\hbar} + \frac{i\mu_{QS} A^* N}{\hbar} \\ \frac{(N - N_0)}{\tau_1} = \frac{i\alpha_x\left(x\rho_{12} - x^*\rho_{12}^*\right) + i\mu_{QS}\left(A\rho_{12} - A^*\rho_{12}^*\right)}{2\hbar} \\ \left(\omega_0^2 - \omega^2 - 2i\omega\gamma\right)x = \chi A \end{cases} \quad (10.2)$$

Our goal is to express ρ_{12}^* and find the effective dielectric constant of CNT layer $\varepsilon_{CNT} = \varepsilon_1 + i\varepsilon_1 = 1 + 4\pi\left(n_{CNT}\frac{\mu_{CNT}\rho_{12}^*}{A} + n_{MM}\frac{qx}{A}\right)$ as a function of all parameters and intensity (the second term in brackets appears due to the MM itself); here n_{CNT} and n_{MM} are the concentrations of CNT and MAs respectively. The imaginary part of the dielectric constant is responsible for the losses and its intensity and frequency dependence could be compared with the experiments.

In order to make expression for the dielectric constant more compact we combine the resonant factors that will often appear in the text introducing the following parameters:

$$\begin{cases} R_x = \omega_0^2 - \omega^2 - 2i\omega\gamma; \quad |R_x|^2 = \left(\omega_0^2 - \omega^2\right)^2 + 4\omega^2\gamma^2 \\ R_\rho = 1 + i(\omega - \omega_{21})\tau_2; \quad |R_\rho|^2 = 1 + (\omega - \omega_{21})^2\tau_2^2 \\ F_1 = \omega_0^2 - \omega^2 - 2\omega\gamma(\omega - \omega_{21})\tau_2 \\ F_2 = \omega_0^2 - \omega^2 + 2\omega\gamma(\omega - \omega_{21})\tau_2 \\ F_3 = \omega_0^2 - \omega^2 \end{cases} \tag{10.3}$$

We also introduce coupling constant $\sigma_{\mathrm{CNT}} = \frac{\alpha_x \chi}{\mu_{\mathrm{CNT}}}$. The expressions containing the non-diagonal element of the density matrix are given by:

$$\begin{cases} \rho_{12} = \frac{-iN\tau_2\mu R_\rho}{\hbar|R_\rho|^2}\left(1 + \frac{\sigma_{\mathrm{CNT}}}{R_x^*}\right)A \\ A\rho_{12} - A^*\rho_{12}^* = \frac{2iN\tau_2\mu|A|^2}{\hbar|R_\rho|^2}\left(1 + \frac{\sigma_{\mathrm{CNT}}F_1}{|R_x|^2}\right) \\ x\rho_{12} - x^*\rho_{12}^* = \frac{i2\chi N\tau_2\mu|A|^2}{\hbar|R_x|^2|R_\rho|^2}\left(\sigma_{\mathrm{CNT}} + F_2\right) \end{cases} \tag{10.4}$$

We finally arrive at a compact expression for the population as a function of the external field:

$$N = \frac{N_0}{1 + \frac{|A|^2}{|A_s|^2}} = \frac{N_0}{1 + S} \tag{10.5}$$

where:

$$\begin{cases} |A_s|^2 = \frac{|A_{s,0}|^2|R_\rho|^2}{1 + \frac{\sigma_{\mathrm{CNT}}(\sigma_{\mathrm{CNT}} + F_3)}{|R_x|^2}}; \qquad |A_{s,0}|^2 = \frac{\hbar^2}{\mu\tau_1\tau_2} \\ S = \frac{|A|^2}{|A_s|^2} = S_0\left[\frac{1 + \frac{\sigma_{\mathrm{CNT}}(\sigma_{\mathrm{CNT}} + F_3)}{|R_x|^2}}{|R_\rho|^2}\right]; \qquad S_0 = \frac{|A|^2}{|A_{s,0}|^2} \end{cases} \tag{10.6}$$

Here S_0 and S account for the saturation and enhanced saturation respectively. Taking into account (10.4) we obtain the following final expression for the dielectric constant for the entire hybrid system of the CNT layer modified in the presence of the MM:

$$\begin{cases} \varepsilon_{CNT} + \varepsilon_x = \varepsilon_1 + i\varepsilon_2 \\ \varepsilon_1 = \varepsilon_{0,1} + \frac{4\pi n_{CNT}\tau_2\mu_{CNT}^2 N_0}{\hbar|R_x|^2(1+S)}\left((\omega-\omega_{21})\tau_2 + \frac{2\omega\gamma\sigma_{CNT}}{|R_x|^2}\right) + \frac{4\pi n_{MM} q\chi F_3}{|R_x|^2} \\ \varepsilon_2 = -\frac{4\pi n_{CNT}\tau_2\mu_{CNT}^2 N_0}{\hbar|R_x|^2(1+S)}\left(1 + \frac{\sigma_{CNT}F_1}{|R_x|^2}\right) + \frac{8\pi q n_{MM}\chi\omega\gamma}{|R_x|^2} \end{cases} \tag{10.7}$$

here ε_{CNT}, ε_x are the contribution of the CNT and plasmonic nanoresonators respectively.

The prediction of our model can be compared with the experimental results [1] in terms of optical losses induced in CNT layer, which are reflected by the imaginary part of its effective dielectric constant (10.7). The relative absorption of light, L_{CNT}, has to be calculated according to the following expression:

$$L_{CNT} = 1 - \exp\left(2i\frac{\omega}{c}\mathrm{Im}\left[\sqrt{\varepsilon_1 + i\varepsilon_2 d}\right]\right) \tag{10.8}$$

which allows us to describe the effect of nonlinearity enhancement quantitatively. Here d is the effective thickness of the CNT layer; positive values of ε_2 correspond to losses, negative—to amplification (the developed model allows us to consider both cases).

Assuming that the imaginary part of ε_{CNT} is smaller compared to the real one, we expand (10.8) into series:

$$L_{CNT} = 1 - \exp\left(2i\frac{\omega}{c}\mathrm{Im}\left[\sqrt{\varepsilon_1 + i\varepsilon_2 d}\right]\right) \approx 1 - \exp\left(-\frac{\omega}{c}\frac{\varepsilon_2}{\sqrt{\varepsilon_1}}d\right) \tag{10.9}$$

In order to be close to the experimental procedure in [1], we considered relative change of absorption due to saturation in terms of the model parameters:

$$\frac{\Delta L_{CNT}}{L_{CNT}} \approx \frac{\omega d}{c}\left(\left.\frac{\varepsilon_2}{\sqrt{\varepsilon_1}}\right|_{|A|^2} - \left.\frac{\varepsilon_2}{\sqrt{\varepsilon_1}}\right|_0\right) \tag{10.10}$$

We also assume for simplicity that the real part of the dielectric constant remains unchanged ($\varepsilon_1|_{|A|^2} = \varepsilon_1|_0 = \varepsilon_{1,0}$):

$$\frac{\Delta L_{CNT}}{L_{CNT}} = \frac{\omega d}{c\sqrt{\varepsilon_{1,0}}}\left(\varepsilon_2\left(|A|^2\right) - \varepsilon_2(0)\right) < 0 \tag{10.11}$$

First, the effect will be evaluated for the case when the resonance frequencies of CNT and MM coincide, namely $\omega = \omega_0 = \omega_{21}$. The frequency dependent coefficients in this case become:

$$\begin{cases} |R_x|^2 = 4\omega^2\gamma^2 \\ |R_\rho|^2 = 1 \\ F_1 = 0 \\ F_2 = 0 \\ F_3 = 0 \\ S = S_0\left[1 + \frac{\sigma_{CNT}^2}{4\omega^2\gamma^2}\right] \end{cases} \tag{10.12}$$

and the relative absorption change (10.12) becomes (no pumping of CNT, $N_0 = -1$):

$$\left(\frac{\Delta L_{CNT}}{L_{CNT}}\right)_{\sigma\neq 0,\,\text{Resonance}} = -\frac{\omega d}{c\sqrt{\varepsilon_1}}\frac{4\pi\tau_2\mu_{CNT}^2}{\hbar}\left(1 - \frac{1}{1+S}\right) \tag{10.13}$$

In order to visualize the effect of the nonlinear response enhancement, it is necessary to introduce a relative absorption change according to the following expression:

$$\frac{\left(\frac{\Delta L_{CNT}}{L_{CNT}}\right)_{\sigma_{CNT}\neq 0,\,\text{Resonance}}}{\left(\frac{\Delta L_{CNT}}{L_{CNT}}\right)_{\sigma_{CNT}=0,\,\text{Resonance}}} = \frac{1 + S_0\left(1 + \frac{\sigma_{CNT}^2}{4\omega^2\gamma^2}\right)}{(1+S_0)\left(1 + \frac{\sigma_{CNT}^2}{4\omega^2\gamma^2}\right)} \tag{10.14}$$

which clearly demonstrates that the increase of the relative absorption change for CNT in the presence of the MM ($\sigma_{CNT} \neq 0$) is bigger than for CNT alone ($\sigma_{CNT} = 0$). The respective dependence is presented in Fig. 10.2 (positive values in Fig. 10.2 correspond to the decrease of absorption).

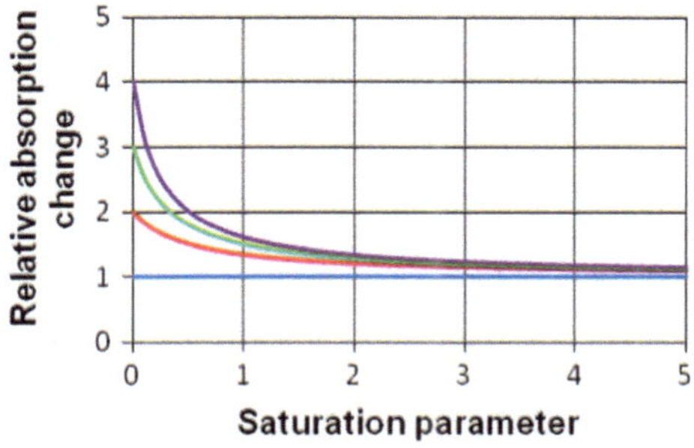

Fig. 10.2 Relative absorption (10.14) as a function of saturation parameter $S_0 = \frac{|A|^2}{|A_{s,0}|^2}$ for different values of coupling $\frac{\sigma_{CNT}^2}{4\omega^2\gamma^2} = 0, 1, 2, 3$ ("0"—blue line, "1"—red line, "2"—green line, "3"—violet line). Note that positive values in Fig. 10.2 correspond to the decrease of absorption [2]

From (10.14) we can conclude the following:

1. The coupling of CNT to the plasmonic nanoresonator enhances the CNT nonlinearity (saturation intensity becomes effectively lower).
2. Losses in the nanoresonator reduce the enhancement effect (higher γ lead to higher saturation intensities).
3. The enhancement is more pronounced in the region of low saturation.

It is rather hard to estimate σ_{CNT}, but even under very conservative assumptions its value can easily become large then $\omega\gamma \sim 10^{28}$ and, consequently the effect of nonlinearity enhancement should be easily observed in the experiments.

It is important to discuss the contribution of the Purcell effect [3] to the nonlinear response of the considered hybrid structure. The Purcell effect appears when the quantum system interacts with the thermostat and is described in the frame of the accepted here approach (see Sect. 9.1.2) by an extra coefficient P in the expressions for the relaxation times $\frac{1}{\tau_1} = \frac{1}{\tau_{1,\mathrm{nr}}} + \frac{P}{\tau_{1,\mathrm{r}}}$, $\frac{1}{\tau_2} = \frac{1}{\tau_1} + \Delta = \frac{1}{\tau_{1,\mathrm{nr}}} + \frac{P}{\tau_{1,\mathrm{r}}} + \Delta$($\tau_{1,\mathrm{nr}}$—is the nonradiative relaxation time, $\tau_{1,\mathrm{r}}$—is the radiative relaxation time, $P > 1$—is the Purcell factor, and Δ—is the extra term, giving difference between τ_1 and τ_2).

The both relaxation times become effectively shorter in the presence of the nanoresonator and thus the Purcell effect causes the effective increase of the saturation intensity—see (10.6). Here the effect of the relaxation time shortening due to Purcell effect is not investigated in details; the mutual interplay between field enhancement and Purcell caused relaxation time shortening requires further investigation.

The following strategy has been adopted in order to assess the validity of the proposed model. First the calculated and measured data for CNT layer alone will be compared and CNT model parameters will be found from fitting the measured data. Then the same procedure will be repeated for the MM without CNT to find the model parameters for the nanoresonator. Finally, we consider coupling between CNT and MM structure and analyse the outcome of the model for different values of the coupling constant σ_{CNT}. Direct comparison between calculated and measured data for the coupled system will be the subject of a separate work.

10.1.2 CNT Alone

It turns out, that experimentally measured absorption spectra of CNT layer exhibit visible asymmetry [1], which can be explained by the fact that the absorption lines of deposited CNTs have different central frequencies. The effect of such inhomogeneous broadening is not included into the density matrix formalism and has to be taken into account by an additional averaging over respective probability distribution function (PDF). We assume in our modelling that PDF of CNT consists of two peaks, with both central frequency and bandwidth treated as the parameters that can be found from fitting the experimental spectrum.

The absorption spectrum in this case was calculated according to the following expressions:

$$\begin{cases} L_{\text{CNT,HOMO}}(\omega,\omega_{21}) = 1 - \exp\left(2i\frac{\omega}{c}\text{Im}\left[\sqrt{\varepsilon_1 + i\varepsilon_2}\right]d\right) \\ \varepsilon_{\text{CNT}} + \varepsilon_x = \varepsilon_1 + i\varepsilon_2 \\ \varepsilon_1 = \varepsilon_{0,1} + \frac{4\pi n_{\text{CNT}}\tau_2^2\mu_{\text{CNT}}^2 N_0(\omega-\omega_{21})}{\hbar\left|R_\rho\right|^2(1+S)} \\ \varepsilon_2 = -\frac{4\pi n_{\text{CNT}}\tau_2^2\mu_{\text{CNT}}^2 N_0}{\hbar\left|R_\rho\right|^2(1+S)} \\ \text{PDF}(\omega_{21}) = C_1\exp\left(-\frac{(\omega-\omega_{21})^2}{2\Delta\omega_1^2}\right) + C_2\exp\left(-\frac{(\omega-\omega_{21})^2}{2\Delta\omega_2^2}\right) \\ L_{\text{CNT, INHOMO}}(\omega) = \int_0^\infty L_{\text{CNT,HOMO}}(\omega,\ \omega_{21}) * \text{PDF}(\omega_{21})\text{d}\omega_{21} \end{cases} \tag{10.15}$$

where the following values of the fitting parameters were used: $\omega_1 = 9.9945\times 10^{14}\,(\text{rad/s}), \omega_2 = 9.2174\times 10^{14}\,(\text{rad/s}), \Delta\omega_1 = 1.65\times 10^{10}\,(\text{rad/s}), \Delta\omega_2 = 5.2\times 10^{10}\,(\text{rad/s})$. The results of the related absorption spectra as a function of wavelength and saturation level are presented in Fig. 10.3a; the respective PDF is given in Fig. 10.3b.

10.1.3 Metamaterial Alone

For retrieving MM parameters we used expressions similar to (10.15) but without invoking the PDF-based averaging procedure:

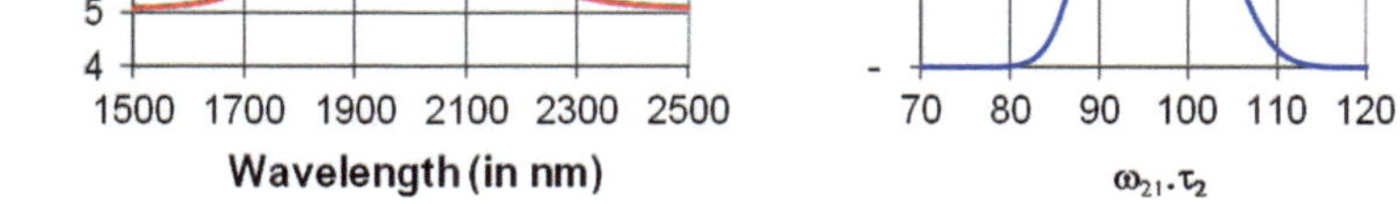

Fig. 10.3 **a** Calculated absorption spectrum of the CNT for different values of saturation parameter $S_0 = \frac{|A|^2}{|A_{s,0}|^2}$. The experimental data coincide one to one with the unsaturated line (blue line) and is not shown separately; **b** the PDF used to calculate the absorption spectrum [2]

$$\begin{cases} L_{\mathrm{MM}}(\omega,\omega_{21}) = 1 - \exp\left(2i\frac{\omega}{c}\mathrm{Im}\left[\sqrt{\varepsilon_1 + i\varepsilon_2}\right]d\right) \\ \varepsilon_{\mathrm{CNT}} + \varepsilon_x = \varepsilon_1 + i\varepsilon_2 \\ \varepsilon_1 = \varepsilon_{0,1} + \frac{4\pi n_{\mathrm{MM}} q\chi\left(\omega_0^2 - \omega\right)}{|R_x|^2} \\ \varepsilon_2 = \frac{8\pi n_{\mathrm{MM}} q\chi\gamma}{|R_x|^2} \end{cases} \quad (10.16)$$

Note that the calculated absorption spectrum of the MM does not match the experimental one reported in [1], which is stipulated by a very simple model for the plasmonic MM, based on a single harmonic oscillator. As it can be seen from (10.16), the spectrum dependence of the absorption is given by the function $|R_x|^2$, and in order to achieve a good match between the observed and calculated MM spectral response it is necessary to either consider a set of several coupled oscillators [4], or just choose an appropriate fitting function for $|R_x|^2$. Here we limit ourselves to demonstrating the adequateness of our very simple model. Correspondingly, in what follows we exploit the calculated absorption spectrum of the MM structure shown in Fig. 10.4. It is worth noting that in the presence of CNT layer the central frequency of the MM absorption peak will be shifted, reflecting the fact that the plasmon resonance frequency depends on the dielectric constant of the environment. This effect is not included in our model itself, but it is taken into account by the fact that the central frequency ω_0 is a fitting parameter.

10.1.4 CNTs Combined with Metamaterial

Here we present the results of modelling the enhancement of CNT nonlinearity based on the values of the fitting parameters retrieved in two previous sections; the only parameter that could not be determined using the experimental data is the coupling constant σ_{CNT} which is the subject of a separate consideration.

We consider two possible realisations of CNT coating: (i) with purely homogeneous distribution of CNT parameters over the MM surface (PDF in Fig. 10.3b

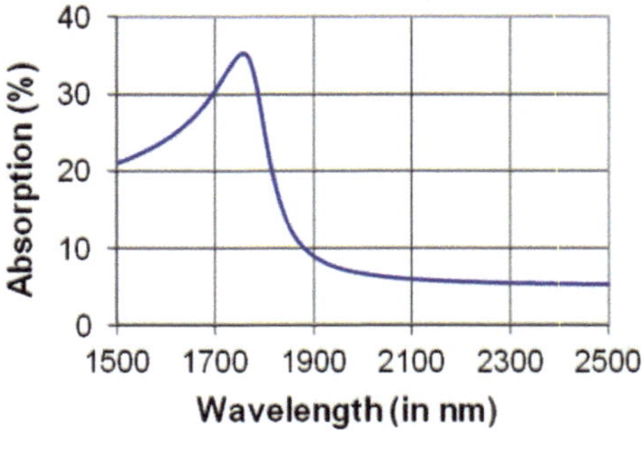

Fig. 10.4 Calculated absorption spectrum of the MM [2]

becomes delta function), and (ii) inhomogeneous, when each CNT has different eigenfrequency and oscillates with eigenphase shifted relative to the incoming field as well as local field of the MM nanoresonator. In homogeneous case we do not take into account spatial distribution of the CNT eigenfrequency in the calculation of absorption, which means that we can simply sum up effective dielectric constants of CNT and MM layers when calculating their combined response. In inhomogeneous case the contribution to the absorption from the CNT molecules becomes not fully coherent and has to be taken into account separately from that of the MM. This can be done by summing up the refractive indexes of CNT and MM instead of their dielectric constants.

As can be seen from Fig. 10.2, the effect of the nonlinearity enhancement is mostly pronounced for small saturation values.

The homogeneous case is described by the following system of equations:

$$\begin{cases} L_{\text{CNT\&MM,HOMO}}(\omega,\omega_{21}) = 1-\exp\left(2i\frac{\omega}{c}\text{Im}\left[\sqrt{\varepsilon_1+i\varepsilon_2}\right]d\right) \\ \varepsilon_{\text{CNT}}+\varepsilon_x = \varepsilon_1+i\varepsilon_2 \\ \varepsilon_1 = \varepsilon_{0,1}+\frac{4\pi n_{\text{CNT}}\tau_2\mu^2N_0}{\hbar|R_x|^2(1+S)}\left((\omega-\omega_{21})\tau_2+\frac{2\omega\gamma\sigma_{\text{CNT}}}{|R_x|^2}\right)+\frac{4\pi n_{\text{MM}}q\chi\left(\omega_0^2-\omega^2\right)}{|R_x|^2} \\ \varepsilon_2 = -\frac{4\pi n_{\text{CNT}}\tau_2\mu^2N_0}{\hbar|R_x|^2(1+S)}\left(1+\frac{\sigma_{\text{CNT}}F_1}{|R_x|^2}\right)+\frac{8\pi q n_{\text{MM}}\chi\omega\gamma}{|R_x|^2} \end{cases} \tag{10.17}$$

while for incoherent case:

$$\begin{cases} L_{\text{CNT\&MM,HOMO}}(\omega,\omega_{21}) = \\ 1-\exp\left(2i\frac{\omega}{c}\text{Im}\left[\sqrt{\varepsilon_{1,\text{CNT\_}A}+i\varepsilon_{2,\text{CNT\_}A}}+\sqrt{\varepsilon_{1,\text{CNT\_}x}+i\varepsilon_{2,\text{CNT\_}x}}+\sqrt{\varepsilon_{1,x}+i\varepsilon_{2,x}}\right]d\right) \\ \varepsilon_{1,\text{CNT\_}A} = \varepsilon_{0,1}+\frac{4\pi n_{\text{CNT}}\tau_2\mu^2N_0}{\hbar|R_x|^2(1+S)}(\omega-\omega_{21})\tau_2 \\ \varepsilon_{2,\text{CNT\_}A} = -\frac{4\pi n_{\text{CNT}}\tau_2\mu^2N_0}{\hbar|R_x|^2(1+S)} \\ \varepsilon_{1,\text{CNT\_}x} = -\frac{4\pi n_{\text{CNT}}\tau_2\mu^2N_0}{\hbar|R_x|^2(1+S)}\frac{2\omega\gamma\sigma_{\text{CNT}}}{|R_x|^2} \\ \varepsilon_{2,\text{CNT\_}x} = -\frac{4\pi n_{\text{CNT}}\tau_2\mu^2N_0}{\hbar|R_x|^2(1+S)}\frac{\sigma_{\text{CNT}}F_1}{|R_x|^2} \\ \varepsilon_{1,x} = \frac{4\pi n_{\text{MM}}q\chi\left(\omega_0^2-\omega^2\right)}{|R_x|^2} \\ \varepsilon_{2,x} = \frac{8\pi q n_{\text{MM}}\chi\omega\gamma}{|R_x|^2} \end{cases} \tag{10.18}$$

Here $\varepsilon_{1,\text{CNT\_}A},\varepsilon_{2,\text{CNT\_}A}$, corresponds to CNT driven by the external field, $\varepsilon_{1,\text{CNT\_}x},\varepsilon_{2,\text{CNT\_}x}$ is the CNT contribution due to the local field of the nanoresonators, and $\varepsilon_{1,x}$, $\varepsilon_{2,x}$ are the effective dielectric constants of the MM alone.

Figure 10.4 presents the results of our calculations for the homogeneous and inhomogeneous cases with the saturation parameter set to 3. For comparison we also present here the absorption of the MM alone (red curve in Fig. 10.5).

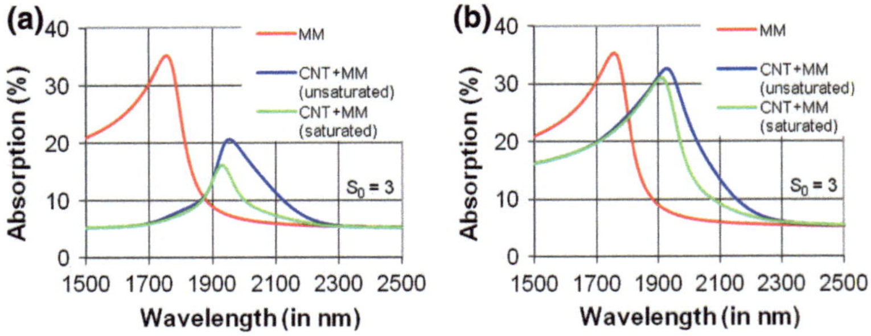

Fig. 10.5 Absorption spectrum of CNTs combined with MM for **a** homogeneous and **b** inhomogeneous cases. Saturation parameter $S_0 = \frac{|A|^2}{|A_{s,0}|^2} = 3$ [2]

The absorption spectra of the CNT-coated MM are red shifted with respect to the uncoated structure due to the fact that CNT effectively increase the density of the dielectric environment for the nanoresonators. In order to demonstrate the effect of the MM on the absorption change due to saturation, we plotted the normalised absorption change $\frac{\Delta L_{\text{CNT\&MM}}}{L_{\text{CNT\&MM}}}$, shown in Fig. 10.6.

Figure 10.6 clearly demonstrates that in homogeneous case the effect appears to be more pronounced, which leads to the general requirement of maximizing the homogeneity of the deposited CNT layer.

10.2 Experimental Investigation of Enhancement of Nonlinear Response in Carbon Nano Tubes (CNT)

10.2.1 Introduction

Nanoscale sized ultrafast optical modulators are key elements for development of integrated all-optical signal processing circuits. Ultrafast modulation of the light

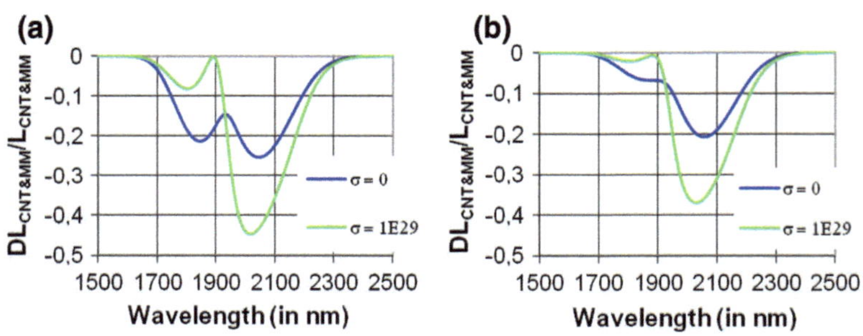

Fig. 10.6 Normalised absorption change spectrum of CNTs combined with MM for **a** homogeneous and **b** inhomogeneous cases. Saturation parameter is $S_0 = \frac{|A|^2}{|A_{s,0}|^2} = 3$ [2]

with light on the nanoscale is a challenging task: optical nonlinearities in conventional nonlinear materials are generally too weak to alter intensity of the light significantly on the sub-wavelength scale [5], semiconductor based devices suffer from slow response time [6] and propagating surface plasmon based structures capable of efficient ultrafast light modulation have μm-scale dimensions [7]. Nanoscale sized devices with improved nonlinear optical properties can be engineered with the use of MMs—artificial media offering wide and rapidly expanding range of new photonic functionalities [8]. Few MM-based structures featuring enhanced all-optical switching properties in THz [9] and near-IR [10, 11] spectral regions have been demonstrated recently. In these studies enhanced switching performance is achieved in MM structures hybridized with active nonlinear media. In such hybrid structures small changes in the refractive index of active medium induced by non-resonant photoexcitation tune MM plasmonic resonance which results in significant modulation of MM transmission near the resonance. Another powerful opportunity for enhancement of nonlinear optical response of MM-based structures, not widely implemented so far, is in exploiting effects of resonant concentration of local fields in the vicinity of MM [12, 13]. Recently we have introduced carbon nanotube MM [1]—hybrid structure of plasmonic MM with semiconducting single-walled carbon nanotubes (CNTs) which employs combination of both mentioned above approaches for enhancement of nonlinear optical response. This is achieved by spectral matching of absorption resonance of nonlinear active medium (CNTs) with MM plasmonic resonance. Nonlinear optical response of the CNT MM is defined by the effect of MM transmission modulation as a result of changes in the refractive index of active medium; however nonlinear response of active medium itself is significantly enhanced by concentration of local fields in MM under the conditions of resonant excitation. We show that this result in the substantial improvement of nonlinear optical performance of the CNT MM in comparison with the other previously reported analogous structures.

Our choice of using semiconducting single-walled CNTs as an active medium was dictated by nanotubes' unique nonlinear optical properties: in addition to high third-order susceptibility at the excitonic absorption resonances CNTs exhibit sub-picosecond recovery time [14, 15] making them extremely promising material for applications in ultrafast all-optical modulators.

10.2.2 Experiment

Plasmonic nanostructure used in the CNT MM is formed by array of asymmetrically split ring resonators and belongs to the class of MMs where weak coupling of the plasmonic excitation mode to the free-space radiation modes creates narrow resonances with asymmetric, Fano-like dispersion [16, 17]. This type of MM resonances features high quality factor and is strongly sensitive to the local environment [17]. High value of quality factor is crucial for efficient concentration of local fields while sensitivity of the MM resonance to the local environment facilitates significant changes of MM transmission in response to small nonlinear changes of refractive index of the CNTs layer. MM structure shown in Fig. 10.1a was fabricated by focused ion beam milling in a 50 nm thick gold film deposited on a 100 nm thick Si_3N_4 membrane. The nanostructure was then covered by $\sim$50 nm thick layer of CNTs using spray-coating technique. We used purified single-walled CNTs with the average diameter of 1.4 nm synthesized by the electric arc discharge method. This method produces natural mixture of CNTs with the 2 to 1 ratio of semiconducting to metallic CNTs. To enable comparison of nonlinear optical response of the CNT MM with the response of CNTs we have fabricated square window in the gold film next to the MM array so the layer of CNTs formed during deposition process had the same thickness on the MM and on the Si_3N_4 membrane window. More details on the CNT MM sample fabrication and characterization by scanning helium ion microscope can be found elsewhere [1]. Spectral position of the plasmonic resonance of uncovered MM λ_p depends on the period of MM D while deposition of the layer of CNTs on the top of MM results in a redshift of plasmonic absorption peak to λ_p, as can be seen from Fig. 10.7b. The structure with the period D = 839 nm where plasmonic resonance in hybrid structure (after CNTs deposition) is spectrally matched to the CNTs excitonic absorption resonance has

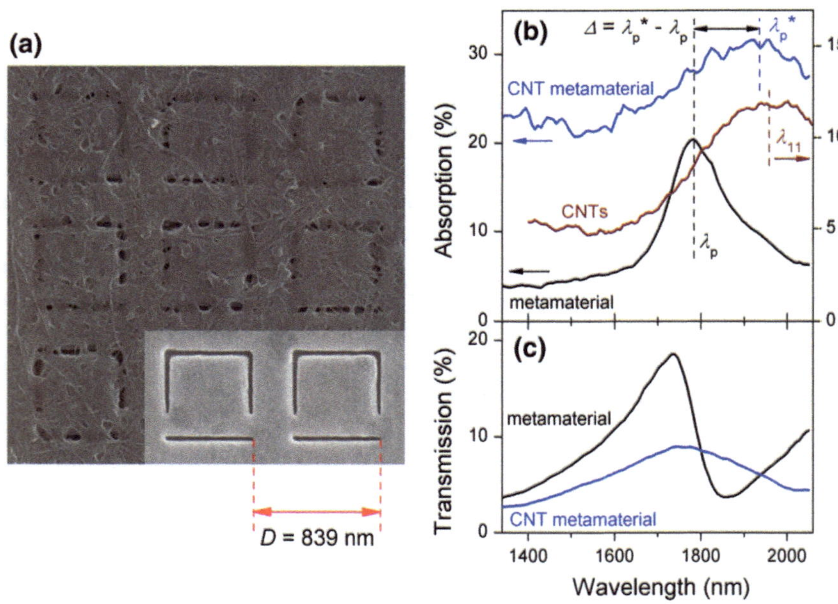

Fig. 10.7 Carbon nanotube MM. **a** Scanning helium ion microscope image of the plasmonic MM covered by CNTs. Inset shows image of two unit cells of uncovered MM. **b** Left scale—plasmonic absorption resonances of uncovered metamaterial (black) and MM covered by CNTs (blue). Right scale—excitonic absorption spectra of the bare CNTs layer (wine). **c** Transmission spectra of uncovered MM (black) and MM covered by CNTs (blue). Presented spectra were measured on CRAIC microspectrometer and correspond to the linear (low intensity) regime [18]

been selected. Nonlinear optical response of the CNT MM was investigated by degenerate (single-colour) non-collinear pump-probe. Both pump and probe beams from tunable 200 fs OPO system pumped by Ti:Sapphire laser operating at 80 MHz were focused to diameters of $\sim 30\ \mu\text{m}$. MM array has smaller size ($22 \times 22\ \mu\text{m}^2$) so it was illuminated uniformly by both pump and probe beams. Intensity of the probe pulse was 10–20 times weaker than the intensity of the pump pulse. Optical response of MM is polarization sensitive and the direction of linear polarization of both pump and probe beams were chosen to excite dark mode resonance in MM.

10.2.3 Results

The nonlinear changes of transmission $\Delta T/T$ of the CNT MM have been retrieved at few delays after the pump pulse (shown in Fig. 10.8). Differential transmission spectrum measured on the bare CNTs layer at the same pump fluency is shown in Fig. 10.8 by open circles for comparison. From experimental data shown in Fig. 10.8 one can see that the nonlinear optical response of the CNT MM has complex sign-changing spectral dispersion, and modulation depth $\Delta T/T$ in the CNT MM is much higher than in the bare CNTs layer.

Observed transient nonlinear dynamics of the CNT MM replicates the dynamics of CNTs: biexponential decay of nonlinear optical response of CNTs was reported in a number of studies where fast component was assigned to the intra-band, and the slow component—to the inter-band carrier relaxation in CNTs [19–21]. For potential applications in all-optical switching it is important that the fast component

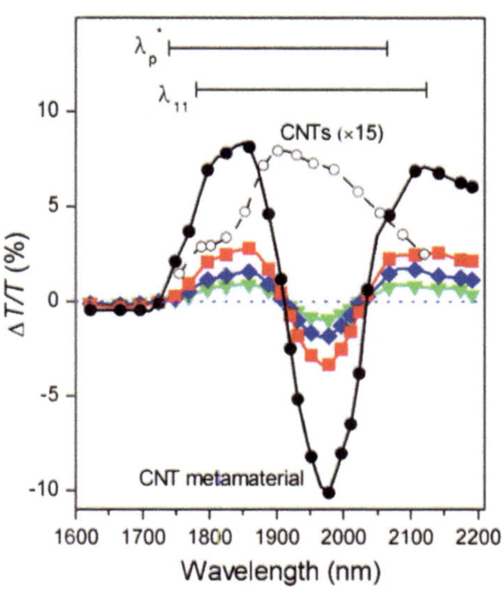

Fig. 10.8 Differential transmission spectra of the CNT MM (solid symbols) taken at 0 (black circles), 0.5 (red squares), 1 (blue diamonds) and 2 ps (green triangles) after the pump pulse. Differential transmission spectrum of the bare CNTs layer measured at 0 ps is shown by open circles. Arrows indicates spectral regions (FWHM) of MM (covered by CNTs) plasmonic (λ_p) and CNTs excitonic (λ_{11}) resonances [18]

provides major contribution to the nonlinear dynamics of both bare CNTs layer and CNT MM.

In contrast to previously reported MM-based optically switching devices [10, 11] in the CNT MM active medium is excited resonantly at the MM plasmonic resonance thus providing opportunity of enhancement of nonlinear optical response of the hybrid structure due to field concentration in plasmonic resonator. To evaluate the effect of the local field enhancement we have compared power dependencies of nonlinear changes of transmission of the CNT MM and the bare CNTs layer. The prime nonlinear optical process in single-walled semiconducting CNTs is saturable absorption which originates from filling of the exited resonant excitonic states and associated with the third-order optical nonlinearities [22] and can be described in a simple two-level saturable absorber model [23].

For the bare CNTs layer from fitting of experimental power dependence saturation fluence was estimated to be 15 μJ/cm^2 at 1960 nm which is in agreement with previously reported saturation fluencies in thin CNTs films ranging from 9.5 μJ/cm^2 [21] to 57 μJ/cm^2 [24]. From the experimental data it can be seen that power dependencies of nonlinear optical response of the CNT MM follows essentially the same saturation law described by (10.6) with lower in compare with bare CNT I_{sat}. Fit of experimental data for the CNT MM at 1820 and 1960 nm gives close saturation fluencies of 5.5 and 5.2 μJ/cm^2, correspondingly, which is ~3 times lower than saturation fluence of the bare CNTs layer measured at 1960 nm. Observed significant decrease of the saturation intensity in the CNT MM as compared to the bare CNTs layer provides experimental evidence for enhancement of nonlinear optical response (see Fig. 10.2) of the layer of CNTs placed in the vicinity of MM due to local field concentration in plasmonic resonator.

10.2.4 Discussion

Nonlinear optical response of the CNT MM can be explained by taking into account interaction between MM plasmonic and CNTs excitonic resonances. Plasmonic resonance of uncovered MM λ_p appears at ~1800 nm as a sharp feature in transmission and corresponding peak in the absorption spectrum (see Fig. 10.7b, c). Deposition of the CNTs layer on MM results in redshift $\Delta\lambda = \lambda_p^* - \lambda_p$ and damping of MM plasmonic resonance (compare spectra for uncovered MM and MM covered by CNTs in Fig. 10.7b, c). Redshift is originated from the real part of refractive index of highly polarizable CNTs while damping is subject to absorption of plasmon evanescent waves in the layer of CNTs and defined by the imaginary part of CNTs' refractive index. We have checked experimentally that the optical response of uncovered MM is linear in the range of wavelengths and fluencies used in our study. Accordingly, nonlinear optical response of the CNT MM is governed by nonlinear processes in the layer of CNTs. Excitonic absorption resonance λ_{11} of CNTs is spectrally overlapped with the plasmonic resonance λ_p^* of MM covered by CNTs. In the nonlinear regime this leads to the two effects: (i) nonlinear optical response of CNTs is enhanced due to resonant concentration of local field in plasmonic resonator, (ii) decrease of absorption (bleaching) in CNTs and associated changes of the real part of refractive index of CNTs leads to recovery of MM plasmonic resonance towards resonance of uncovered MM. This results in the complex spectral dependence of nonlinear changes of transmission of the CNT MM containing spectral regions with positive and negative values of $\Delta T/T$. In the presented in part 10.1 model this effect is not taken into account and does not appear in resulting curves in Fig. 10.6. The experimentally observed spectral dispersion of nonlinear optical response of the CNT MM was reproduced in numerical simulations in COMSOL. To do this we have first calculated linear optical response of uncovered MM using parameters of real MM structure (thicknesses and dielectric constants of gold and Si_3N_4 membrane and geometry of MM slits). Calculated transmission and absorption spectra show good correspondence with the experimental data (compare corresponding curves in Figs. 10.9 and 10.7). To calculate optical response of the hybrid structure 50 nm thick layer of dielectric have been

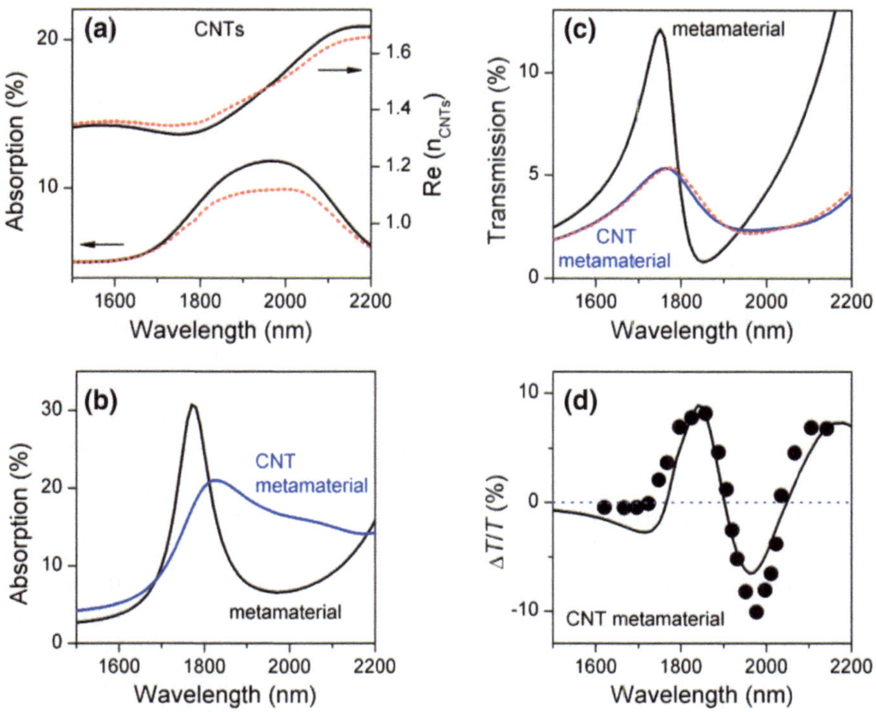

Fig. 10.9 Modelling of nonlinear optical response of the CNT MM. **a** Absorption (left scale) and real part of refractive index (right scale) of the layer of CNTs in the CNT MM at low (black solid line) and high (red dotted line) excitation intensities. **b** Absorption of uncovered MM (black) and the CNT MM (blue) at low intensities. **c** Transmission of uncovered MM (black) and MM covered by CNTs at low (blue solid line) and high (red dotted line) excitation intensities. **d** Comparison of calculated (solid line) and experimental (circles) spectral dispersion of $\Delta T/T$ of the CNT MM [18]

added to simulations representing layer of CNTs on the top of MM. Nonlinear optical response of layer of CNTs which defines nonlinear response of the whole hybrid structure was taken into account in two different spectral dispersions of refractive index of CNT layer corresponding to low (linear limit) and high (non-linear regime) intensities of incident light. Dispersions of the imaginary part of refractive index of CNTs were retrieved from experimental data and dispersions of real part of refractive index were then calculated using Kramers-Kronig relation. For the linear limit (low intensity) experimentally measured absorption spectra of the bare CNTs layer shown on Fig. 10.7b has been used (see also Fig. 10.3). Absorption spectra of the layer of CNTs on plasmonic resonator at high intensity shown in Fig. 10.9a by red dotted line was estimated taking into account linear absorption and differential transmission spectra of the bare CNTs layer (shown in Fig. 10.8), and rescaling factor originating from field enhancement. Resulting transmission spectra of the hybrid structure calculated for low and high incident intensities are shown in Fig. 10.9c by blue solid and red dotted curves, correspondingly. Ratio of these two spectra shown in Fig. 10.9d by solid black line represents simulated nonlinear changes of transmission of the CNT MM. Calculated dispersion of nonlinear optical response of the CNT MM contains spectral regions with positive and negative signs of $\Delta T/T$ and is in excellent agreement with experimental data. We note that two fitting parameters were used in our simulations to achieve good quantitative correspondence with the experimental data. These parameters are average value of the real part of refractive index of CNTs layer across the absorption resonance and rescaling factor defining amplitude of absorption of CNT layer on the plasmonic resonator at high intensity of incident light (in the nonlinear regime). First parameter defines spectral shift and the second one—amplitude of the calculated spectral dependence of $\Delta T/T$ while its shape does not change much with used small variations of fitting parameters.

Discussed above considerations and results of numerical simulations refer to steady state corresponding to the continuous pumping of the CNT MM. To understand experimentally observed nonlinear dynamics it is important to point out the fact that the recombination of excited carriers in CNTs is essentially non-radiative.

From microscope image shown in Fig. 10.7a it is seen that most of the CNTs in the hybrid structure are aggregated in bundles. It is well established that while in the individual (suspended) CNTs efficient band gap photoluminescence (PL) can be observed, in bundled CNTs PL appears to be strongly quenched [25]. The reason for PL quenching in CNTs bundles is suggested to be inter-tube energy transfer from the semiconducting to the metallic nanotubes within each bundle, followed by rapid non-radiative carrier cooling in the metallic nanotubes [25]. It has been shown recently that in a hybrid structure where MM plasmonic resonance is coupled to optically active resonances in PbS quantum dots cavity quantum electrodynamics Purcell effect leads to multifold enhancement of PL intensity [26] and, presumably, to the acceleration of recombination rate in quantum dots. In contrast the rate of non-radiative recombination in the CNTs layer in the CNT MM is not affected by the presence of MM nano-resonator despite the strong resonant coupling between CNTs excitonic and MM plasmonic resonances in the hybrid structure. In the coupled system of the CNT MM energy can be transferred from plasmonic resonator to CNTs. Manifestation of this energy transfer is the observed decrease of saturation intensity of nonlinear optical response in the CNT MM as a result of additional pumping of CNTs layer by local field concentrated in MM. Energy transfer in opposite direction (from CNTs to plasmonic resonator) is negligible as the excitation created in CNTs rapidly dissipates to the thermostat nonradiatively. Non-radiative relaxation of excitation in the layer of CNTs on the timescale of few hundreds of fs however is still much slower than the period of plasmonic oscillations in MM (few fs—corresponding to the light period of the laser pulse) and decay of plasmonic oscillations (few tens of fs as can be estimated from quality factor of plasmonic resonator: $Q \approx 5.5$ for MM covered by CNTs). It is thereby assumed that the relaxation of excitation in CNTs layer adiabatically tunes plasmonic resonance and, as a result, nonlinear dynamics of the CNT MM replicates the dynamics in bare CNTs layer as it is observed in the experiment. We note that from the point of view of applications for ultrafast light modulators bundling of CNTs and quenching of PL can be considered as an advantage. Indeed, reported radiative lifetimes in individual CNTs are typically lying in the range from 10 to 100 ps [19, 20, 27–29] while in bundled CNTs nonradiative relaxation occurs on sub-picosecond scale [14, 15, 19, 21, 23]. Finally, we compare modulation characteristics of the CNT MM with the characteristics of analogous nanoscale sized ultrafast optical switching structures reported previously—Ag/α-Si/Ag fishnet nanostructures [10, 11] and nanostructured gold medium [13]. First, we note that all of the structures (including CNT MM) have comparable thicknesses (100–200 nm) and transmission levels at the wavelength of modulated light (5–15%). As modulation depth depends strongly on the pump fluence we compare pump fluencies required to induce some fixed change of transmission. In Ag/α-Si/Ag nanostructures and the nanostructured gold medium close pump fluencies of 100 μJ/cm^2 [10, 11] and 80 μJ/cm^2 [13], correspondingly, required to induce modulation depth $\Delta T/T = 10\%$. In the CNT MM the same switching ratio is achieved with the order of magnitude lower pump fluence of just 10 μJ/cm^2. Such dramatic decrease of switching fluence is achieved due to: (i) use of CNTs as nonlinear medium where third-order nonlinear susceptibility is significantly enhanced due to quantum confinement of electron-hole motion on one-dimensional space [30] and (ii) concentration of local fields in the plasmonic resonator resulting in additional pumping of nonlinear medium. With increasing pump fluence modulation depth in Ag/α-Si/Ag nanostructures and the nanostructured gold medium increases nearly linearly reaching maximum reported value $\Delta T/T = 70\%$ at 900 μJ/cm^2 [11]. In the CNT MM we observed saturation of the modulation depth with the maximum value estimated as $\Delta T/T = 15\%$. Although saturation of the modulation depth limits switching performance of CNT MM we believe it can be significantly improved by optimizing thickness of CNTs layer and, especially, by using selected semiconducting CNTs [31]. Indeed, value of $\Delta T/T$ in CNT MM is limited by non-saturable absorption in CNTs. Metallic CNTs in the natural mixture of CNTs used in this study do not contribute to the nonlinear optical response but induce additional losses. Accordingly, removing metallic CNTs from the mixture can significantly improve switching characteristics of the CNT MM. Recovery time of the CNT MM (<500 fs) defining switching speed is slightly shorter than recovery times of Ag/α-Si/Ag nanostructures (750 fs [10] and 600 fs [11]) but significantly longer than response time of the nanostructured gold medium ($\sim$40 fs [13]). When comparing nonlinear characteristics of discussed above nanostructures it is important to note that Ag/α-Si/Ag nanostructures can be switched only with the visible pump pulse, necessary to photoexcite carriers above the α-Si band gap, while changes of transmission are probed in the near-IR. This eliminates possibility of using Ag/α-Si/Ag nanostructures as nonlinear optical devices working with the single laser beam. On the contrary, nanostructured gold medium can be used as very efficient optical limiter or saturable absorber, but it's capabilities of light with light modulation are limited by requirement for pump and probe beams to be coherent (this requirement originates from the two-photon nonlinear absorption process employed in the nanostructured gold medium [13]). In contrast, CNT MM can be both pumped and probed with either coherent or frequency shifted beams in the near-IR and it can also work as efficient nonlinear device with a single beam as it was shown in our previous study [1]. From this comparison one can see that optical switching performance of CNT MM has number of significant advantages in comparison with previously reported devices with analogous functionalities.

10.3 Conclusion

An analytical model for describing complex dynamics of a hybrid system consisting of interacting classical and quantum resonant structures has been developed. An application of this model for enhancement of saturation nonlinearity in the system of coupled CNT and MM has been demonstrated. The model adequately reproduces absorption spectrum of the CNT layer alone taking into account inhomogeneous distribution of the nanotubes' absorption lines. The model clearly demonstrates the effect of saturation nonlinearity enhancement due to the presence of the MM, and the dependence of the effect on homogeneity of the CNT layer. It has to be pointed out again that in the presented in part 10.1 analytical model the effect of MM resonance position change power dependence $\omega_0(S)$ is not taken into account and does not appear in the resulting curves in Fig. 10.6 (compare with Figs. 10.8 and 10.9d).

It has been experimentally demonstrated that the CNT MM possesses exceptional nonlinear optical properties in the near-infrared part of the spectrum—high modulation depth on the nanoscale length at low switching power and ultrafast relaxation time, exceeding characteristics of previously reported materials. The CNT MM is spectrally flexible: while in the present study optical switching in the spectral region around 1900 nm has been demonstrated, resonant nonlinear properties of the CNT MM can be easily tuned in the spectral range of 1–2 μm with CNTs of different diameter [32] and appropriately scaling the MM, thus covering the entire second and third optical telecom spectral windows.

References

1. A. Nikolaenko, F. Angelis, S. Boden, N. Papasimakis, P. Ashburn, E. Fabrizio, N. Zheludev, Carbon nanotubes in a photonic metamaterials. PRL **104**, 153902 (2010)
2. A. Chipouline, S. Sugavanam, V.A. Fedotov, A.E. Nikolaenko, Analytical model for active metamaterials with quantum ingredients. J. Opt. **14**, 114005 (2012)
3. A.F. Koenderink, On the use of Purcell factors for plasmon antennas. Opt. Lett. **35**, 4208 (2010)
4. J. Petschulat, C. Menzel, A. Chipouline, C. Rockstuhl, A. Tünnermann, F. Lederer, T. Pertsch, Multipole approach to metamaterials. Phys. Rev. B **78**, 043811 (2008)
5. N. Zheludev, Nonlinear optics on the nanoscale. Contemp. Phys. **43**, 365 (2002)
6. V. Almeida, C. Barrios, R. Panepucci, M. Lipson, All-optical control of light on a silicon chip. Nature **431**, 1081 (2004)
7. K. MacDonald, Z. Samson, M. Stockman, N. Zheludev, Ultrafast active plasmonics. Nature Photon. **3**, 55–58 (2009)
8. N. Zheludev, The road ahead for metamaterials. Science **328**, 582 (2010)
9. W. Padilla, A. Taylor, C. Highstrete, M. Lee, R. Averitt, Dynamical electric and magnetic metamaterial response at terahertz frequencies. PRL **96**, 107401 (2006)
10. D.J. Cho, W. Wu, E. Ponizovskaya, P. Chaturvedi, A.M. Bratkovsky, S.-Y. Wang, X. Zhang, F. Wang, Y.R. Shen, Ultrafast modulation of optical metamaterials. Opt. Express **17**, 17652 (2009)
11. K.M. Dani, Z. Ku, P.C. Upadhya, R.P. Prasankumar, A.J. Taylor, S.R.J. Brueck, Ultrafast nonlinear optical spectroscopy of a dual-band negative index metamaterial all-optical switching device. Opt. Express **19**, 3973 (2011)
12. J. Pendry, A. Holden, D. Robbins, W. Stewart, IEEE Trans. Microw. Theory Tech. **47**, 2075 (1999)
13. M. Ren, B. Jia, J.-Y. Ou, E. Plum, J. Zhang, K.F. MacDonald, A. Nikolaenko, J. Xu, M. Gu, N. Zheludev, Nanostructured plasmonic medium for terahertz bandwidth all-optical switching. Adv. Mater. Published on-line, http://onlinelibrary.wiley.com/doi/10.1002/adma.201103162/abstract
14. Y.-C. Chen, N.R. Raravikar, L.S. Schadler, P.M. Ajayan, Y.-P. Zhao, T.-M. Lu, G.-C. Wang, X.-C. Zhang, Ultrafast optical switching properties of single-wall carbon nanotube polymer composites at 1.55 μm. APL **81**, 975–977 (2002)
15. S. Tatsuura, M. Furuki, Y. Sato, I. Iwasa, M. Tian, H. Mitsu, Semiconductor carbon nanotubes as ultrafast switching materials for optical telecommunications. Adv. Mater. **15**, 534–537 (2003)
16. V.A. Fedotov, M. Rose, S.L. Prosvirnin, N. Papasimakis, N.I. Zheludev, Sharp trapped-mode resonances in planar metamaterials with a broken structural symmetry. PRL **99**, 147401 (2007)
17. B. Luk'yanchuk, N. Zheludev, S. Maier, N. Halas, P. Nordlander, H. Giessen, C. Chong, The Fano resonance in plasmonic nanostructures and metamaterials. Nat. Mater. **9**, 707–715 (2010)
18. A. Nikolaenko, N. Papasimakis, A. Chipouline, F. De Angelis, E. Di Fabrizio, N. Zheludev, THz bandwidth optical switching with carbon nanotube metamaterial. Opt. Express **20**(6), 6068 (2012)
19. L. Huang, H.N. Pedrosa, T.D. Krauss, Ultrafast ground-state recovery of single-walled carbon nanotubes. PRL **93**, 017403 (2004)
20. G. Ostojic, S. Zaric, J. Kono, M. Strano, V. Moore, R. Hauge, R. Smalley, Interband recombination dynamics in resonantly excited single-walled carbon nanotubes. PRL **92**, 117402 (2004)
21. W.B. Cho, J.H. Yim, S.Y. Choi, S. Lee, A. Schmidt, G. Steinmeyer, U. Griebner, V. Petrov, D.-I. Yeom, K. Kim, F. Rotermund, Boosting the non linear optical response of carbon

nanotube saturable absorbers for broadband mode-locking of bulk lasers. Adv. Funct. Mater. **20**, 1937 (2010)

22. T. Hasan, Z. Sun, F. Wang, F. Bonaccorso, P.H. Tan, A.G. Rozhin, A.C. Ferrari, Nanotube-polymer composites for ultrafast photonics. Adv. Mater. **21**, 3874 (2009)
23. R.W. Boyd, *Nonlinear Optics*, 2nd edn. (Academic Press, 2003)
24. K.H. Fong, K. Kikuchi, C.S. Goh, S.Y. Set, R. Grange, M. Haiml, A. Schlatter, U. Keller, Solid-state Er:Yb:glass laser mode-locked by using single-wall carbon nanotube thin film. Opt. Lett. **32**, 38–40 (2007)
25. M. O'Connell, S. Bachilo, C. Huffman, V. Moore, M. Strano, E. Haroz, K. Rialon, P.J. Boul, W. Noon, C. Kittrell, J. Ma, R. Hauge, R. Weisman, R. Smalley, Band gap fluorescence from individual single-walled carbon nanotubes. Science **297**, 593–596 (2002)
26. K. Tanaka, E. Plum, J.Y. Ou, T. Uchino, N. Zheludev, Multi-fold enhancement of quantum dot luminescence in a plasmonic metamaterial. PRL **105**, 227403 (2010)
27. F. Wang, G. Dukovic, L. Brus, T. Heinz, Time-resolved fluorescence of carbon nanotubes and its implication for radiative lifetimes. PRL **92**, 177401 (2004)
28. S. Reich, M. Dworzak, A. Hoffmann, C. Thomsen, M. Strano, Excited-state carrier lifetime in singlewalled carbon nanotubes. Phys. Rev. B **71**, 033402 (2005)
29. P. Avouris, M. Freitag, V. Perebeinos, Carbon-nanotube photonics and optoelectronics. Nat. Photon. **2**, 341–350 (2008)
30. A. Maeda, S. Matsumoto, H. Kishida, T. Takenobu, Y. Iwasa, M. Shiraishi, M. Ata, H. Okamoto, Large optical nonlinearity of femiconducting single-walled carbon nanotubes under resonant excitations. PRL **94**, 047404 (2005)
31. M.C. Hersam, Progress towards monodisperse single-walled carbon nanotubes. Nat. Nanotech. **3**, 387 (2008)
32. H. Kataura, Y. Kumazawa, Y. Maniwa, I. Umezu, S. Suzuki, Y. Ohtsuka, Y. Achiba, Optical properties of single-wall carbon nanotubes. Synth. Met. **103**, 2555 (1999)

Chapter 11
Application of the Model of "Quantum" Metamaterials: Regular and Stochastic Dynamics of Nanolaser (Spaser)

11.1 Introduction

One of the main drawbacks of plasmonic nanostructures, restricting their potential application, is the intrinsic (ohmic) losses caused by the interaction of the free electrons of the metal with thermostat (irreversible losses) and radiative losses. The more localized light is to the metal surface, the more concentrated the plasmonic field fraction is inside the metal resulting in the appearance of higher dissipative losses [1]. Passive losses as a limiting factor was pointed out a rather long time ago [2, 3] and only recently new materials have been suggested in order to mitigate the losses [4–7]. Nevertheless, keeping in mind metal as a main candidate for the plasmonic components, the only way to compensate the losses is to use optically active materials in combination with the nanostructures [8–12].

The nanoresonator changes the radiative properties of the quantum system coupled to the nanoresonator [13, 14] and can cause both enhancement [15, 16] or inhibition [17] of spontaneous emission. Nevertheless, in the case of regular dynamics of a nanolaser, spontaneous emission does not affect the dynamics when operating well above threshold; influence of the Purcell effect on the stimulated emission process is undistinguishable from the field enhancement effect and can be taken into account by appropriate choice of the phenomenological coefficients in the model.

The nanolaser dynamics is based on energy transfer from the excited quantum emitter to the plasmons, and therefore depends strongly on the positioning of the emitters near the nanoresonator [18]. For example, an appropriate positioning of the emitters can enhance generation of bright and suppress generation of dark modes, and vice versa [18, 19]. The emitters appear to be coupled with the plasmonic modes from ones side and with the far field (radiative) modes from another one. Moreover, the radiative losses can exceed dissipative ones by a factor of two [20, 21]. In the model developed here, the radiative losses are included in the damping coefficient γ for a plasmonic mode. An appropriate positioning of the emitters can

A. Chipouline and F. Küppers, *Optical Metamaterials: Qualitative Models*, Springer Series in Optical Sciences 211, https://doi.org/10.1007/978-3-319-77520-3_11

redistribute energy transfer in favor of a plasmonic mode which is more effective near sharp angles by a factor of $(kr)^3$, where k is the wave vector and r is the curvature of the shape of the nanoresonator [22–24]. Nevertheless, positioning the emitters too close to the metallic surface can cause quenching of inversion and so must be additionally avoided [25].

The principles of nanolaser design is suggested and developed in [12, 26–28] and was later experimentally realized in various different configurations [29–38]. First observed in gold nanoparticles (NPs) coated by dye-doped dielectric shells [26], spasing action was reported in hybrid plasmonic waveguides [31], semiconductor quantum dots on metal film [39, 40], plasmonic nanocavities and nanocavity arrays [41–46], and metallic NPs and nanorods [47, 48] and recently was studied in graphene-based structures [49]. Recent achievements in this area are summarized in several review articles [8, 18, 50].

The small spaser size well below the diffraction limit gives rise to numerous promising applications, e.g., in sensing [45] or medical diagnostics [48]. However, most experimental realizations of spaser-based nanolasers were carried out in relatively large systems, while only a handful of experiments reported spasing action in small systems with overall size below 50 nm [30, 48].

Theoretical models of the nanolaser can be approximately divided by fully numerical [51, 52] and semi-analytical [28, 53, 54], with the model developed here belonging to the latter approach. In both versions, the quantum dynamics of the emitters is described by the density matrix method [55] adopted for two, three, or four level schemes. The main difference is in the description of the plasmonic oscillations. In order to make the model treatable (at least to some extent) analytically, the plasmon dynamics has to be reduced to some version of the harmonic oscillator equation, which finally results in the well-known point-like dynamic laser model [56]. This model has been used many times for investigation of the laser dynamics as a self-oscillating system, and as a modeling task for various problems of nonlinear (including stochastic) dynamics, stability analysis, etc. [56–58]. To some extent, this model describes especially well the nanolaser due to the monomode (or two-mode) oscillation [59], while the usual "macro" laser tends to operate in multimode regime. Nevertheless, there are several new tasks, which have not been previously addressed, or have not been addressed fully or in a consistent manner.

The standard threshold condition for gain coupled to a resonance mode [56] needs to be modified in realistic plasmonic systems [28, 60]. For example, it has long been known that fluorescence of a molecule placed sufficiently close to a metal surface is quenched due to the Ohmic losses in the metal [61, 62]. During the past decade, numerous experiments (see [25–51] in [63]) reported fluorescence enhancement by the resonant dipole surface plasmon mode in spherical metal NP that was followed by quenching due to coupling to nonresonant modes as the molecules moved closer to the NP surface [64–66]. Another important factor is the direct dipole dipole interactions between gain molecules which causes random Coulomb shifts of molecules' excitation energies and therefore could lead to the system dephasing [67–69].

It should be emphasized, that the rigorous numerical calculations undoubtedly provide results closer to the experimentally realized data, but at the same time hide the physics of the problem. In order to understand necessity of an analytical treatment, consider the problem of instability of the nanolaser operation under the action of an external field. In case of the numerical approach there is no way to subdivide fields [52], generated by the plasmons and incoming one, while analytical modeling provides clear qualitative explanation of appearance of the unstable regimes [54].

A semi-classical model for plasmonic nanolaser, developed earlier in Chap. 10 (10.15), (10.21) is now used to investigate the regular [without stochastic terms in (10.15) and (10.21)] dynamics (Sect. 11.2), a rigorous multimode model is developed combining quantum dynamics with Green function technique (Sect. 11.3), and the problem of bandwidth of the nanolaser is considered in Sect. 11.4.

There are several options for the nanolaser model. Basically, any laser model consists of the resonator model and the model for active media. Each of them (the resonator and active media models) has two options, namely: single/double or multimode resonator model, and classical or quantum model for active media, see Fig. 11.1. Single mode can be used to describe a wire-like resonator, double mode can be used to describe a coupled double-wire system. These two options (single or double) consists of one or two coupled equations for the resonator dynamics; anyway, either single or double mode resonator models are mathematically the same one. The double mode model is necessary to describe magnetic mode and consequently take into account the magnetic response. The multimode model is

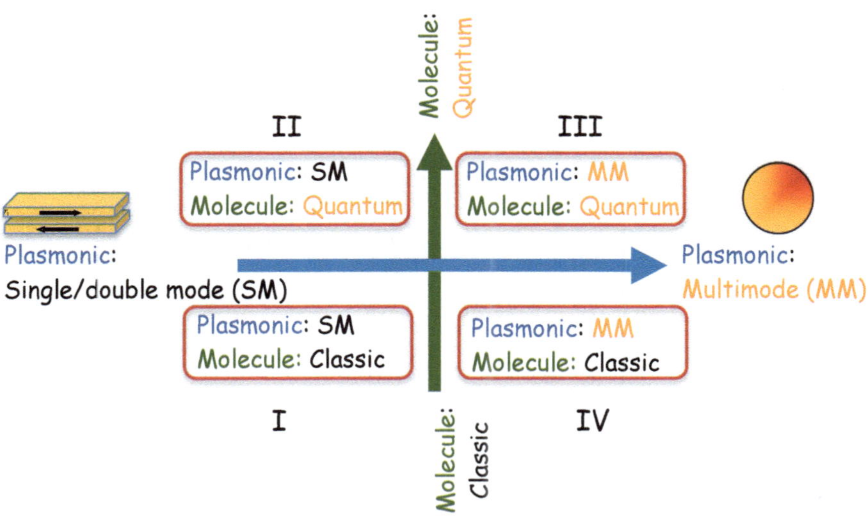

Fig. 11.1 Four possible versions for the model of nanolaser: two options for the resonator modeling (single or multimode resonator), and two options for the active media (classical or quantum)

based on Mie approach (for example, this approach is perfectly suited for the symmetric systems e.g. nano spheres or nano cylinders.

Version I and IV are used very rear due to the fact that the optical amplification (modelling of active media) cannot be described in the frame of classical approach. Nevertheless, some authors still use a harmonic oscillator equation with the positive losses to describe dynamics of an active media.

Version II is the mostly widely used, this approach is assumed in this chapter excepting Sect. 11.3, where a multipole model (plus interaction between active molecules) is presented.

11.2 Regular Spaser Dynamics

As it has been mentioned above, the elaborated set of (10.15) is well known in laser dynamics and has been widely used to model different effects in nonlinear dynamics. System (10.15) in stationary operation mode (after long enough time after inversion is switched on):

$$\begin{cases} \tilde{\rho}_{12} = \frac{i\alpha_x \tau_2 \tilde{x}^* N}{\hbar R_\rho} \\ N - N_0 = \frac{i\alpha_x \tau_1 \left(\tilde{x}\tilde{\rho}_{12} - \tilde{x}^* \tilde{\rho}_{12}^*\right)}{2\hbar} \\ \tilde{x} = \frac{\alpha_\rho \tilde{\rho}_{12}^*}{R_x} \\ R_\rho = (1 + i(\omega - \omega_{21})\tau_2) \\ R_x = \left(\omega_0^2 - \omega^2 - 2i\omega\gamma\right) \end{cases} \tag{11.1}$$

possesses an analytical stable-state solution:

$$\begin{cases} \omega_G = \frac{\omega_{21}\gamma\tau_2 + \sqrt{(\omega_{21}\gamma\tau_2)^2 + \omega_0^2(1 + 2\gamma\tau_2)}}{(1 + 2\gamma\tau_2)} \\ N_{STAT} = \frac{\hbar\left((\omega_G - \omega_{21})\left(\omega_0^2 - \omega_G^2\right)\tau_2 + 2\gamma\omega_G\right)}{2\alpha_x \alpha_\rho \tau_2} \\ |\tilde{x}|_{STAT}^2 = \frac{(N_0 - N_{STAT})}{N_{STAT}} |x|_S^2 \\ |x|_S^2 = \frac{\hbar^2\left(1 + (\omega_G - \omega_{21})^2 \tau_2^2\right)}{\alpha_x^2 \tau_1 \tau_2} \\ |\tilde{\rho}_{12}|_{STAT}^2 = \frac{\tau_2}{\tau_1} N_{STAT}(N_0 - N_{STAT}) \end{cases} \tag{11.2}$$

Physically it means, that the spaser adjusts its own generation frequency ω_G according to the parameters of the system (but NOT according to the pump), and finally stabilizes itself by adjusting oscillation amplitude $|\tilde{x}|_{STAT}^2$. The pump threshold is given by evident requirement $|\tilde{x}|_{STAT}^2 > 0$:

$$N_{0,\text{threshold}} > N_{\text{STAT}} \Rightarrow N_0 > \frac{\hbar\left((\omega_{\text{G}} - \omega_{21})\left(\omega_0^2 - \omega_{\text{G}}^2\right)\tau_2 + 2\gamma\omega_{\text{G}}\right)}{2\alpha_x\alpha_\rho\tau_2} \tag{11.3}$$

In the case of full resonance $\omega_{\text{G}} = \omega_0 = \omega_{12}$ stationary parameters of generation become:

$$\begin{cases} \omega_{\text{G}} = \omega_0 = \omega_{21} \\ N_{\text{STAT}} = \frac{\hbar\gamma\omega_0}{\alpha_x\alpha_\rho\tau_2} \\ |\tilde{x}|^2_{\text{STAT}} = \frac{\alpha_x\alpha_\rho\tau_2}{\hbar\gamma\omega_0}\left(N_0 - \frac{\hbar\gamma\omega_0}{\alpha_x\alpha_\rho\tau_2}\right)|x|^2_{\text{S}} \\ |x|^2_{\text{S}} = \frac{\hbar^2}{4\alpha_x^2\tau_1\tau_2} \\ |\tilde{\rho}_{12}|^2_{\text{STAT}} = \frac{\hbar\gamma\omega_0}{\alpha_x\alpha_\rho\tau_1}\left(N_0 - \frac{\hbar\gamma\omega_0}{\alpha_x\alpha_\rho\tau_2}\right) \end{cases} \tag{11.4}$$

To investigate transient processes of the dynamical transition to the stationary regime, numerical methods must be used to solve the system (10.15). A solution of this system (without stochastic terms) is presented in Fig. 11.2.

More complicated regular dynamics demonstrate multipole nanolaser (multipole spaser) action. As expected, in this case both bright and dark oscillation modes can be excited. Nevertheless, excitation of dark modes is significantly affected by the placement of the emitters around the nanoresonators, for example around double wires structure. It is intuitively clear, that in order to generate an anisymmetric (dark) mode the symmetry of the structure has to be broken, and not all variants of the emitter positioning will be appropriate for the dark mode generation. Three different layouts of the emitters in the vicinity of the nanoresonators are shown in Fig. 11.3, where gold strips represent nanowires and red circles represent emitters with the resonant wavelengths corresponding to bright or dark modes respectively. The layouts shown in Fig. 11.3a, b do not possess the dark (antisymmetric) modes, provided there are no some external factors breaking the symmetry.

The statement above can be rather straightforwardly proven in the framework of the model developed here. Following the same logic as in the elaboration of (10.21), for the three layouts in Fig. 11.3 the respective system of equations are:

$$\begin{cases} \frac{\mathrm{d}\tilde{\rho}_{12}^{(1)}}{\mathrm{d}t} + \tilde{\rho}_{12}^{(1)}\left(\frac{1}{\tau_2} + i(\omega - \omega_{21})\right) = \frac{i\alpha_x(m_{\text{s}} + m_{\text{a}})^* N^{(1)}}{4\hbar} \\ \frac{\mathrm{d}N^{(1)}}{\mathrm{d}t} + \frac{\left(N^{(1)} - N_0\right)}{\tau_1} = \frac{i\alpha_x\left((m_{\text{s}} + m_{\text{a}})\tilde{\rho}_{12}^{(1)} - (m_{\text{s}} + m_{\text{a}})^*\tilde{\rho}_{12}^{(1)*}\right)}{2\hbar} \\ \frac{\mathrm{d}\tilde{\rho}_{12}^{(2)}}{\mathrm{d}t} + \tilde{\rho}_{12}^{(2)}\left(\frac{1}{\tau_2} + i(\omega - \omega_{21})\right) = \frac{i\alpha_x(m_{\text{s}} - m_{\text{a}})^* N^{(2)}}{4\hbar} \\ \frac{\mathrm{d}N^{(2)}}{\mathrm{d}t} + \frac{\left(N^{(2)} - N_0\right)}{\tau_1} = \frac{i\alpha_x\left((m_{\text{s}} - m_{\text{a}})\tilde{\rho}_{12}^{(2)} - (m_{\text{s}} - m_{\text{a}})^*\tilde{\rho}_{12}^{(2)*}\right)}{2\hbar} \\ 2(\gamma - i\omega)\frac{\mathrm{d}m_{\text{s}}}{\mathrm{d}t} = 2\alpha_\rho\left(\rho_{12}^{(1)} + \rho_{12}^{(2)}\right) - \left(\omega_0^2 - \omega^2 - 2i\gamma\omega - \sigma\right)m_{\text{s}} \\ 2(\gamma - i\omega)\frac{\mathrm{d}m_{\text{a}}}{\mathrm{d}t} = 2\alpha_\rho\left(\rho_{12}^{(1)} - \rho_{12}^{(2)}\right) - \left(\omega_0^2 - \omega^2 - 2i\gamma\omega - \sigma\right)m_{\text{a}} \end{cases} \tag{11.5a}$$

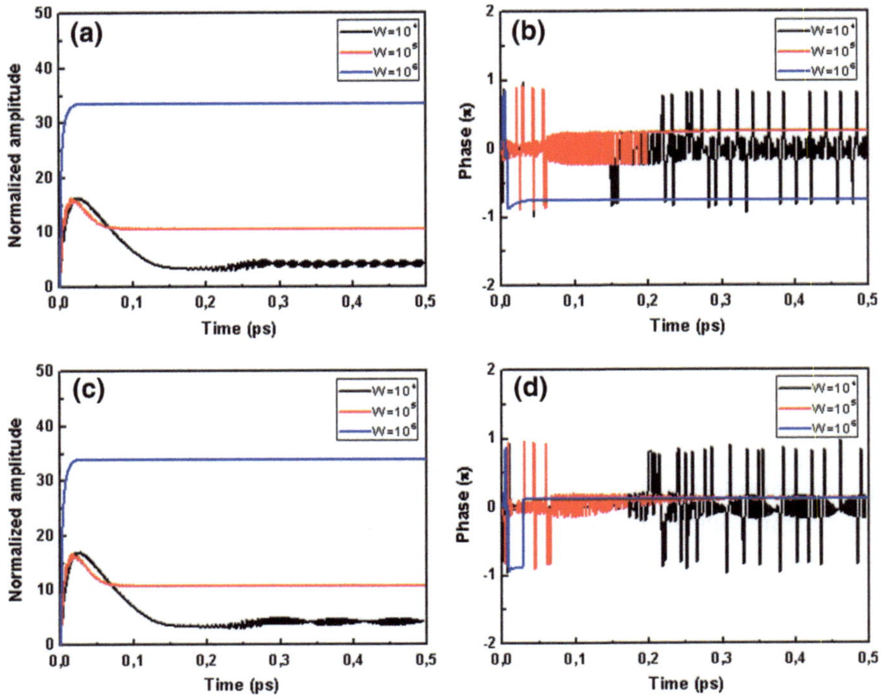

Fig. 11.2 Transient processes at the switching on the generation of the spaser for **a**, **b** resonant case $\omega_G = \omega_0 = \omega_{12}$ and **c**, **d** non resonant case $\omega_{12} < \omega_G < \omega_p$. Parameters are chosen to provide stable generation ($N_0 > N_{0,\text{threshold}}$); dark, red, and blue lines represent dynamics for increased pump. Phase of the oscillation is calculated with respect to one of the non-diagonal elements of the density matrix ρ_{12}

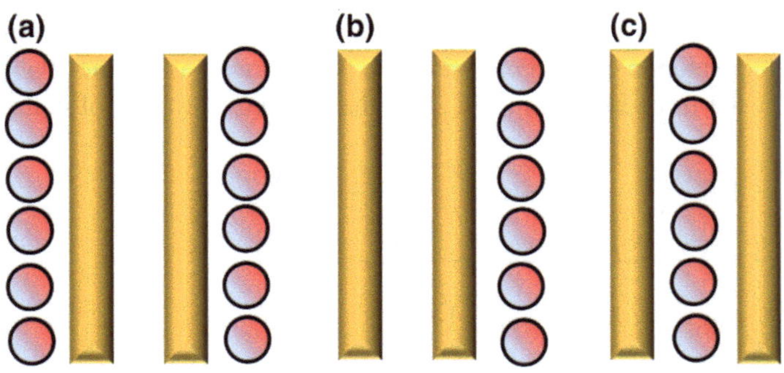

Fig. 11.3 Three possible emitter positioning around the nanoresonators. Red circles represent emitters with resonant wavelengths corresponding to dark (antisymmetric) or bright (symmetric) resonances respectively. Only one layout **b** appears to be fully compatible with the dark mode oscillation

$$\begin{cases} \frac{d\tilde{\rho}_{12}}{dt} + \tilde{\rho}_{12}\left(\frac{1}{\tau_2} + i(\omega - \omega_{21})\right) = \frac{i\alpha_x\left(m_s^* + m_a^*\right)N}{2\hbar} \\ \frac{dN}{dt} + \frac{(N - N_0)}{\tau_1} = \frac{i\alpha_x\left(\left(m_s + m_a\right)\tilde{\rho}_{12} - \left(m_s^* + m_a^*\right)\tilde{\rho}_{12}^*\right)}{4\hbar} \\ 2(\gamma - i\omega)\frac{dm_s}{dt} + \left(\omega_0^2 - \omega^2 - 2i\omega\gamma + \sigma\right)m_s = \alpha_\rho\tilde{\rho}_{12}^* \\ 2(\gamma - i\omega)\frac{dm_a}{dt} + \left(\omega_0^2 - \omega^2 - 2i\omega\gamma - \sigma\right)m_a = \alpha_\rho\tilde{\rho}_{12}^* \end{cases} \quad (11.5b)$$

$$\begin{cases} \frac{d\rho_{12}^{(1)}}{dt} + \tilde{\rho}_{12}^{(1)}\left(\frac{1}{\tau_2} + i(\omega - \omega_{21})\right) = \frac{i\alpha_x m_s^* N^{(1)}}{2\hbar} \\ \frac{dN}{dt} + \frac{\left(N^{(1)} - N_0\right)}{\tau_1} = \frac{i\alpha_x\left(m_s\tilde{\rho}_{12}^{(1)} - m_s^*\tilde{\rho}_{12}^{(1)*}\right)}{\hbar} \\ 2(\gamma - i\omega)\frac{dm_s}{dt} = 4\alpha_\rho\rho_{12}^{(1)} - \left(\omega_0^2 - \omega^2 - 2i\gamma\omega - \sigma\right)m_s \\ 2(\gamma - i\omega)\frac{dm_a}{dt} = -\left(\omega_0^2 - \omega^2 - 2i\gamma\omega - \sigma\right)m_a \end{cases} \quad (11.5c)$$

It can be seen for the case of (11.5c) that the antisymmetric mode is not excited and in the case of (11.5a) the antisymmetric mode is strongly suppressed provided $\rho_{12}^{(1)} \approx \rho_{12}^{(2)}$. Note that both cases (11.5a) and (11.5c) (and layouts in Fig. 11.2a, c respectively) can most easily be realized by a hypothetical bulk MM, where the layers of the emitters are placed between two sets of nanoresonators on either side. This results in the suppression of dark mode generation and consequently an inhibition of the magnetic response of the MMs [70]. In this case the only mechanism braking symmetry is the retardation of the propagating wave between the nanowires [are not taken into consideration in (11.5a–11.5c)] [71]. It is worth noting that to the best of our knowledge, the conclusion regarding the necessity of asymmetric emitter's placement has not been mentioned previously in the literature. On the other hand, the layout in Fig. 11.3b is the typical format for samples with the nanoresonators on the substrate, where the metallic surfaces are functionalized with the emitters, or a polymer emitter solution, spin coated on the top of the metallic nanoresonators. The stationary state of (11.5b) is:

$$\begin{cases} \tilde{\rho}_{12} = \frac{i\alpha_x\tau_2\left(m_s^* + m_a^*\right)N}{2\hbar R_\rho} \\ N - N_0 = \frac{i\alpha_x\tau_1\left(\left(m_s + m_a\right)\tilde{\rho}_{12} - \left(m_s^* + m_a^*\right)\tilde{\rho}_{12}^*\right)}{4\hbar} \\ m_s = \frac{\alpha_\rho\tilde{\rho}_{12}^*}{R_{x,s}} \\ m_a = \frac{\alpha_\rho\tilde{\rho}_{12}^*}{R_{x,a}} \\ R_\rho = \left(1 + i(\omega - \omega_{21})\tau_2\right) \\ R_{x,s} = \left(\omega_0^2 - \omega^2 - 2i\omega\gamma + \sigma\right) \\ R_{x,a} = \left(\omega_0^2 - \omega^2 - 2i\omega\gamma - \sigma\right) \end{cases} \quad (11.6)$$

$$\begin{cases} \mathrm{Re}\left[\frac{R_\rho R^*_{x,s}R^*_{x,a}}{R^*_x}\right] = 0 \Rightarrow \omega_G \\ N_{STAT} = \frac{-i\hbar}{\alpha_x\alpha_\rho\tau_2}\left[\frac{R_\rho R^*_{x,s}R^*_{x,a}}{R^*_x}\right]\Big|_{\omega=\omega_G} \\ |x|^2_S = \frac{\hbar^2\left(1+(\omega_G-\omega_{21})^2\tau_2^2\right)}{\alpha_x^2\tau_1\tau_2} \\ |\tilde{\rho}_{12}|^2_{STAT} = \frac{\tau_2}{\tau_1}N_{STAT}(N_0 - N_{STAT}) \\ |\overline{m}_a + \overline{m}_s|^2_{STAT} = \frac{(N_0-N_{STAT})}{N_{STAT}}|x|^2_S \\ |\overline{m}_s|^2_{STAT} = \frac{|\tilde{\rho}_{12}|^2_{STAT}\alpha_\rho^2}{|R_{x,s}|^2} \\ |\overline{m}_a|^2_{STAT} = \frac{|\tilde{\rho}_{12}|^2_{STAT}\alpha_\rho^2}{|R_{x,a}|^2} \end{cases} \quad (11.7)$$

It is interesting to note that the relation between bright and dark mode intensities $\frac{|\overline{m}_s|^2_{STAT}}{|\overline{m}_a|^2_{STAT}} = \frac{|R_{x,a}|^2}{|R_{x,s}|^2}$ does not depend on the pump since the phase of the oscillations of the both modes are undetermined as in the case of the dipole-like spaser. Dynamics of the multipole spaser, including the transient part, is shown in Fig. 11.4.

Dynamics of the dipole-like spaser under the influence of an external field requires deeper consideration. In this case, competition between its own dynamics and the driving force, due to an external field, can result in unstable operation [54]. It has to be emphasized, that this case corresponds to hypothetically realizable MMs. This MM consists of MAs, considered in this chapter, where an external wave, propagating in the MMs plays a role of the external driving force.

This effect appears only when coupling between emitters and MAs is significant; otherwise the plane wave will be just amplified by the free emitters, which are not coupled to the MAs and no instabilities will be observed [72]. A model which does not distinguish between gains provided by free and coupled emitters can lead to wrong conclusions about the inability of loss compensation without crossing the threshold for the spaser generation [28]. In order to investigate the effect of competition between the generated fields and externally driving forces, a system (10.14) without stochastic terms is used for the dipole-like MA:

$$\begin{cases} \frac{d\tilde{\rho}_{12}}{dt} + \tilde{\rho}_{12}\left(\frac{1}{\tau_2} + i(\omega - \omega_{21})\right) = \frac{i\alpha_x\tilde{x}^*N}{2\hbar} + \frac{i\mu_{QS}A^*N}{2\hbar} \\ \frac{dN}{dt} + \frac{(N-N_0)}{\tau_1} = \frac{i\alpha_x\left(\tilde{x}\tilde{\rho}_{12}-\tilde{x}^*\tilde{\rho}^*_{12}\right) + i\mu_{QS}\left(A\tilde{\rho}_{12}-A^*\tilde{\rho}^*_{12}\right)}{\hbar} \\ 2(\gamma - i\omega)\frac{d\tilde{x}}{dt} + \left(\omega_0^2 - \omega^2 - 2i\omega\gamma\right)\tilde{x} = 2\alpha_\rho\tilde{\rho}^*_{12} + \chi A \end{cases} \quad (11.8)$$

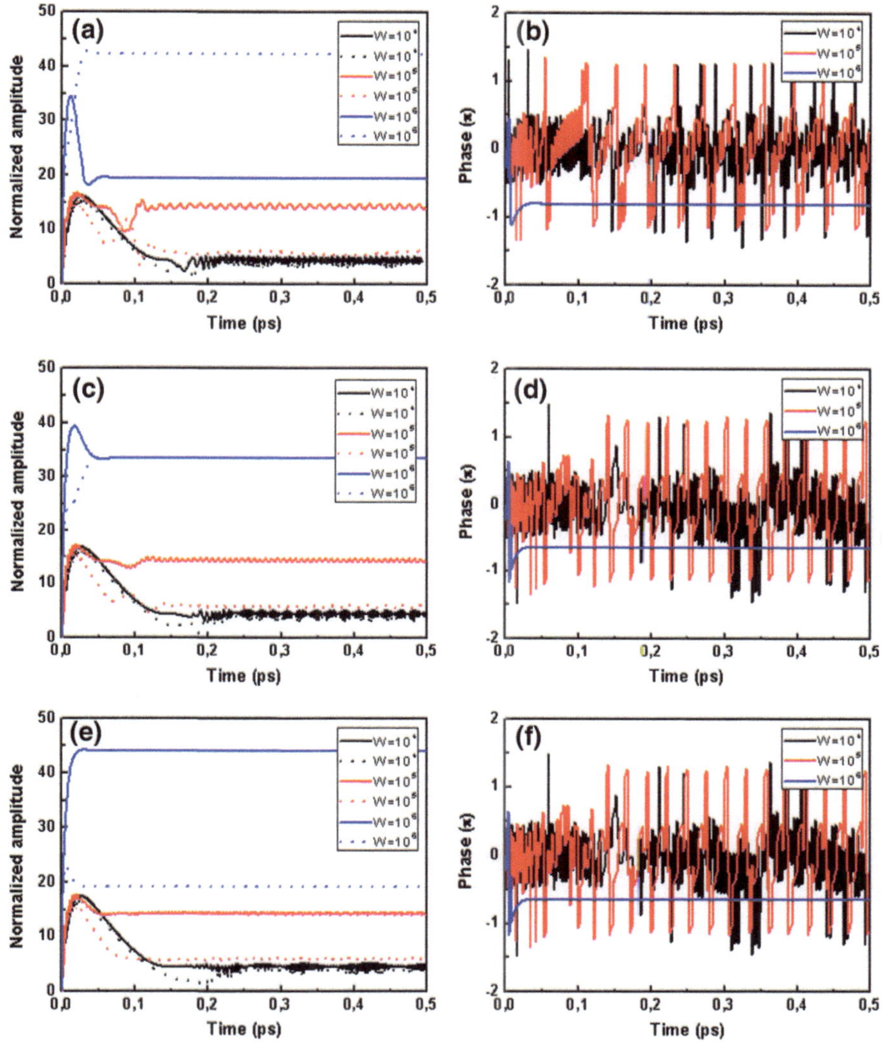

Fig. 11.4 Transient processes at the switching on the generation of the multipole spaser (solid—symmetric and dotted—asymmetric modes) at the center luminescent of the emitters coinciding with **a**, **b** center frequency of a single oscillator $\omega_0 = \omega_{12} = \omega_G$, **c**, **d** symmetric mode $\omega_p < \omega_G < \omega_s = \sqrt{\omega_0^2 + \sigma} = \omega_{21}$, and **e**, **f** asymmetric mode $\omega_a = \sqrt{\omega_0^2 - \sigma} = \omega_{21} < \omega_G < \omega_p$. Parameters are chosen to provide stable generation ($N_0 > N_{0,\text{threshold}}$); black, red, and blue lines represent dynamics for increased pump respectively $N_0(\text{black}) < N_0(\text{red}) < N_0(\text{blue})$. Phase of the oscillation is calculated with respect to one for non-diagonal element of the density matrix ρ_{12}

For the stationary state operation it becomes:

$$\begin{cases} \tilde{\rho}_{12} = \frac{iN\tau_2}{2\hbar R_\rho}\left(\alpha_x \tilde{x}^* + \mu_{QS}A^*\right) \\ N - N_0 = \frac{i\tau_1}{\hbar}\left(\alpha_x\left(\tilde{x}\tilde{\rho}_{12} - \tilde{x}^*\tilde{\rho}_{12}^*\right) + \mu_{QS}\left(A\tilde{\rho}_{12} - A^*\tilde{\rho}_{12}^*\right)\right) \\ \tilde{x} = \frac{\left(2\alpha_\rho\tilde{\rho}_{12}^* + \chi A\right)}{R_x} \\ R_\rho = \left(1 + i(\omega - \omega_{21})\tau_2\right) \\ R_x = \left(\omega_0^2 - \omega^2 - 2i\omega\gamma\right) \end{cases} \tag{11.9}$$

In (11.9) the external field A is assumed to be given and serves as an external driving force for the spaser, which reduces the problem to the well-known one of laser synchronization by an external field [58].

A qualitative analysis of the dynamics follows shortly. It is clear from (11.4) that the spaser itself does not set the phase of its own oscillations, and even a small external resonant wave can potentially force the spaser to generate in phase with the external field, provided the intensity of the external field does not affect the dynamics. As the intensity of the external field increases, the dynamics of the spaser itself starts to compete with the dynamics caused by the external driving forces and instability occurs. Note that if the frequency of the external field does not coincide with that of the generated field, then the spaser frequency shifts to the frequency of the external field. If the external field intensity becomes high enough to overcome the spaser dynamics, stable generation appears again, as described in [54].

From the math's point of view, system (11.9) possesses several solutions for the same set of parameters resulting in instability and in contrast to (11.4), gives a solution for the phase as well [(11.4) gives solutions for the respective amplitudes only]. The stability diagram, for the case of resonant operation $\omega_0 = \omega_{12}$ is presented in Fig. 11.4. Note, that it is supposed that the spaser oscillates at the frequency of the external field, which is chosen to coincide with both resonant frequencies $\omega = \omega_0 = \omega_{12}$. It can be seen, that stable operation is possible in two regions of parameters (below the surface in Fig. 11.5a and above the surface in Fig. 11.5b), while between the surfaces laser operation is unstable.

To the best of my knowledge, the similar stability analysis remains to be done for the multipole spaser with the external field (11.10). Here the initial results of the analysis are given, but the problem requires further investigation. Basically, in the case of generation of a purely bright or purely dark mode, system (11.10) is equivalent to (11.8) for dipole-like MA. When the both modes are excited at the same time, the competition between them adds more complexity to the dynamics and the stability analysis becomes non trivial. Full stability analysis of the multipole spaser is not included in this work and will be presented elsewhere, in the future. Preliminary result of the stability diagram is given in Fig. 11.6.

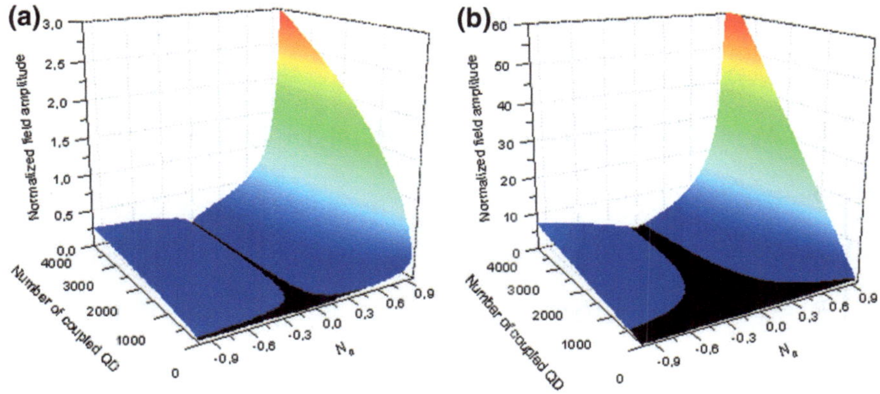

Fig. 11.5 Stability diagram of the dipole-like spaser driven by an external resonant field $\omega = \omega_0 = \omega_{12}$. The surfaces in **a** and **b** are separated due to the different range values. The stability region in **a** is below the surface, while stability region in **b** is above the surface; the instability region is between the surfaces (**a**) and (**b**)

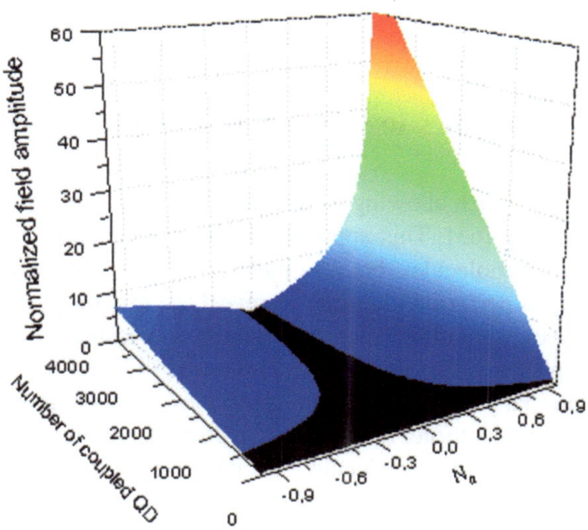

Fig. 11.6 Stability diagram of the multipole spaser driven by an external field which is chosen to be in resonance with eigen plasmonic frequency: $\omega_{\text{as}} < \omega_{\text{external}} < \omega_0 = \omega_{21} < \omega_{\text{s}}$

$$\begin{cases} \frac{d\tilde{\rho}_{12}}{dt} + \tilde{\rho}_{12}\left(\frac{1}{\tau_2} + i(\omega - \omega_{21})\right) = \frac{i\alpha_x\left(m_{\text{s}}^* + m_{\text{a}}^*\right)N}{2\hbar} + \frac{i\mu_{\text{QS}}A^*N}{2\hbar} \\ \frac{dN}{dt} + \frac{(N - N_0)}{\tau_1} = \frac{i\alpha_x\left((m_{\text{s}} + m_{\text{a}})\tilde{\rho}_{12} - \left(m_{\text{s}}^* + m_{\text{a}}^*\right)\tilde{\rho}_{12}^*\right) + 2i\mu_{\text{QS}}\left(A\tilde{\rho}_{12} - A^*\tilde{\rho}_{12}^*\right)}{4\hbar} \\ 2(\gamma - i\omega)\frac{dm_{\text{s}}}{dt} + \left(\omega_0^2 - \omega^2 - 2i\omega\gamma + \sigma\right)m_{\text{s}} = \alpha_\rho\tilde{\rho}_{12}^* \\ 2(\gamma - i\omega)\frac{dm_{\text{a}}}{dt} + \left(\omega_0^2 - \omega^2 - 2i\omega\gamma - \sigma\right)m_{\text{a}} = \alpha_\rho\tilde{\rho}_{12}^* \end{cases} \tag{11.10}$$

Stable-state system for the multipole spaser is:

$$\begin{cases} \tilde{\rho}_{12} = \frac{iN}{2\hbar R_\rho}\left(\alpha_x\left(m_{\rm s}^* + m_{\rm a}^*\right) + \mu_{\rm QS}A^*\right) \\ N - N_0 = \frac{i\tau_1}{4\hbar}\left(\alpha_x\left((m_{\rm s} + m_{\rm a})\tilde{\rho}_{12} - \left(m_{\rm s}^* + m_{\rm a}^*\right)\tilde{\rho}_{12}^*\right) + 2i\mu_{\rm QS}\left(A\tilde{\rho}_{12} - A^*\tilde{\rho}_{12}^*\right)\right) \\ m_{\rm s} = \frac{\alpha_\rho \tilde{\rho}_{12}^*}{R_{x,{\rm s}}} \\ m_{\rm a} = \frac{\alpha_\rho \tilde{\rho}_{12}^*}{R_{x,{\rm a}}} \\ R_\rho = (1 + i(\omega - \omega_{21})\tau_2) \\ R_{x,{\rm s}} = \left(\omega_0^2 - \omega^2 - 2i\omega\gamma + \sigma\right) \\ R_{x,{\rm a}} = \left(\omega_0^2 - \omega^2 - 2i\omega\gamma - \sigma\right) \end{cases} \tag{11.11}$$

It turns out, that for the case of the multipole spaser the stability region is above the surface, while under the surface all solutions are unstable. In contrast to the dipole spaser, only one surface appears as a result of the stability analysis.

11.3 Spaser Dynamics in Case of Multimode Generation

In this chapter an approach allowing us to consider multimode resonator rigorously is presented following [63]. This approach includes interaction not only between the modes and active molecules, but also interaction between the active molecules. A numerical study is performed in order to reveal the role of quenching and direct interactions between gain molecules in reaching the lasing threshold for small spherical Nano spheres (NPs) with a metal core and a doped dielectric shell. It is shown that for the large number of gain molecules, the coupling to nonresonant modes plays no significant role. In contrast, the direct dipole-dipole interactions, by causing random shifts in gain molecules' excitation energies, can hinder reaching the lasing threshold in small NP-based spasers. A semiclassical approach that combines Maxwell-Bloch equations with the Green's function formalism is used to derive the threshold condition in terms of exact system eigenstates, which were found numerically.

A composite spherical NP with a metallic core of radius R_c and dielectric shell of thickness h is considered. The shell is doped with M active molecules at random positions $\mathbf{r}_j$ (see inset in Fig. 11.7).

For small NPs one can use the quasistatic approximation for electromagnetic fields. Within the semiclassical approach, the gain molecules are described by pumped two-level systems, with resonant frequency ω_{21}, while electromagnetic fields are treated classically. Each molecule is characterized by the polarization and population inversion N_j is the density matrix for the jth molecule. In the rotating

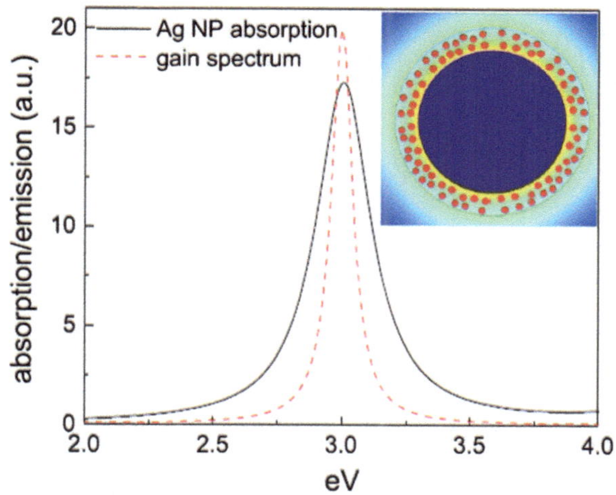

Fig. 11.7 Normalized spectra for a spherical Ag NP with radius R = 5 nm and gain molecule with maximum tuned to plasmon resonance. Inset: Schematics of a composite NP with Ag core and dielectric shell doped with M active molecules [63]

wave approximation, the steady-state molecule dynamics is described by optical Bloch equations (see Chap. 9):

$$\begin{cases} \frac{d\rho_{12}}{dt} + \rho_{12}\left(\frac{1}{\tau_2} + i(\omega - \omega_{21})\right) = \frac{i\mu A^* N}{\hbar} \\ \frac{dN}{dt} + \frac{(N-N_0)}{\tau_1} = \frac{i\mu\left(A\rho_{12} - A^*\rho_{12}^*\right)}{2\hbar} \end{cases} \tag{11.12}$$

Here $A(\vec{r}_j)$ is the slow amplitude of the local field at the point of the jth molecule. The local field $A(\vec{r}_j)$ is created by all molecular dipoles in the presence of a NP and satisfies the Maxwell equation

$$\Delta A(\vec{r}_j) + \frac{\omega^2}{c^2}\varepsilon(\vec{r},\omega)A(\vec{r}_j) = \frac{4\pi\omega^2\mu}{c^2}\sum_j \rho_j\delta(\vec{r} - \vec{r}_j) \tag{11.13}$$

where $\varepsilon(\vec{r},\omega)$ is the local dielectric function given by metal, shell, and outside dielectric functions in the corresponding regions and $\mu\rho_j$ is the molecule dipole moment. The solution of (11.13) has the form:

$$A(\vec{r}) = A_0(\vec{r}) + \frac{4\pi\omega^2\mu}{c^2}\sum_j G(\vec{r},\vec{r}_j)\rho_j \tag{11.14}$$

where $A_0(\vec{r})$ is a solution of the homogeneous part of (11.13) (i.e., in the absence of molecules) and $G(\vec{r},\vec{r}_j)$ is the Green's dyadic in the presence of a NP. After expressing the polarization in terms of local fields using (11.12) and then eliminating the local fields using (11.14), system (11.12) takes the form

$$\sum_{k=1}^{M}\left[\left(\frac{1}{\tau_2}+i(\omega-\omega_{21})\right)\delta_{jk}-N_jD_{jk}\right]\rho_k=\mu A_0(\vec{r}_j)$$
$$N_j-N_0+4\tau_1\mathrm{Im}\sum_{k=1}^{M}\left[\rho_j^*D_{jk}\rho_k\right]=\frac{4\mu\tau_1}{\hbar}\mathrm{Im}\left[\rho_jA_0^*(\vec{r}_j)\right] \tag{11.15}$$

where δ_{jk} and $D_{jk}(\omega)$ are, respectively, the Kronecker symbol and frequency-dependent coupling matrix in the configuration space given by

$$D_{jk}=\frac{4\pi\omega^2\mu^2}{c^2\hbar}G(\vec{r},\vec{r}_j) \tag{11.16}$$

Equations (11.15) and (11.16) constitute the model for active molecules near a plasmonic NP. For a sufficiently high pump rate (sufficiently high N_0), spasing action is possible provided that losses are compensated [12, 26]. The collective system eigenstates defined by the homogeneous part of system (11.15) is of interest,

$$\sum_{k=1}^{M}\left[\left(\frac{1}{\tau_2}+i(\omega-\omega_{21})\right)\delta_{jk}-N_jD_{jk}\right]\rho_k=0$$
$$N_j-N_0+4\tau_1\mathrm{Im}\sum_{k=1}^{M}\left[\rho_j^*D_{jk}\rho_k\right]=0 \tag{11.17}$$

Following the procedure employed previously for studying plasmon-mediated cooperative emission (Pustovit 2009), [73] eigenstates $|J\rangle$ of the coupling matrix D are introduced as

$$D|J\rangle=\Lambda_j|J\rangle,\quad \Lambda_j=\Lambda_j'+i\Lambda_j'' \tag{11.18}$$

where Λ_j' and Λ_j'' are, respectively, real and imaginary parts of system eigenvalues $\Lambda_j'+i\Lambda_j''$ which represent the frequency shift and decay rate of an eigenstate $|J\rangle$. We now introduce *collective* variables for polarization and population inversion as

$$\rho_j=\sum_{k=1}^{M}\langle\bar{J}\mid j\rangle\rho_j,$$
$$N_{JJ'}=\sum_{k=1}^{M}\langle\bar{J}\mid j\rangle N_j\langle j\mid J'\rangle \tag{11.19}$$

where, to ensure the orthonormality, we used the eigenstates $|\bar{J}\rangle$ of complex-conjugate matrix D_{jk} corresponding to the advanced Green's function of (11.13). Multiplying the first equation of system (11.17) by $\langle\bar{J}\mid.j\rangle$ and then summing both equations over j, system (11.17) in the basis of collective eigenstates takes the form

$$\sum_{J'=1}^{M}\left[\left(\frac{1}{\tau_2}+i(\omega-\omega_{21})\right)\delta_{JJ'}-N_{JJ'}\Lambda_{J'}\right]\rho_{J'}=0$$
$$N_{\sum 0}-N+4\tau_1\mathrm{Im}\sum_{J=1}^{M}\left[\Lambda_J''|\rho_J|^2\right]=0 \tag{11.20}$$

where $N=\sum_{j=1}^{M}N_j$ is the ensemble population inversion and $N_{\sum 0}=N_0M$. The mixing of collective states J through $N_{JJ'}$ originates from the inhomogeneity of the N_j distribution for individual molecules. In the following, we assume that, for a sufficiently large ensemble, this inhomogeneity is weak and adopt $N_{JJ'}=\delta_{JJ'}\overline{N}$, where $\overline{N}=N/M$ is the *average* population inversion per molecule. Note that, in this approximation, the individual molecule polarizations ρ_j are still random due to the molecules' spatial distribution. The first equation of system (11.19) then yields the characteristic equation for each state

$$\left(\frac{1}{\tau_2}+i(\omega_g-\omega_{21})\right)-\overline{N}\Lambda_J(\omega_g)=0 \tag{11.21}$$

implying that each eigenstate acquires self-energy $\overline{N}\Lambda_J(\omega_g)$ due to the interactions of molecules with the NP and each other, ω_g is the laser frequency generation. The resonance frequency of mode J is determined by the real part of (11.21)

$$\omega_g=\omega_{21}+\overline{N}\Lambda_J(\omega_g) \tag{11.22}$$

while its imaginary part,

$$\overline{N}\tau_2\Lambda_J''(\omega_g)=1 \tag{11.23}$$

determines $\overline{N}$ and, in fact, represents the lasing threshold condition. Eliminating $\overline{N}$ we obtain the equation for resonance frequency ω_g,

$$\tau_2(\omega_g-\omega_{21})=\Lambda_J''(\omega_g)/\Lambda_J''(\omega_g) \tag{11.24}$$

Equations (11.22)–(11.24) are valid for any plasmonic system with weak inhomogeneity of gain population inversion. For the spherical core-shell NP that we consider, the plasmon modes are characterized by angular momentum l and by well separated frequencies ω_l. However, each system eigenstate $|J\rangle$ contains, in general, contributions from all l since NP spherical symmetry is broken down by the random distribution of molecules within the shell. In order to establish the relation of our model to a conventional spaser description [12, 26, 27] let us assume for now a largely homogeneous spatial distribution of molecules in the shell and disregard the effects of direct dipole-dipole interactions. This could be considered one extreme of real systems where dyes do not interact due to mutual orientation and distribution.

In this case, the eigenstates $|J\rangle$ are dominated by molecules' coupling with the lth plasmon mode and can be labeled as Λ_J. Assume now that gain excitation energy is close to some lth plasmon energy, $\omega_g \approx \omega_l$. In this case, for small overall system size, there is a (2l + 1)-fold degenerate eigenstate of matrix (11.16) which scales linearly with the number of molecules as

$$\Lambda_J \sim M\lambda_l \tag{11.25}$$

where λ_l is the *single-molecule* self-energy (Pustovit 2009, 2015) [73]. For example, the single-molecule self-energy λ_l due to the nearfield coupling to the dipole (l = 1) plasmon mode is given by Pustovit (2009, 2015), [73] (also see below)

$$\lambda_l = \frac{4\mu^2}{\hbar}\frac{\alpha_1(\omega)}{r^6} \tag{11.26}$$

where r is the average distance to the NP center and $\alpha_1(\omega)$ is NP dipole polarizability (for simplicity, we assumed normal dipole orientation relative to the NP surface). Near the plasmon resonance $\omega \sim \omega_p$, the NP polarizability can be approximated as

$$\alpha_1(\omega) = \frac{R^3\omega_p}{\omega_p - \omega - i/\tau_p} \tag{11.27}$$

where τ_p is the plasmon lifetime and R is the overall NP size. Then (11.24) yields the standard expression for resonance frequency [12, 26, 27]

$$\omega_g = \frac{\omega_p\tau_p + \omega_{21}\tau_2}{\tau_p + \tau_2} \tag{11.28}$$

For exact molecule-plasmon resonance, $\omega_p = \omega_{21}$, the solution of (11.24) is $\omega_g = \omega_p = \omega_{21}$ (i.e., there is no frequency shift), and we have $\alpha_1'' \sim R^3Q$, where $Q = \omega_p\tau_p$ is the plasmon quality factor. Then, for $r \sim R$, (11.23) takes the form

$$\frac{\mu^2\tau_2}{\hbar R^3}NQ \sim 1 \tag{11.29}$$

For small NPs, the local fields penetrate the entire system volume, i.e., $V_m \sim R^3$, conditions (11.29) coincide with one from [27] $\frac{\mu^2\tau_2}{\hbar V_m}NQ \sim 1$. For general gain distribution in the shell, each of the exact system eigenstates contains a contribution from nonresonant plasmon modes. For a single fluorescing molecule coupled to a dipole plasmon mode, the high-l modes' contribution leads to fluorescence quenching if the molecule is sufficiently close to the metal surface (see [25–51] in [63]). At the same time, the role of direct dipole-dipole interactions between gain molecules confined in a small volume may be significant as well due to large

Coulomb shifts of molecules' excitation frequencies [67–69]. Both the mode-mixing and direct-coupling effects can be incorporated on an equal footing within our approach through the corresponding terms in the matrix (11.16). The detailed description of the performed numerical calculations can be found in [63]. Here only final results are presented.

In Fig. 11.8 normalized resonance frequency shift $\Delta\omega\tau_2 = (\omega_g - \omega_{21})\tau_2$ and threshold population inversion per molecule $n_s = N_s/M$ are shown as a function of shell thickness h for M = 600 gain molecules randomly distributed in the shell on top of an R_c = 5 nm Ag core. All curves are plotted for distances larger than 0.5 nm in order to minimize the nonlocal effects [66]. The gain frequency ω_{21} was chosen to coincide with the dipole plasmon frequency $\omega_p \approx 3.0$ eV for the parameters chosen. In the single-mode case (l = 1) and in the absence of direct dipole-dipole coupling, the calculated ω is nearly vanishing, in agreement with (11.36), while n_s increases with h before reaching its maximum value n_s = 1 at h $\approx$ 2.35 nm. This threshold behavior is consistent with condition (11.29) as the latter implies the increase of N with mode volume until the full population inversion N = M is reached, which, in the case of low gain molecule number M = 600, takes place for a relatively small shell thickness. Very similar results are obtained for higher number of modes l (up to l = 50) are incorporated in the coupling matrix. Neither ω nor n_s shows significant deviations from the l = 1 curves except for an unrealistically small shell thickness below 0.5 nm (not shown here).

This behavior should be contrasted with the single-molecule case, where the molecule decay into high-l modes leads to fluorescence quenching at several-nanometer distances from the NP surface (see [25–51] in [63]). A similar quenching effect was demonstrated in cooperative emission of a relatively small number (M < 100) of dyes (Pustovit 2009), [73]. For larger ensembles, however, the quenching effects apparently become insignificant due to the effective restoration of spherical symmetry, which inhibits the mode mixing.

Turning the direct dipole-dipole interactions between gain molecules to a maximum has a dramatic effect on both resonance frequency and threshold

Fig. 11.8 **a** Spasing threshold n_s, where the hatched region represents the gain condition and the spasing region is shaded gray, and **b** normalized frequency shift $\Delta\omega\tau_2 = (\omega_g - \omega_{21})\tau_2$ for M = 600 molecules with $\omega_p = \omega_{21}$ are plotted versus shell thickness h with and without direct coupling for the dipole (l = 1) plasmon mode and for up to l = 50 modes included [63]

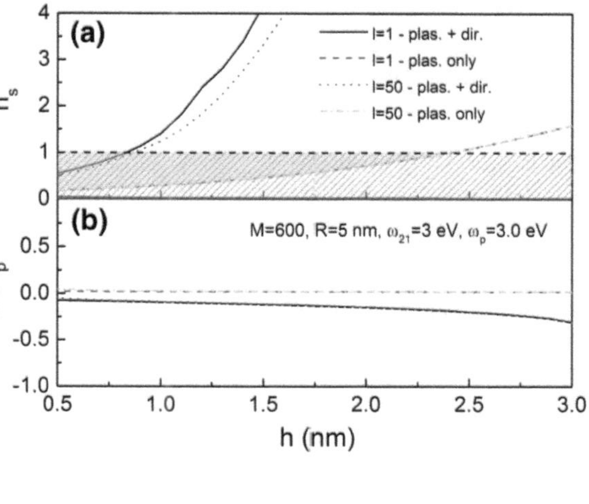

population inversion. The resonance frequency exhibits negative shift relative to the plasmon frequency, whose amplitude increases with h. The overall negative sign of $\Delta\omega\tau_2 = (\omega_g - \omega_{21})\tau_2$ is due to the normal orientation of molecule dipoles relative to the NP surface, while the increase of absolute value of $\Delta\omega\tau_2 = (\omega_g - \omega_{21})\tau_2$ with h is due to reduced plasmonic contribution. Note that real systems would lie somewhere between the non-interacting case and this maximum dipole-dipole interaction case where the choice of the molecules' normal dipole orientation may overestimate $\Delta\omega\tau_2 = (\omega_g - \omega_{21})\tau_2$ compared to more realistic random orientations. Even so, the new resonance frequency lies well within the plasmon spectral band (i.e. $\Delta\omega\tau_2 \ll 1$). At the same time, the maximal threshold value $n_s = N_s/M = 1$ is reached at about $h = 1$ nm, indicating that, in the presence of direct coupling between gain molecules, the dependence (11.37) is no longer valid. Note that here the mode mixing has a somewhat larger effect than in the absence of direct coupling, presumably due to the violation of spherical symmetry by much stronger interactions between closely spaced molecules.

Figure 11.9, the calculations are repeated for a larger number of gain molecules, $M = 1000$, which show two notable differences from the $M = 600$ case. In the absence of direct coupling between gain molecules, the maximal threshold value $n_s = 1$ is reached at larger shell thickness values. However, when the direct coupling is turned on, the maximal threshold value is reached at a *smaller* value of $h \approx 0.75$ nm, which must be attributed to stronger dipole-dipole interactions for higher gain densities.

At the same time, the effect of mode mixing in the n_s dependence on h becomes more pronounced, which is also related to stronger interactions between more closely spaced molecules that can effectively break spherical symmetry in a larger system. The major effect of direct dipole-dipole interactions is the *random* Coulomb shift of gain molecules' excitation frequencies, which may lead to the detuning between individual gain molecules and surface plasmon resonance. Note that the *average* negative shift that is due to normal orientation of molecular dipoles can be compensated by changing the gain molecules' excitation frequency.

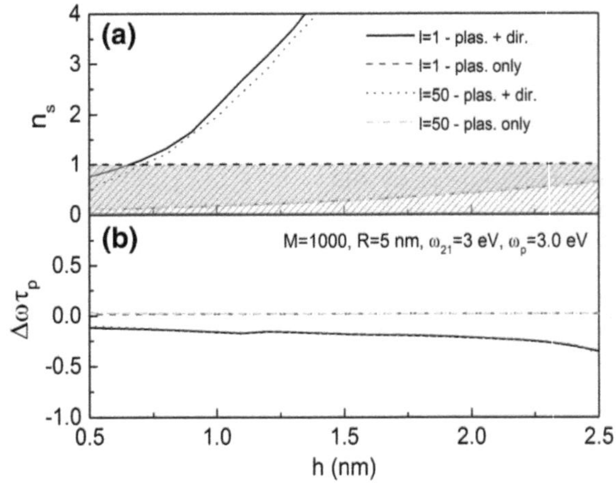

Fig. 11.9 **a** Spasing threshold n_s, where the hatched region represents the gain condition and the spasing region is shaded gray, and **b** normalized frequency shift $\Delta\omega\tau_2 = (\omega_g - \omega_{21})\tau_2$ for $M = 1000$ molecules with $\omega_p = \omega_{21}$ are plotted versus shell thickness h with and without direct coupling for the dipole ($l = 1$) plasmon mode and for up to $l = 50$ modes included [63]

In Figs. 11.10 and 11.11 we show calculated ω and *ns* for both redshifted (ω_{21} = 2.95 eV) and blue shifted (ω_{21} = 3.05 eV) gain frequencies relative to the SP resonance at 3.0 eV. As expected, for ω_0 = 2.95 eV, the average shift of ω is strongly reduced, while it increases for ω_0 = 3.05 eV [see Figs. 11.10b and 11.11b]. However, the maximal threshold value $n_s = 1$ is now reached for even smaller shell thickness $h <$ 0.5 nm [see Figs. 11.10a and 11.11a], indicating that the loss of coherence is caused by the *fluctuations* of gain excitation energies.

In summary, it was found that for sufficiently large (~1000) gain molecule numbers, the quenching is negligibly small and a single-mode approximation should work well for realistic systems. In contrast, the direct dipole dipole interactions causes random Coulomb shifts of gain molecules' excitation frequencies, and may lead to system dephasing and hinder reaching the spasing threshold in small systems. These two regimes serve as edges to an identified parameter window in which spasing can likely be achieved.

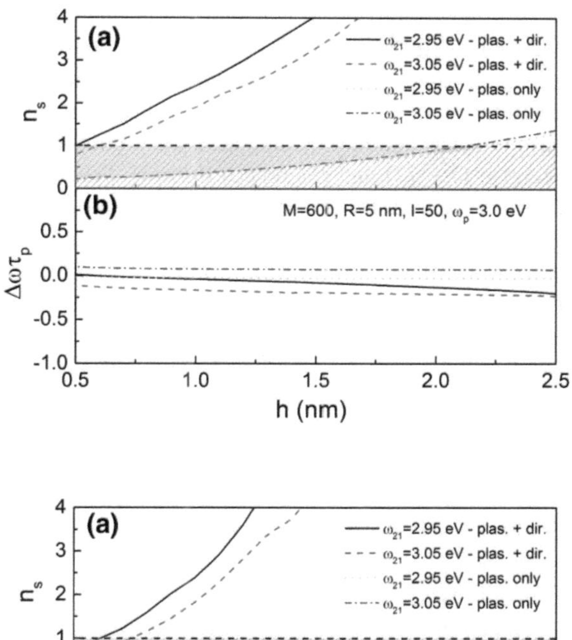

Fig. 11.10 **a** Spasing threshold n_s, where the hatched region represents the gain condition and the spasing region is shaded gray, and **b** frequency shift $\Delta\omega\tau_2 = (\omega_g - \omega_{21})\tau_2$ for M = 600 molecules are plotted versus shell thickness h for gain spectral bands centered at 2.95 and 3.05 eV [63]

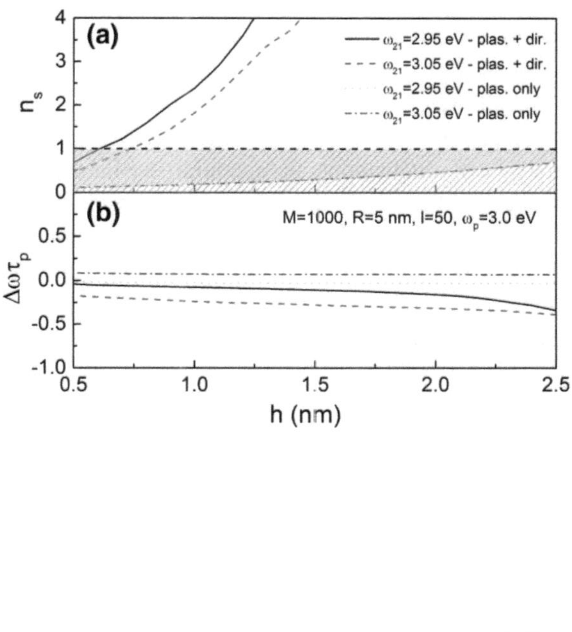

Fig. 11.11 **a** Spasing threshold n_s, where the hatched region represents the gain condition and the spasing region is shaded gray, and **b** frequency shift $\Delta\omega\tau_2 = (\omega_g - \omega_{21})\tau_2$ for M = 1000 molecules are plotted versus shell thickness h for gain spectral bands centered at 2.95 and 3.05 eV [63]

11.4 Stochastic Properties of Spasers

As it was mentioned in the introduction of this chapter, one of the areas to be developed for spasers is the theory of stochastic laser fluctuation, which should describe the linewidth of the spaser oscillations. A short plasmon relaxation time in combination with the same order of magnitude phase relaxation time of the quantum dots (main candidate for the emitters for the spaser) prohibits use of the well-established Schawlow-Townes expression [74].

A simple method for calculating the linewidth of sources of stimulated coherent radiation originating from spontaneous noise and thermal fluctuations has been proposed in the framework of the two-level model of the active medium. An expression for the spectral width has been derived beyond the adiabatic approximation, under an arbitrary relation between relaxation times in a self-sustained oscillating system and taking into account the finite spectral bandwidth of spontaneous radiation. The result is applicable for a wide class of coherent radiation sources.

This new approach to calculating the natural line width of sources of induced electromagnetic radiation is without any constraint on the relation between the relaxation times of a self-sustained oscillation system. The spectral width of quantum oscillators is an important characteristic for their application. For this reason, it was actively investigated since putting the maser into operation [74, 75] (see also [76, 77] and references therein). When optical range quantum oscillators (lasers) appeared, the theory of their fluctuations was newly developed [78, 79]. In contrast to masers, where thermal fluctuations dominate, the main source of noise in lasers is spontaneous emission [56, 57, 80]. Furthermore, the relations between the relaxation times of physical parameters in masers and lasers emitting in different ranges of electromagnetic waves are different. A quantum oscillator is a system of two coupled resonance circuits whose quality factors are usually very different. For this reason, the theoretical analysis of their fluctuations was performed in the adiabatic approximation, where the variation of rapidly relaxed parameters is considered quasistatically. At the same time, losses in circuits can be comparable to those in nanolasers, which are currently attracting particular attention (see, e.g., [12, 29, 30, 59]). For example, losses of metallic nanocavities, due primarily to ohmic losses, lead to large damping coefficients $\gamma \sim 10^{13}\ \mathrm{s}^{-1}$ [71] (see below).

In this work, the theory of fluctuations of radiation of quantum oscillators is developed for an arbitrary relation between the relaxation times. A consistent analysis of fluctuations in quantum oscillators should be based on a quantum mechanical theory. At the same time, the semiclassical approach used in this work gives correct results for the linewidth of laser radiation when describing an active medium by negative losses with saturation [57, 80] and with the use of the quantum calculation results for sources of fluctuations [81–83]. Let us consider a single mode quantum oscillator with an active medium, having the homogeneously broadened line of a resonance transition, described by (10.11) without an external field $E_{\mathrm{ext}} = 0$. In the semiclassical approach, the generation process is described by the following system of coupled equations (see, e.g., [82, 84, 85]; note, that another set of variables in contrast with the ones in Chap. 10 is used here):

$$\begin{cases} \frac{d^2P}{dt^2} + 2\gamma_2\frac{dP}{dt} + \omega_0^2 P = -\frac{2|d|\omega_0 EN}{\hbar} \\ \frac{dN}{dt} + \gamma_2(N - N_0) = -\frac{2E}{\hbar\omega_0}\frac{d\mathbf{P}}{dt} \\ \frac{d^2E}{dt^2} + 2\gamma\frac{dE}{dt} + \omega_0^2 E = -4\pi\frac{q_{el}}{m_{el}}\frac{d^2P}{dt^2} \end{cases} \tag{11.30}$$

Here, $P(t)$ is the polarization of the active medium; $N(t)$ is the difference between the populations of working levels; N_0 is the equilibrium value in the absence of radiation; $E(t)$ is the electric field strength in a cavity; d is the matrix element of the dipole moment of the working transition; $\gamma_2 = 1/T_2$ and $\gamma_1 = 1/T_1$, where T_2 and T_1 are the relaxation times for the polarization and populations of the levels, respectively; and the coefficient γ takes into account losses in the cavity. To simplify the subsequent analysis, we accept that the resonance frequency of the two-level transition ω_{21} coincides with the cavity frequency ω_c, i.e., $\omega_{21} = \omega_c = \omega_0$. The solution of (11.30) is sought in the form:

$$\begin{cases} P(t) = \frac{1}{2}(p(t)\exp(i\omega_0 t) + p(t)^*\exp(-i\omega_0 t)) \\ E(t) = \frac{1}{2}(A(t)\exp(i\omega_0 t) + A(t)^*\exp(-i\omega_0 t)) \end{cases} \tag{11.31}$$

Note, that in contrast to the previous chapters, signs of the oscillating terms are chosen to be negative for the complex conjugated parts. In the approximations of "slowly varying complex amplitudes" and "rotating wave", we obtain the system of reduced equations:

$$\begin{cases} \frac{dp}{dt} + \gamma_2 p - \frac{i|d|AN}{\hbar} = \xi_{sp}(t) \\ \frac{dN}{dt} + \gamma_1(N - N_0) = \frac{i(A^*p - Ap^*)}{2\hbar} \\ \frac{dA}{dt} + \gamma A + 2\pi\omega_0 p = \xi_T(t) \end{cases} \tag{11.32}$$

where a dot means time derivative. Equations (11.32) are supplemented by Langevin fluctuation sources $\xi_{sp}(t)$ and $\xi_T(t)$, which are due to the spontaneous emission of atoms (molecules) of the active medium and thermal noise of the cavity and, as was mentioned above, can be consistently obtained from quantum analysis (see, e.g., [56, 82, 83]). The fluctuation sources $\xi_{sp}(t)$ and $\xi_T(t)$ are statistically independent. As usual, the random process $\xi_T(t)$ is considered as delta correlated:

$$\langle\xi_{sp}(t)\rangle = 0, \langle\xi_{sp}(t)\xi_{sp}(t+\tau)\rangle = \gamma_2 D_{sp}\exp[-\gamma_2|\tau|] \tag{11.33}$$

and the process $\xi_{sp}(t)$ is taken in the form of "color" noise with the correlation time $\tau_c = 1/\gamma_2$:

$$\langle \xi_{\mathrm{T}}(t) \rangle = 0, \langle \xi_{\mathrm{T}}(t)\xi_{\mathrm{T}}(t+\tau) \rangle = 2D_{\mathrm{T}}\delta(\tau) \tag{11.34}$$

This expression makes it possible to take into account a finite band of the noise due to spontaneous emission. In the limit $\gamma_2 \to \infty\,(T_2 \to 0)$, the correlation function behaves as $2D_{\mathrm{sp}}\delta(t_2 - t_1)$. The coefficients D_{sp} and D_{T} will be discussed below. To solve the formulated problem, we need the following additional equation [easily obtained from (11.32)]:

$$\frac{\mathrm{d}}{\mathrm{d}t}\Sigma(t) + (\gamma + \gamma_2)\Sigma(t) = \eta(t) \tag{11.35}$$

where

$$\Sigma(t) = p(t)A^*(t) + p^*(t)A(t) \tag{11.36}$$

and

$$\eta(t) = p(t)\xi_{\mathrm{T}}^* + A^*(t)\xi_{\mathrm{sp}}(t) + k.c. \tag{11.37}$$

is a random function. In this case, $\langle \eta(t) \rangle = 0$, because fluctuations $p(t)$ and $A(t)$ only depend directly on $\xi_{sp}(t)$ and $\xi_T(t)$, respectively. The correlation function of the process $\eta(t)$ is given by the expression:

$$\langle \eta(t)\eta(t+\tau) \rangle = 4D_{\mathrm{T}}|p(t)|^2\delta(\tau) + 2\gamma_2 D_{\mathrm{sp}}|A(t)|^2\exp[-\gamma_2|\tau|] \tag{11.38}$$

Let us represent the complex amplitude of the field in the form of the amplitude and phase, $A(t) = |A(t)|\exp(i\phi(t))$, and express the phase as:

$$\phi(t) = \frac{1}{i2}\ln\left(\frac{A(t)}{A^*(t)}\right) \tag{11.39}$$

Correspondingly, in contrast to traditional approaches, we describe the temporal dynamics of the phase by the equation:

$$\frac{\mathrm{d}\phi}{\mathrm{d}t} = \frac{\dot{A}A^* - A\dot{A}^*}{i2|A|^2} \tag{11.40}$$

The above threshold generation regime, characterized by small fluctuations of the amplitude, is of special interest for applications. In this case, amplitude fluctuations in the denominator of (11.40) can be neglected; i.e., $|A|^2 = |A_{\mathrm{st}}|^2$ (A_{st} is the steady state amplitude of oscillations) can be assumed.

The substitution of expression (11.32) into (11.40) yields:

$$\frac{\mathrm{d}}{\mathrm{d}t}\phi(t) = -\frac{1}{2|A_{\mathrm{st}}|^2}[2\pi\Sigma(t) + i\zeta_{\mathrm{T}}(t)] \tag{11.41}$$

In this equation, $\zeta_T(t)$ can be represented in the form:

$$\sigma_T(t) = A^*(t)\xi_T(t) - A(t)\xi_T^*(t) = A_{nc}^*(t)\xi_T(t) - A_{nc}(t)\xi_T^*(t) \tag{11.42}$$

where $A(t)_{nc}$ is the random part of $A(t)$ that does not correlate with $\xi_T(t)$ and, therefore, $\langle\varsigma_T(t)\rangle = 0$. The change of $A(t)$ to $A(t)_{nc}$, made in (11.32), does not affect the statistical characteristics of the process $\zeta_T(t)$. Indeed, since the process $\xi_T(t)$ is delta correlated (correlation time $\tau_{\xi_E} \to 0$), only the increment $\delta A(t)$ obtained by the function $A(t)$ at times immediately preceding the time t can correlate with this process. Hence, in the time interval Δt satisfying the conditions $\tau_{\xi_E} \ll \Delta t \ll \gamma^{-1}$, the function $A(t)$ can be represented in the form $A(t) = A(t - \Delta t)_{nc} + \delta A(t)$, where the correlated part is given by the expression:

$$\delta A(t) = \int_{t-\Delta t}^{t} \xi(t')\mathrm{d}t' \tag{11.43}$$

and $A(t - \Delta t)_{nc}$ can be changed to $A(t)_{nc}$. Taking into account that the change in $A(t)_{nc}$ throughout the interval Δt is small. This approach is justified in, e.g., [86]. Bearing in mind (11.32), we obtain:

$$\langle\delta A(t)\xi_T^*(t)\rangle = \langle\delta A^*(t)\xi_T(t)\rangle = \int_{t-\Delta t}^{t} \langle\xi_T^*(t)\xi_T(t')\rangle\mathrm{d}t' = D_T \tag{11.44}$$

Therefore, the correlated part of the amplitude $A(t)$ does not affect the statistics of the process $\zeta_T(t)$ with the correlation function:

$$\langle\varsigma_T(t)\varsigma_T(t_1)\rangle = -4|A_{st}|^2 D_T\delta(t_1 - t) \tag{11.45}$$

According to (11.41) and (11.35), $\langle\dot{\varphi}(t)\rangle = 0$ and the phase increment $\Delta\varphi(\tau)$ and its $\langle\Delta\varphi^2(\tau)\rangle$ dispersion in the time interval τ are determined by the expressions:

$$\Delta\varphi(\tau) = \int_{t-\tau}^{t} \dot{\varphi}(t')\mathrm{d}t' \tag{11.46}$$

$$\langle(\Delta\varphi(\tau))^2\rangle = \int\int_{t-\tau}^{t} R_{\dot{\varphi}}(t'' - t')\mathrm{d}t''\mathrm{d}t', R_{\dot{\varphi}}(t'' - t') = \langle\varphi(t'')\varphi(t')\rangle \tag{11.47}$$

To find the correlation function $R_{\dot{\varphi}}(t'' - t')$, we use the Wiener–Khinchin theorem:

$$R_{\dot{\varphi}}(\tau) = \int_{-\infty}^{+\infty} S_{\dot{\varphi}}(\Omega)\exp(i\Omega\tau)\mathrm{d}\tau \tag{11.48}$$

where $S_{\dot{\varphi}}(\Omega)$ is the spectral density of fluctuations of the frequency deviation $\delta\omega(t)$ = (t). The Fourier spectrum $\dot{\varphi}(\Omega)$ is obtained by means of the Fourier transform of (11.35) and (11.41). Taking into account the statistical stationarity of the random processes $\xi_{\mathrm{sp}}(t)$ and $\xi_{\mathrm{T}}(t)$, we can obtain the expression $\langle\dot{\varphi}(\Omega_1)\dot{\varphi}(\Omega)\rangle = S_{\dot{\varphi}}(\Omega)\delta(\Omega_1 - \Omega)$. The substitution of the expression for $S_{\dot{\varphi}}(\Omega)$ into (11.48) and (11.48) into (11.47) gives the following expression for the dispersion of the phase increment:

$$\begin{aligned}\left\langle(\Delta\varphi(\tau))^2\right\rangle &= \frac{D_{\mathrm{T}}}{|A|^2}|\tau| + \frac{\gamma^2 - 2\gamma\Gamma}{\Gamma^3|A|^2}D_{\mathrm{T}}F(\Gamma|\tau|)\\ &+ \frac{(2\pi\omega_0\gamma_2)^2}{|A|^2(\Gamma^2 - \gamma_2^2)}D_{\mathrm{sp}}\left\{\gamma_2^{-3}F(\gamma_2|\tau|) - \Gamma^{-3}F(\Gamma|\tau|)\right\}\end{aligned} \tag{11.49}$$

Here, $\Gamma = \gamma\ + \gamma_2$ and:

$$F(x) = \mathrm{e}^{-x} + x - 1 \tag{11.50}$$

In (11.49) and below, the subscript "st" in the parameters A, p, and N is omitted in order to simplify expressions. Recall that the steady state values of these parameters are determined in (11.32), disregarding fluctuations. Moreover, p in (11.49) is expressed in terms of A from time independent (11.32). The spectral density of the field is determined by the expression:

$$S_{\mathrm{E}}(\omega) = \frac{|A|^2}{4\pi}\int \exp\left[-0,5 < (\Delta\varphi(\tau))^2 > \ - i(\omega - \omega_0)\tau\right]\mathrm{d}\tau \tag{11.51}$$

and can be generally calculated only numerically; analytical results can be obtained only in particular cases. The correlation time is related to the width of the field spectrum $\Delta\omega_{\mathrm{osc}}$ as $\tau_{\mathrm{c}}^{(\mathrm{E})} \approx (\Delta\omega_{\mathrm{osc}})^{-1}$.

Letting $\gamma_2\tau_{\mathrm{cor}}^{(E)} \gg 1$,; i.e., $\gamma_2 \gg (\Delta\omega_{\mathrm{osc}})^{-1}$, then $\langle(\Delta\varphi(\tau))^2\rangle = D_{\varphi}|\tau|$. In this case, the radiation spectrum has a Lorentzian shape and the FWHM is given by the expression:

$$\Delta\omega_{\mathrm{osc}} = D_{\varphi} = \frac{\gamma_2^2}{(\gamma + \gamma_2)^2|A|^2}\left(D_{\mathrm{T}} + (2\pi\omega_0/\gamma_2)^2 D_{\mathrm{sp}}\right) \tag{11.52}$$

Expression (11.52) was derived without any assumptions regarding the relations between the relaxation times associated with the coefficients γ, γ_1, and γ_2. Consequently, (11.52) for the natural linewidth of the radiation spectrum of quantum oscillators is general and applicable to a wide class of quantum oscillators whose dynamics can be described by the two level model of the active medium. However, it should be noted that the source of noise $\xi_{sp}(t)$, under the accepted condition, is considered as delta correlated: This assumption may be incorrect for fluctuations of the polarization, if the radiation spectral width $\Delta\omega_{osc}$ is comparable with $1/\gamma_2$. If this is the case then, (11.49) should be used. The corresponding effect can be demonstrated with the following example. Let $\Gamma|\tau| \gg 1$, but $\gamma_2|\tau| < 1$. Then:

$$\begin{aligned}\left\langle (\Delta\varphi(\tau))^2 \right\rangle = & \frac{\gamma_2^2}{(\gamma+\gamma_2)^2|A|^2}\left(D_{\mathrm{T}} + (2\pi\omega_0/\gamma_2)^2 D_{\mathrm{sp}}\right)|\tau| \\ & - \frac{(1-0.5\gamma_2|\tau|)(2\pi\omega_0)^2}{(\Gamma^2-\gamma_2^2)|A|^2} D_{\mathrm{sp}}|\tau|\end{aligned} \tag{11.53}$$

It can easily be seen that the last term in (11.53), in comparison to (11.52), leads both to a decrease in the width of the spectrum and modification of its shape. The coefficient DT associated with thermal noise can be determined using the fluctuation–dissipation theorem or the principle of equidistribution of energy with respect to degrees of freedom [57, 80]. The average thermal energy of the mode in the cavity near the frequency ω_0 is:

$$\langle w(\omega_0) \rangle = \hbar\omega_0\bar{n} \tag{11.54}$$

$$\bar{n} = 1/[\exp(\hbar\omega_0/k_{\mathrm{B}}T) - 1] \tag{11.55}$$

where $\bar{n}$ is the average number of thermal photons in the mode, k_{B} is the Boltzmann constant and T is the absolute temperature. Note that the energy of zero oscillations is disregarded in (11.55); i.e., $\bar{n}$ is used instead of $\bar{n}+0.5$. According to (11.32), the spectral density of fluctuations of the amplitude $A(t)$ for a passive cavity, i.e., when $p(t) = 0$, is given by the expression:

$$S_A(\Omega) = \frac{D_{\mathrm{T}}}{\pi\left(\gamma^2+\Omega^2\right)} \tag{11.56}$$

According to this expression, the thermal energy of the mode is:

$$\langle w(\omega_0) \rangle = \frac{D_{\mathrm{T}}}{2\pi}\int_{-\infty}^{\infty}\frac{d\Omega}{\left(\gamma^2+\Omega^2\right)} = \frac{D_{\mathrm{T}}}{2\gamma} \tag{11.57}$$

Comparing (11.57) and (11.46), we obtain:

$$D_{\mathrm{T}} = 2\gamma\hbar\omega_0\bar{n} \tag{11.58}$$

The problem of determining the coefficient D_{sp}, which depends on the characteristics of the two level medium (consisting of atoms or molecules) and takes into account spontaneous emission, is more complicated. The corresponding quantum calculations of this coefficient can be found, e.g., in [56, 82, 83]. At the same time, the form of (11.42) means we avoid this problem. Under the condition $\gamma_2 \gg \gamma$ (solid state lasers), (11.42) is reduced to the expression known as the Schawlow–Townes formula. In terms of measurable parameters, this expression has the form [57, 80]:

$$\Delta\omega_{osc}^{ST} = \frac{2\gamma^2\hbar\omega_0}{P_0}\left[\langle n\rangle + \frac{N_2}{N_2 - (g_2/g_1)N_1}\right] \tag{11.59}$$

Here, $P_0 = 0.5\gamma|A|^2$ is the power of radiation of the oscillator, including the power at the output of the cavity and the power spent on the compensation of losses; $\gamma = 0.5\Delta\omega_{res}$ ($\Delta\omega_{res}$ is the pass band of the passive cavity); $N = N_2 - (g_2/g_1)N_1$ is the difference between the populations between the upper N_2 and lower N_1 levels near the generation threshold; and g_2 and g_1 are the degeneracy orders of the respective levels. Note that the first term in (11.59) is related to coefficient (11.58). Thus, (11.42) can be rewritten in the form:

$$\Delta\omega_{osc} = 2\left(\frac{\gamma\gamma_2}{\gamma+\gamma_2}\right)^2\frac{\hbar\omega_0}{P_0}\left[\langle n\rangle + \frac{N_2}{N_2 - (g_2/g_1)N_1}\right] \tag{11.60}$$

Formula (11.60) generalizes (11.59) for the case of an arbitrary relation between the relaxation parameters γ and γ_2. Similar expressions have been found for the case of so called bad-cavity laser [87–89]. In terms of the quality factor of the cavity $Q_{res} = 2\omega_0/\gamma$ and the "quality factor of the levels of the transition" $Q_{lev} = 2\omega_0/\gamma_2$, (11.50) can be written as:

$$\Delta\omega_{osc} = \frac{2\hbar\omega_0^3}{(Q_{res} + Q_{lev})P_0}\left[\langle n\rangle + \frac{N_2}{N_2 - (g_2/g_1)N_1}\right] \tag{11.61}$$

According to this expression, the width of the radiation line depends on the sum of the quality factors of the cavity and the particular two level transition.

Let us illustrate the application of these results. For a molecular oscillator (maser) based on ammonia molecules [84], the excitation frequency $\nu_0 = \omega_0/2\pi = 2.4 \times 1010$ Hz (($\hbar\omega_0 = 1.5 \times 10^{-16}$ erg), $\gamma_2 = 6.6 \times 10^3\ \text{s}^{-1}$, and $\gamma = 6 \times 10^6\ \text{s}^{-1}$ ($\gamma_2 \ll \gamma$). Even at the temperature $T = 300$ K, the energy is $k_B T = 4 \times 10^{-14}$ erg $\gg \hbar\omega_0$. Therefore, $\hbar\omega_0\bar{n} = k_B T$ and the last term in (11.61) can be neglected. As a result, the following known result is obtained for the molecular oscillator [75, 76]:

$$\Delta\omega_{osc} = \frac{2\gamma_2^2 k_B T}{P_0} \tag{11.62}$$

In this example, the spectral linewidth depends on thermal fluctuations and the parameter γ_2, which is determined by how long the molecules remain in the cavity [75, 76]. The situation in lasers is opposite; $\gamma_2 \gg \gamma$ and $\hbar\omega_0 \gg k_B T$, where the cavity parameter γ and spontaneous radiation [the last term in (11.59)] play the main roles. The approach developed in this work for calculating phase fluctuations of radiation provides a unified analysis of the spectral linewidth of quantum oscillators with various dependencies between the relaxation times. Expressions (11.43) and (11.50) are also applicable for the calculation of the spectral linewidth when the coefficients γ and γ_2 are comparable, for example, for $\Delta\omega_{osc}$ a model of a nanolaser using the adiabatic approximation cannot be used. At this current time, there is no experimental data that can verify or disprove our results. Although the authors of [30] observed the narrowing of the excitation band spectrum of the nanolaser to 2–5 nm, this effect cannot with certainty be attributed to laser emission. To conclude, we note that the method developed in this work can also be used to calculate the radiation spectrum of optical parametric oscillators.

Depending on the relation between the pumping parameter N_0 and the steady-state value of the population inversion N_{th} one can consider three operation regimes of the nanolaser: (1) the below threshold regime $N_0 < N_{th}$, (2) the near-threshold regime $N_0 \sim N_{th}$ and (3) the case of the above-threshold regime $N_0 > N_{th}$. The generalized Schawlow-Townes-like formula works only for the above-threshold regime (see [90]), so that the numerical analysis in the other two cases, which can be realised in the experiments, is of primary interest.

The numerical study of the radiation linewidth of a stochastic system (1) was conducted by the second order Runge–Kutta stochastic method [91]. For this purpose the phase shift $\Delta\phi(\tau) = \phi(t) - \phi(t-\tau)$ was considered, where $\phi(t)$ is the phase of the slowly-varying component of the amplitude, appearing due to the stochastic forces.

The above-threshold regime of the laser is the most convenient one, since the linewidth is predominantly determined by the phase fluctuations, while the amplitude ones are well-supressed. The power spectrum of the radiation is obtained through the Fourier transform of the autocorrelation function: $B(\tau) = \langle A^*(t)A(t-\tau)\rangle \sim \exp(-\frac{1}{2}\langle(\Delta\phi(\tau))^2\rangle)$. In Fig. 11.12 the numerical and analytical calculations of the

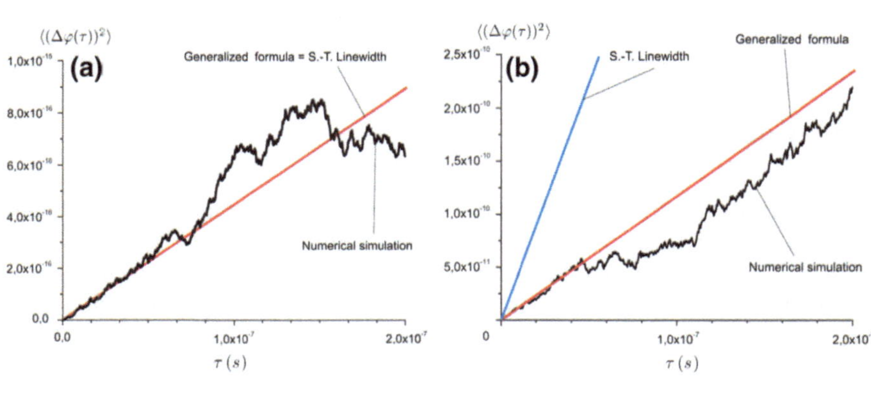

Fig. 11.12 Analytical (red and blue curves) and numerical results (black curves) of the variance $\left\langle (\Delta\phi(\tau))^2 \right\rangle$ as a function of time delay. The curves are plotted for the following parameters: $\lambda = 0.532$ μm, $|d| = 2.5 \times 10^{-17}$ esu, $\gamma_1 = 10^9\,\text{s}^{-1}$, $\gamma_2 = 10^{13}\,\text{s}^{-1}$, $N_0 = 10^{17}\,\text{cm}^{-3}$, $D_{SP} = 2 \times 10^{-7}$. The resonator field attenuation rates are: **a** $\gamma = 10^7\,\text{s}^{-1}$ which corresponds to ordinary laser with high quality resonator, and **b** $\gamma = 10^{13}\,\text{s}^{-1}$ which corresponds to nanoresonator

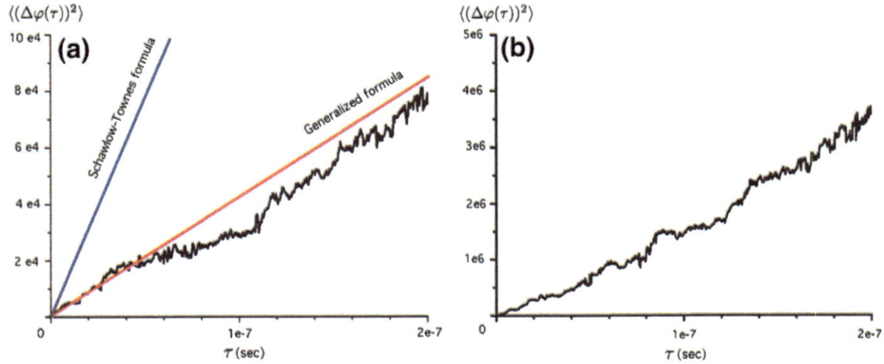

Fig. 11.13 Analytical (red and blue curves) and numerical results (black curves) of the variance $\left\langle (\Delta\phi(\tau))^2 \right\rangle$ as a function of time delay for the case of the nanolaser $\gamma = 10^{13}\,\mathrm{s}^{-1}$. The curves are plotted for the following parameters: $\lambda = 0.532$ μm, $|d| = 2.5 \times 10^{-17}$ esu, $\gamma_1 = 10^9\,\mathrm{s}^{-1}$, $\gamma_2 = 10^{13}\,\mathrm{s}^{-1}$, $D_{SP} = 2 \times 10^{-7}$. The pumping parameter corresponds to: **a** below threshold regime and **b** above but near threshold regime (in this case analytical approximation is no more valid). The y scales are different in both figures [90]

variance of the phase shift $\Delta\phi(\tau)$ is presented. Figure 11.12a corresponds to the ordinary laser with $\gamma \ll \gamma_1 \ll \gamma_2$, while Fig. 11.12b—for the nanolaser with $\gamma \sim \gamma_2 \gg \gamma_1$. As can be seen from Fig. 11.12, the analytical and numerical results are in good agreement; the standard ST expression for the nanolaser overestimates significantly the more correct results.

Expression (11.49) has been elaborated under approximation of no amplitude fluctuations, which corresponds to the laser generation well above threshold. On the other hand, it would be interesting to compare the results of our model with the rigorous calculations and Schawlow-Towns expression. The comparison is presented in Fig. 11.13.

11.5 Conclusion

The coupled dynamics model of the plasmonic nanoresonator and quantum emitter has been applied to describe regular and stochastic properties of the nanolaser (spaser). The dynamics has been considered in the case of an applied resonant external field and without, for both the dipole-like and quadrupole-like MAs. The case of multimode spaser generation has been considered. The regions for stable and unstable dynamics have been found and in the framework of the same model, the problem of the bandwidth of the spaser generation has been considered. It has been found, that the standard Schawlow-Towns expression significantly overestimates the bandwidth and cannot be used; a modified expression has been suggested. The analytical expression is shown to approximate closely the rigorous solution of the stochastic dynamics even for the operation mode just above the threshold, where the amplitude fluctuations are not a priori supposed to be negligible. This proves again, that the phase fluctuations play a decisive role for the spaser generation bandwidth.

References

1. S. Maier, Plasmonic field enhancement and SERS in the effective mode volume picture. Opt. Express **14**, 1957 (2006)
2. A.D. Boardman, *Electromagnetic Surface Modes* (Wiley, New York, 1982)
3. H. Raether, *Surface Plasmons* (Springer, New York, 1988)
4. C. Soukoulis, M. Wegener, Optical metamaterials—more bulky and less lossy. Science **330**, 1633 (2010)
5. A. Boltasseva, H. Atwater, Low-loss plasmonic metamaterials. Science **331**, 290 (2011)
6. S. Anlage, The physics and applications of superconducting metamaterials. J. Opt. **13**, 024001 (2011)
7. H.-T. Chen et al., Tuning the resonance in high-temperature superconducting terahertz metamaterials. PRL **105**, 247402 (2010)
8. P. Berini, I. De Leon, Surface plasmon–polariton amplifiers and lasers. Nat. Photon. **6**, 16 (2011)
9. C. Soukoulis, M. Wegener, Past achievements and future challenges in the development of three-dimensional photonic metamaterials. Nat. Photon. **5**, 523–530 (2011)
10. W.L. Barnes, Fluorescence near interfaces: the role of photonic mode density. J. Mod. Opt. **45**, 661 (1998)
11. S. Ramakrishna, J. Pendry, Removal of absorption and increase in resolution in a near-field lens via optical gain. Phys. Rev. B **67**, 201101(R) (2003)
12. D.J. Bergman, M.I. Stockman, PRL **90**, 027402 (2003)
13. E.M. Purcell, Phys. Rev. **69**, 681 (1946)
14. A.F. Koenderink, On the use of Purcell factors for plasmon antennas. Opt. Lett. **35**, 4208 (2010)
15. N. Blombergen, R. Pound, Phys. Rev. **95**, 8 (1954)
16. M. Strandberg, Phys. Rev. **106**, 617 (1957)
17. F. Bunkin, A. Oraevsky, Izv. Vuzov. Radiophysika **2**(2), 181 (1959)
18. O. Hess, J.B. Pendry, S.A. Maier, R.F. Oulton, J.M. Hamm, K.L. Tsakmakidis, Active nanoplasmonic metamaterials. Nat. Mater. **11**, 573 (2012)
19. N. Liu, L. Langguth, J.K.T. Weiss, M. Fleischhauer, T. Pfau, H. Giessen, Nat. Mater. **8**, 758 (2009)
20. C. Rockstuhl et al., Resonances of split-ring resonator metamaterials in the near infrared. Appl. Phys. B **84**, 219–227 (2006)
21. M. Husnik et al., Absolute extinction cross-section of individual magnetic split-ring resonators. Nat. Photon. **2**, 614 (2008)
22. D.E. Chang, A.S. Sorensen, P.R. Hemmer, M.D. Lukin, PRL **97**, 053002 (2006)
23. D. Martin-Cano, L. Martin-Moreno, F. Garcia-Vidal, E. Moreno, Nano Lett. **10**, 3129 (2010)
24. V.V. Klimov, Nanoplasmonics. Phys. Usp. **51**(8), 839–844 (2008)
25. V.V. Klimov, *Nanoplasmonika*. ISBN 978-5-9221-1205-5 (2010) (in Russian)
26. M. Stockman, Spaser explained. Nat. Photonics **2**, 327 (2008)
27. M. Stockman, The spaser as a nanoscale quantum generator and amplifier. J. Opt. **12**, 024004 (2010)
28. M. Stockman, Spaser action, loss-compensation, and stability in plasmonic systems with gain. PRL **106**, 156802 (2011)
29. N. Zheludev, S. Prosvirin, N. Papasimakis, V. Fedotov, Lasing spaser. Nat. Photonics **2**, 351 (2008)

30. M. Noginov, G. Zhu, A. Belgrave, R. Bakker, V. Shalaev, E. Narimanov, S. Stout, E. Herz, T. Suteewong, U. Wiesner, Demonstration of a spaser-based nanolaser. Nature **460**, 1110 (2009)
31. R. Oulton, V. Sorger, T. Zentgraf, R.-M. Ma, C. Gladden, L. Dai, G. Bartal, X. Zhang, Plasmon lasers at deep subwavelength scale. Nature **461**, 629 (2009)
32. M. Hill, M. Marell, E. Leong, B. Smalbrugge, Y. Zhu, M. Sun, P. van Veldhoven, E. Jan Geluk, F. Karouta, Y.-S. Oei, R. Nötzel, C.-Z. Ning, M. Smit, Lasing in metal-insulator-metal sub-wavelength plasmonic waveguides. Opt. Express **17**, 11107 (2009)
33. Z. Zhu, H. Liu, S. Wang, T. Li, J. Cao, W. Ye, X. Yuan, S. Zhu, Optically pumped nanolaser based on two magnetic plasmon resonance modes. APL **94**, 103106 (2009)
34. A. Banerjee, R. Li, H. Grebel, Surface plasmon lasers with quantum dots as gain media. APL **95**, 251106 (2009)
35. M. Nezhad, A. Simic, O. Bondarenko, B. Slutsky, A. Mizrahi, L. Feng, V. Lomakin, Y. Fainman, Room-temperature subwavelength metallo-dielectric lasers. Nat. Photon. **4**, 395 (2010)
36. R.-M. Ma, R. Oulton, V. Sorger, G. Bartal, X. Zhang, Room-temperature sub-diffraction-limited plasmon laser by total internal reflection. Nat. Mater. **10**, 110 (2011)
37. R.A. Flynn, C.S. Kim, I. Vurgaftman, M. Kim, J.R. Meyer, A.J. Mäkinen, K. Bussmann, L. Cheng, F.-S. Choa, J.P. Long, A room-temperature semiconductor spaser operating near 1.5 μm. Opt. Express **19**, 8954 (2011)
38. C.-Y. Wu, C.-T. Kuo, C.-Y. Wang, C.-L. He, M.-H. Lin, H. Ahn, S. Gwo, Plasmonic green nanolaser based on a metal-oxide-semiconductor structure. Nano Lett. **11**, 4256 (2011)
39. E. Plum, V.A. Fedotov, P. Kuo, D.P. Tsai, N.I. Zheludev, Opt. Express **17**, 8548 (2009)
40. Y. Lu, C.-Y. Wang, J. Kim, H.-Y. Chen, M.-Y. Lu, Y.-C. Chen, W.-H. Chang, L.-J. Chen, M.I. Stockman, C.-K. Shih, S. Gwo, Nano Lett. **14**, 4381 (2014)
41. R. Ma, R. Oulton, V. Sorger, G. Bartal, X. Zhang, Nat. Mater. **10**, 110 (2010)
42. K. Ding, Z.C. Liu, L.J. Yin, M.T. Hill, M.J.H. Marell, P.J. van Veldhoven, R. Nöetzel, C.Z. van Ning, Phys. Rev. B **85**, 041301(R) (2012)
43. Y.-J. Lu, J. Kim, H.-Y. Chen, C.I. Wu, N. Dabidian, C.E. Sanders, C.-Y. Wang, M.-Y. Lu, B.-H. Li, X. Qiu, W.-H. Chang, L.-J. Chen, G. Shvets, C.-K. Shih, S. Gwo, Science **337**, 450 (2012)
44. W. Zhou, M. Dridi, J.Y. Suh, C.H. Kim, D.T. Co, M.R. Wasielewski, G.C. Schatz, T.W. Odom, Nat. Nanotechnol. **8**, 506 (2013)
45. R.-M. Ma, S. Ota, Y. Li, S. Yang, X. Zhang, Nat. Nanotechnol. **9**, 600 (2014)
46. A. Yang, T.B. Hoang, M. Dridi, C. Deeb, M.H. Mikkelsen, G.C. Schatz, T.W. Odom, Nat. Commun. **6**, 6939 (2015)
47. X. Meng, A.V. Kildishev, K. Fujita, K. Tanaka, V.M. Shalaev, Nano Lett. **13**, 4106 (2013)
48. E.I. Galanzha, R. Weingold, D.A. Nedosekin, M. Sarimollaoglu, A.S. Kuchyanov, R.G. Parkhomenko, A.I. Plekhanov, M.I. Stockman, V.P. Zharov, arXiv:1501.00342
49. V. Apalkov, M.I. Stockman, Light Sci. Appl. **3**, e191 (2014)
50. Y. Yin, T. Qiu, J. Li, P. Chu, Plasmonic nano-lasers. Nano Energy **1**, 25 (2012)
51. J.A. Gordon, R.W. Ziolkowski, The design and simulated performance of a coated nano-particle laser. Opt. Express **15**, 2622 (2007)
52. S. Wuestner, A. Pusch, K. Tsakmakidis, J. Hamm, O. Hess, Gain and plasmon dynamics in active negative-index metamaterials. Phil. Trans. R. Soc. A **369**, 3525 (2011)
53. A. Sarychev, G. Tartakovsky, Magnetic plasmonic metamaterials in actively pumped host medium and plasmonic nanolaser, Phys. Rev. B **75**, 085436 (2007)
54. E. Andrianov, A. Pukhov, A. Dorofeenko, A. Vinogradov, A. Lisyansky, Forced synchronization of spaser by an external optical wave. Opt. Express **19**, 24849 (2011)
55. V.M. Fain, *Quantum Radio Physics, Vol. 1: Photons and Nonlinear Media* (Sovetskoe Radio, Moscow, 1972) (in Russian)
56. G. Haken, *Laser Light Dynamics* (North Holland, Amsterdam, 1985)
57. S. Akhmanov, Y. D'yakov, A. Chirkin, *Introduction to Statistical Radio Physics and Optics* (Nauka, Moscow, 1981) (in Russian)

58. A. Pikovsky, M. Rosenblum, J. Kurths, *Synchronization. A Universal Concept in Nonlinear Sciences* (Cambridge University Press, Cambridge, 2001)
59. I. Protsenko, A. Uskov, O. Zaimidoroga et al., Phys. Rev. A **71**, 063812 (2005)
60. N. Arnold, B. Ding, C. Hrelescu, T.A. Klar, Beilstein J. Nanotechnol. **4**, 974 (2013)
61. R.R. Chance, A. Prock, R. Silbey, in *Advances in Chemical Physics*, vol. 37, ed. by I. Prigogine, S.A. Rice (Wiley, Hoboken, 1978)
62. H. Metiu, Prog. Surf. Sci. **17**, 153 (1984)
63. V. Pustovit, A. Urbas, A. Chipouline, T. Shahbazyan, Coulomb and quenching effects in small nanoparticle-based spasers. PRB (2016)
64. J. Gersten, A. Nitzan, Spectroscopic properties of molecules interacting with small dielectric particles. J. Chem. Phys. **75**(3), 1139 (1981)
65. R. Ruppin, J. Chem. Phys. **76**(4) (1982)
66. V.N. Pustovit, T.V. Shahbazyan, J. Chem. Phys. **136**, 204701 (2012)
67. N.E. Rehler, J.H. Eberly, Phys. Rev. A **3**, 1735 (1971)
68. R. Friedberg, S.R. Hartmann, Phys. Rev. A **10**, 1728 (1974)
69. B. Coffey, R. Friedberg, Phys. Rev. A **17**, 1033 (1978)
70. S. Wuestner, J.M. Hamm, A. Pusch, F. Renn, K. Tsakmakidis, O. Hess, Control and dynamic competition of bright and dark lasing states in active nanoplasmonic metamaterials. Phys. Rev. B **85**, 201406(R) (2012)
71. J. Petschulat, C. Menzel, A. Chipouline, C. Rockstuhl, A. Tünnermann, F. Lederer, T. Pertsch, Multipole approach to metamaterials. Phys. Rev. B **78**, 043811 (2008)
72. S. Xiao, V. Drachev, A. Kildishev, X. Ni, U. Chettiar, H.-K. Yuan, V. Shalaev, Loss-free and active optical negative-index metamaterials. Nat. Lett. **466**, 735 (2010)
73. V. Pustovit, T. Shahbazyan, Phys. Rev. B **82**, 075429 (2010)
74. A. Schawlow, C. Townes, Phys. Rev. **112**, 1940 (1958)
75. V.S. Troitskii, Zh. Eksp. Teor. Fiz. **34**, 390 (1958) [Sov. Phys. JETP **7**, 271 (1958)]. Radiotekhn. Elektron. **3**, 1298, (1958)
76. J. Singer, *Masers* (Wiley, New York, 1959)
77. A. Malakhov, *Fluctuations in Self Oscillatory Systems* (Nauka, Moscow, 1968) (in Russian)
78. F. Arecchi, M. Scully, H. Haken, W. Weidlich, *Quantum Fluctuations of Laser Emission* (Mir, Moscow, 1974) (in Russian)
79. M. Lax, in *Statistical Physics, Phase Transitions and Superfluidity*, vol. 271, ed. by M. Chrétien, E.P. Gross, S. Deser (Gordon and Breach, New York, 1968)
80. A. Yariv, *Quantum Electronics*, 2nd edn. (Wiley, New York, 1975)
81. Y. Klimontovich (ed.), *Wave and Fluctuation Processes in Lasers* (Nauka, Moscow, 1974) (in Russian)
82. A. Oraevsky, J. Opt. Soc. Am. B **5**, 933 (1988)
83. M. Scully, M. Zubairy, *Quantum Optics* (Cambridge University Press, Cambridge, 1997)
84. G. Strakhovskii, A. Uspenskii, *Fundamentals of Quantum Electronics* (Vysshaia Shkola, Moscow, 1973) (in Russian)
85. R. Pantell, H. Puthoff, *Fundamentals of Quantum Electronics* (Wiley, New York, 1969)
86. S. Akhmanov, Y. D'yakov, A. Chirkin, *Statistical Radiophysics and Optics. Random Oscillations and Waves in Linear Systems* (Fizmatlit, Moscow, 2010) (in Russian)
87. S. Kuppens, M. van Exter, J. Woerdman, Quantum limited linewidth of a bad-cavity laser. PRL **72**, 3815 (1994)
88. A.Z. Khoury, M.I. Kolobov, L. Davidovich, Quantum-limited linewidth of a bad-cavity laser with inhomogeneous broadening. Phys. Rev. A **53**, 1120 (1996)
89. M. Exter, S. Kuppens, J. Woerdman, Theory for the linewidth of a bad-cavity laser. Phys. Rev. A **51**, 809 (1995)
90. A.S. Chirkin, A.V. Chipouline, Generalized expression for the natural width of the radiation spectrum of quantum oscillators. JETP Lett. **93**, 114 (2011)
91. P.E. Kloeden, E. Platen, *Numerical Solution of Stochastic Differential Equations (Stochastic Modelling and Applied Probability)* (Springer, Berlin, 2011)

Chapter 12
Plane Wave Propagation in Metamaterials with Gain

12.1 Introduction and General Approach

An overview of the interaction of the plasmonic nanostructures and quantum systems has been presented in the introductions to Chap. 10 and this chapter. In this chapter peculiarities appearing in the description of the plane wave propagation in MMs will be underlined, and then developed using the model discussed in Chap. 10.

As mentioned previously, the mitigation of the optical losses in MMs could be potentially achieved by a combination of lower loss materials [1, 2] and by providing gain by doping of the MMs with optically active emitters—see recent reviews [3, 4]. It has been shown experimentally that this form of loss compensation does not prohibit the negative index property of the MM [5]. In addition, the coupling with optically active emitters can compensate losses of the plasmonic components making them feasible for telecom applications [6–9]. Along with the plasmonic waveguides, other active components like modulators and switchers form a full-scale nomenclature for application in the next generation signal processing devices [10].

Several types of theoretical models have been suggested in order to describe gain processes in MMs and plasmonic waveguides. Analytical or semi-analytical models [11, 12] used the density matrix approach for the quantum dynamics description from the very beginning, but to the best of my knowledge have not been combined with the multipole approach [13] and consequently do not provide an adequate enough platform to investigate the properties of MMs fully. Instead, the vast majority of the publications have utilized the computational approach [4, 14] which gives results close to the experimentally realizable data. Unfortunately, in some cases the numerical approach cannot subdivide different physical effects and as a result hides or limits any physical insight. For example, for the case of competition between spaser eigen generation and dynamics driven by an external field the

A. Chipouline and F. Küppers, *Optical Metamaterials: Qualitative Models*, Springer Series in Optical Sciences 211, https://doi.org/10.1007/978-3-319-77520-3_12

numerical approaches mix all fields (generated by MAs and the external one) in one and thus far cannot demonstrate instability effect in the form given by analytical analysis [15].

In this chapter the multipole approach, in combination with the density matric formalism is used for establishing of the model for MMs with gain. This approach allows us to investigate analytically or semi-analytically the interplay between gain and magnetic properties of the MMs, the influence of internally unstable operation mode for spasers (MAs coupled with emitters, which the MMs consist of) on the propagation characteristics, and finally to optimize MM design. Moreover, the presented model is in line with the previously presented approach (actually is its natural extension on the problem of plane wave propagation) and from the other side pretty clear and observable, which makes the model a perfect platform for various university courses (see also Chap. 14).

In view of the results achieved in previous chapters it becomes clear that the process of the loss compensation depends significantly on the degree of coupling of the emitters and MAs. It is clear, that in a real experimental realization the MM will contain both types of emitters, coupled and uncoupled, which has to be incorporated into the model. Here, we assume that total averaged concentration of the emitters (number of the emitters in an elementary volume, containing one MA) is kept constant for all cases considered below. In contrast, the particular emitters coupled to the MAs δ can vary from zero ($\delta = 0$, no coupled emitters) to one ($\delta = 1$, all emitters are coupled); both of them contribute to the polarizability:

$$\begin{cases} \frac{\partial^2 E_x}{\partial y^2} + \frac{\omega^2}{c^2}\left(E_x + 4\pi P_x(y, \rho_{12}, \omega)\right) + \frac{i4\pi\omega}{c}\frac{\partial M_z(y,\rho_{12},\omega)}{\partial y} = 0 \\ P_x(y, \rho_{12}, \omega) = \eta q m_{\mathrm{s}} - \frac{\partial Q_{xy}}{\partial y} + \eta_{\mathrm{QS}}(1-\delta)\mu_{\mathrm{QS}}\tilde{\rho}^*_{12,\mathrm{un}} + \eta_{\mathrm{QS}}\delta\mu_{\mathrm{QS}}\tilde{\rho}^*_{12,\mathrm{c}} \\ Q_{xy}(y, \rho_{12}, \omega) = \eta q y_1 m_{\mathrm{a}} \\ M_z(y, \rho_{12}, \omega) = \frac{i\omega\eta q y_1}{c} m_{\mathrm{a}} \\ \frac{\mathrm{d}\tilde{\rho}_{12,\mathrm{un}}}{\mathrm{d}t} + \tilde{\rho}_{12,\mathrm{un}}\left(\frac{1}{\tau_2} + i(\omega - \omega_{21})\right) = \frac{i\mu_{\mathrm{QS}}E^*_{x,1}N_{\mathrm{un}}}{\hbar} + \xi_\rho \\ \frac{\mathrm{d}N_{\mathrm{un}}}{\mathrm{d}t} + \frac{(N_{\mathrm{un}} - N_0)}{\tau_1} = \frac{i\mu_{\mathrm{QS}}\left(E_{x,1}\tilde{\rho}_{12,\mathrm{un}} - E^*_{x,1}\tilde{\rho}^*_{12,\mathrm{un}}\right)}{2\hbar} \\ \frac{\mathrm{d}\tilde{\rho}_{12,\mathrm{c}}}{\mathrm{d}t} + \tilde{\rho}_{12,\mathrm{c}}\left(\frac{1}{\tau_2} + i(\omega - \omega_{21})\right) = \frac{i\alpha_x\left(m^*_{\mathrm{s}} + m^*_{\mathrm{a}}\right)N}{2\hbar} + \frac{i\mu_{\mathrm{QS}}E^*_{x,1}N_{\mathrm{c}}}{\hbar} + \xi_\rho \\ \frac{\mathrm{d}N_{\mathrm{c}}}{\mathrm{d}t} + \frac{(N_{\mathrm{c}} - N_0)}{\tau_1} = \frac{i\alpha_x\left((m_{\mathrm{s}} + m_{\mathrm{a}})\tilde{\rho}_{12,\mathrm{c}} - \left(m^*_{\mathrm{s}} + m^*_{\mathrm{a}}\right)\tilde{\rho}^*_{12,\mathrm{c}}\right) + 2i\mu_{\mathrm{QS}}\left(E_{x,1}\tilde{\rho}_{12,\mathrm{c}} - E^*_{x,1}\tilde{\rho}^*_{12,\mathrm{c}}\right)}{4\hbar} \\ 2(\gamma - i\omega)\frac{\mathrm{d}m_{\mathrm{s}}}{\mathrm{d}t} + \left(\omega_0^2 - \omega^2 - 2i\omega\gamma + \sigma\right)m_{\mathrm{s}} = \delta\alpha_\rho\tilde{\rho}^*_{12,\mathrm{c}} + \chi\left(E_{x,1} + E_{x,2}\right) + \xi_{\mathrm{ms}} \\ 2(\gamma - i\omega)\frac{\mathrm{d}m_{\mathrm{a}}}{\mathrm{d}t} + \left(\omega_0^2 - \omega^2 - 2i\omega\gamma - \sigma\right)m_{\mathrm{a}} = \delta\alpha_\rho\tilde{\rho}^*_{12,\mathrm{c}} + \chi\left(E_{x,1} - E_{x,2}\right) + \xi_{\mathrm{ma}} \end{cases} \tag{12.1}$$

Here $\tilde{\rho}_{12,\mathrm{un}}$ and N_{un} describe dynamics of the uncoupled emitters, and $\tilde{\rho}_{12,\mathrm{un}}$ and N_{un} are responsible for the coupled emitter dynamics respectively. For the case of simplest dipole-like MA system (12.1) becomes:

$$\begin{cases} \frac{\partial^2 E_x}{\partial y^2} - \frac{\omega^2}{c^2}\left(E_x + 4\pi P_x(y, \rho_{12}, \omega)\right) = 0 \\ P_x(y, \rho_{12}, \omega) = \eta q m_s + \eta_{QS}(1-\delta)\mu_{QS}\tilde{\rho}^*_{12,un} + \eta_{QS}\delta\mu_{QS}\tilde{\rho}^*_{12,c} \\ \frac{d\tilde{\rho}_{12,un}}{dt} + \tilde{\rho}_{12,un}\left(\frac{1}{\tau_2} + i(\omega - \omega_{21})\right) = \frac{i\mu_{QS}E^*_x N_{un}}{\hbar} + \xi_\rho \\ \frac{dN_{un}}{dt} + \frac{(N_{un} - N_0)}{\tau_1} = \frac{i\mu_{QS}\left(E_x\tilde{\rho}_{12,un} - E^*_x\tilde{\rho}^*_{12,un}\right)}{2\hbar} \\ \frac{d\tilde{\rho}_{12,c}}{dt} + \tilde{\rho}_{12,c}\left(\frac{1}{\tau_2} + i(\omega - \omega_{21})\right) = \frac{i\alpha_x m^*_s N}{2\hbar} + \frac{i\mu_{QS}E^*_x N_c}{\hbar} + \xi_\rho \\ \frac{dN_c}{dt} + \frac{(N_c - N_0)}{\tau_1} = \frac{i\alpha_x\left(m_s\tilde{\rho}_{12,c} - m^*_s\tilde{\rho}^*_{12,c}\right) + 2i\mu_{QS}\left(E_x\tilde{\rho}_{12,c} - E^*_x\tilde{\rho}^*_{12,c}\right)}{4\hbar} \\ 2(\gamma - i\omega)\frac{dm_s}{dt} + \left(\omega_0^2 - \omega^2 - 2i\omega\gamma + \sigma\right)m_s = \delta\alpha_\rho\tilde{\rho}^*_{12,c} + 2\chi E_x + \xi_{ms} \end{cases} \tag{12.2}$$

When all emitters are placed between MAs and are free of coupling ($\delta = 0$), the problem is reduced to the propagation of a plane wave with loss and gain, where gain dynamics and loss (plasmonic) dynamics are separated from each other:

$$\begin{cases} \frac{\partial^2 E_x}{\partial y^2} + \frac{\omega^2}{c^2}\left(E_x + 4\pi P_x(y, \rho_{12}, \omega)\right) + \frac{i4\pi\omega}{c}\frac{\partial M_z(y, \rho_{12}, \omega)}{\partial y} = 0 \\ P_x(y, \rho_{12}, \omega) = \eta q m_s - \frac{\partial Q_{xy}}{\partial y} + \eta_{QS}\mu_{QS}\tilde{\rho}^*_{12,un} \\ Q_{xy}(y, \rho_{12}, \omega) = \eta q y_1 m_a \\ M_z(y, \rho_{12}, \omega) = \frac{i\omega\eta q y_1}{c} m_a \\ \frac{d\tilde{\rho}_{12,un}}{dt} + \tilde{\rho}_{12,1}\left(\frac{1}{\tau_2} + i(\omega - \omega_{21})\right) = \frac{i\mu_{QS}E^*_{x,1}N_{un}}{\hbar} + \xi_\rho \\ \frac{dN_{un}}{dt} + \frac{(N_{un} - N_0)}{\tau_1} = \frac{i\mu_{QS}\left(E_{x,1}\tilde{\rho}_{12,un} - E^*_{x,1}\tilde{\rho}^*_{12,un}\right)}{2\hbar} \\ 2(\gamma - i\omega)\frac{dm_s}{dt} + \left(\omega_0^2 - \omega^2 - 2i\omega\gamma + \sigma\right)m_s = \chi\left(E_{x,1} + E_{x,2}\right) + \xi_{ms} \\ 2(\gamma - i\omega)\frac{dm_a}{dt} + \left(\omega_0^2 - \omega^2 - 2i\omega\gamma - \sigma\right)m_a = \chi\left(E_{x,1} - E_{x,2}\right) + \xi_{ma} \end{cases} \tag{12.3}$$

and no new physical phenomena are expected.

12.2 Propagation of a Plane Wave in a Metamaterial with Dipole-Like Metaatoms

12.2.1 Master System of Equations

First, the case of dipole-like MAs (12.2) is considered in the case of CW operation without noise terms, namely:

$$\begin{cases} \frac{\partial^2 E_x}{\partial y^2} + \frac{\omega^2}{c^2}(E_x + 4\pi P_x(y, \rho_{12}, \omega)) = 0 \\ P_x(y, \rho_{12}, \omega) = \eta q m_s + \eta_{QS}(1-\delta)\mu_{QS}\tilde{\rho}^*_{12,un} + \eta_{QS}\delta\mu_{QS}\tilde{\rho}^*_{12,c} \\ \tilde{\rho}_{12,un} = \frac{i\mu_{QS}E_x^* N_{un}}{2\hbar R_\rho} \\ N_{un} - N_0 = \frac{i\tau_1\mu_{QS}\left(E_x\tilde{\rho}_{12,un} - E_x^*\tilde{\rho}^*_{12,un}\right)}{4\hbar} \\ \tilde{\rho}_{12,c} = \frac{iN_c}{2\hbar R_\rho}\left(\alpha_x m_s^* + 2\mu_{QS}E_x^*\right) \\ N_c - N_0 = \frac{i\tau_1}{4\hbar}\left(\alpha_x\left(m_s\tilde{\rho}_{12,c} - m_s^*\tilde{\rho}^*_{12,c}\right) + 2\mu_{QS}\left(E_x\tilde{\rho}_{12,c} - E_x^*\tilde{\rho}^*_{12,c}\right)\right) \\ m_s = \frac{\delta\alpha_\rho\tilde{\rho}^*_{12,c} + 2\chi E_x}{R_s} \\ R_\rho = (1 + i(\omega - \omega_{21})\tau_2) \\ R_s = \left(\omega_0^2 - \omega^2 - 2i\omega\gamma + \sigma\right) \end{cases} \tag{12.4}$$

The Helmholz propagation equation for the electric field can be transformed using Slowly Varying Approximation (SVA):

$$E_x(y) = A_x(y)\exp(iky) + \frac{\omega^2}{c^2}(E_x(y,\omega) + 4\pi P_x(y,\omega)) = 0$$
$$\Rightarrow \frac{\partial^2 E_x(y,\omega)}{\partial y^2} = \frac{\partial^2 A_x(y,\omega)}{\partial y^2} + 2ik\frac{\partial A_x(y,\omega)}{\partial y} - k^2 A_x \approx 2ik\frac{\partial A_x(y,\omega)}{\partial y} - k^2 A_x \tag{12.5}$$

Note, that all equations for $\tilde{\rho}_{12,un}$, $\tilde{\rho}^*_{12,c}$, N_{un}, N_c, and m_s remain invariant under transformation:

$$\begin{cases} E_x \to A_x\exp(iky) \\ \tilde{\rho}_{12,un} \to \tilde{\rho}_{12,un}\exp(-iky) \\ \tilde{\rho}_{12,c} \to \tilde{\rho}_{12,c}\exp(-iky) \\ m_s \to m_s\exp(iky) \end{cases} \tag{12.6}$$

Combination of (12.5) and (12.6) finally results in:

$$\begin{cases} ik\frac{\partial A_x(y,\omega)}{\partial y} + 2\pi\frac{\omega^2}{c^2}\left(\delta\left(q\eta\frac{\alpha_\rho}{R_s} + \eta_\rho\mu_{QS}\right)\rho^*_{12,c}(\omega, A_x) + \eta_\rho\mu_{QS}(1-\delta)\rho^*_{12,un}(\omega, A_x)\right) = 0 \\ \tilde{\rho}_{12,un} = \frac{i\mu_{QS}A_x^* N_{un}}{2\hbar R_\rho} \\ N_{un} - N_0 = \frac{i\tau_1\mu_{QS}\left(A_x\tilde{\rho}_{12,un} - A_x^*\tilde{\rho}^*_{12,un}\right)}{4\hbar} \\ \tilde{\rho}_{12,c} = \frac{iN_c}{2\hbar R_\rho}\left(\alpha_x m_s^* + 2\mu_{QS}A_x^*\right) \\ N_c - N_0 = \frac{i\tau_1}{4\hbar}\left(\alpha_x\left(m_s\tilde{\rho}_{12,c} - m_s^*\tilde{\rho}^*_{12,c}\right) + 2\mu_{QS}\left(A_x\tilde{\rho}_{12,c} - A_x^*\tilde{\rho}^*_{12,c}\right)\right) \\ m_s = \frac{\delta\alpha_\rho\tilde{\rho}^*_{12,c} + 2\chi A_x}{R_x} \\ R_\rho = (1 + i(\omega - \omega_{21})\tau_2) \\ R_s = \left(\omega_0^2 - \omega^2 - 2i\omega\gamma + \sigma\right) \end{cases} \tag{12.7}$$

where the following substitution has been made $k^2 = \frac{\omega^2}{c^2}\left(1 + 4\pi q \eta \frac{\chi}{R_s}\right)$. System (12.7) is the master system of equations to describe propagation of the plane wave in a MM with MAs, consisting of plasmonic nanoresonators coupled with QS and free QS embedded in the matrix of the MM itself. In the case of Quantum Dots, for example, as QS, (12.7) describes loss compensation and respective optical properties of the MM.

Solutions of (12.7), especially transition region, can only be found numerically. Nevertheless, even before the solutions are demonstrated, one can qualitatively predict different operation modes using the results obtained in Chap. 11, where stable and unstable operation modes of the spaser have been found and analyzed.

12.2.2 Loss Compensation by Completely Uncoupled QS ($\delta = 0$)

In this case (12.7) becomes:

$$\begin{cases} ik\frac{\partial A_x(y,\omega)}{\partial y} + 2\pi\frac{\omega^2}{c^2}\eta_\rho \mu_{QS}\rho^*_{12,\mathrm{un}}(\omega, A_x) = 0 \\ \tilde{\rho}_{12,\mathrm{un}} = \frac{i\mu_{QS}A^*_x N_{\mathrm{un}}}{2\hbar R_\rho} \\ N_{\mathrm{un}} - N_0 = \frac{i\tau_1\mu_{QS}\left(A_x\tilde{\rho}_{12,\mathrm{un}} - A^*_x\tilde{\rho}^*_{12,\mathrm{un}}\right)}{4\hbar} \\ m_s = \frac{2\chi A_x}{R_x} \\ R_\rho = (1 + i(\omega - \omega_{21})\tau_2) \\ R_s = \left(\omega_0^2 - \omega^2 - 2i\omega\gamma + \sigma\right) \end{cases} \quad (12.8)$$

Solution of (12.7) describes rather trivial loss compensation in a media with dipoles and will not be considered here; more information about this kind of problems can be found in any textbook about optical amplifiers, see for example [16]. Under appropriate choice of parameters (QS concentration and pump) plane wave stabilizes own intensity by saturation and propagates stably.

The results of the solution of the propagation (12.8) are presented in Fig. 12.1, where amplitude A_x of the propagating wave is presented in color grade as a function of frequency and propagation length.

The parameters of the MM (nanoresonator concentration, QS concentration, pump etc.) have been chosen in order to reach full compensation, so that amount of gain provided to the plane wave is enough to compensate the losses. The stable propagation can be achieved around the central gain frequency of the QS, which is chosen the same as the resonant frequency of the nanoresonators.

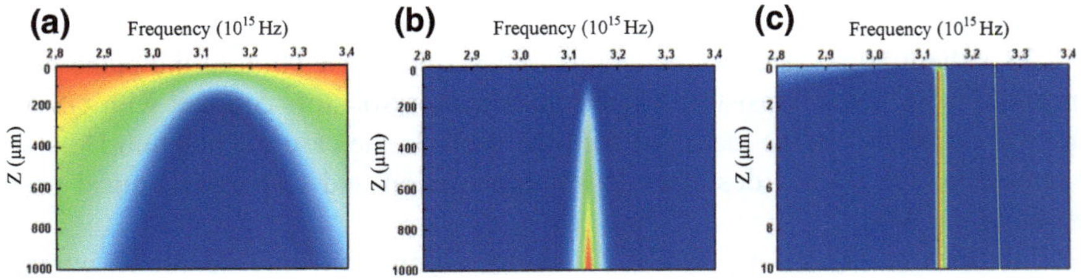

Fig. 12.1 **a** Propagation of the plane wave in the case of MM with dipole-like MAs without QS, blue color corresponds to lower amplitudes. Peak of the losses corresponds to the resonance frequency for a single nanoresonator $\omega_0 = 3.14 \times 10^{15}$ Hz. **b** Propagation of the plane wave in media filled by pumped QS with the resonant frequency $\omega_{21} = 3.14 \times 10^{15}$ Hz. **c** Propagation of the plane wave in the case of completely uncoupled QS $\delta = 0$ in MM with dipole-like MAs, $\omega_0 = \omega_{21} = 3.14 \times 10^{15}$ Hz. Parameters (QS concentration and pump) are chosen in order to provide full loss compensation, which is achieved around QS peak gain frequency

12.2.3 Loss Compensation by Completely Coupled QS ($\delta = 1$)

In this case (12.7) is reduced to the system, which describes the situation considered, for example, in [12]:

$$\begin{cases} ik\frac{\partial A_x(y,\omega)}{\partial y} + 2\pi\frac{\omega^2}{c^2}\left(q\eta\frac{2\alpha_\rho}{R_x} + \eta_\rho\mu_{\mathrm{QS}}\right)\rho^*_{12,\mathrm{c}}(\omega, A_x) = 0 \\ \tilde{\rho}_{12,\mathrm{c}} = \frac{iN_\mathrm{c}}{2\hbar R_\rho}\left(\alpha_x m^*_\mathrm{s} + 2\mu_{\mathrm{QS}}A^*_x\right) \\ N_\mathrm{c} - N_0 = \frac{i\tau_1}{4\hbar}\left(\alpha_x\left(m_\mathrm{s}\tilde{\rho}_{12,\mathrm{c}} - m^*_\mathrm{s}\tilde{\rho}^*_{12,\mathrm{c}}\right) + 2\mu_{\mathrm{QS}}\left(A_x\tilde{\rho}_{12,\mathrm{c}} - A^*_x\tilde{\rho}^*_{12,\mathrm{c}}\right)\right) \\ m_\mathrm{s} = \frac{\delta\alpha_\rho\tilde{\rho}^*_{12,\mathrm{c}} + 2\chi A_x}{R_x} \\ R_\rho = (1 + i(\omega - \omega_{21})\tau_2) \\ R_\mathrm{s} = \left(\omega_0^2 - \omega^2 - 2i\omega\gamma + \sigma\right) \end{cases} \tag{12.9}$$

In this case full loss compensation coincides with the threshold of the spaser generation—which means, that the spaser's own dynamics will be affected by the external wave, resulting in potentially unstable operation [15], see also Chap. 11. To this extent, the results about instability of operation of the MM under full loss compensation correspond to the conclusion of [12]. Nevertheless, stable operation is still possible (see Chap. 11) for not completely coupled case. On the other hand, it is hard to believe that under real experimental situations all emitters appear to be coupled with the plasmonic nanoresonators. Therefore, an experimental realization of the case $\delta = 1$ and respective instability prognoses remain rather exotic ones.

The solutions of propagation (12.9) are presented in Fig. 12.2. In this case peak luminescence coincided (as before) with the nanoresonator frequency $\omega_{21} = \omega_0$ and the complex structure of the spectrum is caused by strong coupling between the QS and the nanoresonators. It should be emphasized that for the same parameters, but without coupling the amplitude of the propagating wave cannot be stabilized for any

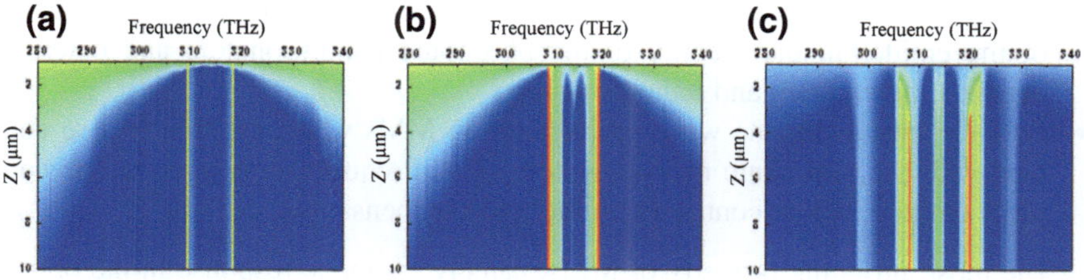

Fig. 12.2 Propagation of the plane wave in the case of totally coupled QDs in MM with dipole-like MAs. The amount of QD is fixed, $\omega_{12} = \omega_0 = 314\,\text{THz}$, N_0 (the pump rate) is varied 0.93 0.97 0.98 for (**a**), (**b**), and (**c**) respectively. One can see that there are certain frequencies where losses are totally compensated and stable propagation is possible. Amount of such frequencies increases as the pump rate increases

frequency. It proves in turn that the coupling between the QS and the nanoresonator reduces significantly an amount of QS necessary for the full loss compensation and is crucial for the experimentally achievable full loss compensation in MMs.

12.2.4 Loss Compensation by Partially Coupled QS $(0 < \delta < 1)$

It has been shown that the coupling of plasmonic nanoresonators to QSs increases amplification and offers a preferable realization of loss compensation [17]. As just discussed, a fully coupled realization tends to result in unstable operation [15]. In the intermediate case (12.7) the number of coupled emitters is almost sufficient to cross the generation threshold. The rest of losses will be compensated by free emitters. This seems to be perfect solution and sets a clear strategy for the design of the MMs with fully compensated losses. The solution, qualitatively looks similar to the one presented in Fig. 12.2 and is not shown here. The main difference is in the wider region of parameters for stable propagation (the parameters correspond to the stability region in terms of Chap. 11). The full analysis of stability of the plane wave propagation in MMs with gain has not been published and is a subject for further investigation.

12.3 Propagation of Plane Wave in Metamaterial with Quadrupole-Like Metaatoms

12.3.1 Master System of Equations

In summary, consideration of dipole-like MAs and the respective MMs reveal mainly two new physical phenomena, namely:

- Enhancement of the amplification processes due to the appropriate positioning of the emitter in the vicinity of the plasmonic nanoresonators and coupling between the emitters and nanoresonators;
- Possibly unstable plane wave propagation in MMs where all emitters are coupled with the plasmonic nanoresonators, which requires non zero free (uncoupled) emitters which contribute to the loss compensation.

The most intriguing property of MMs, namely magnetic response in the optical frequency domain, cannot be considered in the framework of the dipole-like MAs and requires a full consideration of multipoles, namely magnetic dipole and quadrupole terms. In such a case, the system (12.1) will be reduced to the system of equation for SVA in analogy with the consideration of the dipole-like MAs. Firstly, system (12.1) for CW operation and without stochastic terms becomes:

$$\begin{cases} \frac{\partial^2 E_x}{\partial y^2} + \frac{\omega^2}{c^2}\left(E_x + 4\pi P_x(y, \rho_{12}, \omega)\right) + \frac{i4\pi\omega}{c}\frac{\partial M_z(y,\rho_{12},\omega)}{\partial y} = 0 \\ P_x(y, \rho_{12}, \omega) = \eta q m_\mathrm{s} - \frac{\partial Q_{xy}}{\partial y} + \eta_\mathrm{QS}(1-\delta)\mu_\mathrm{QS}\tilde{\rho}^*_{12,\mathrm{un}} + \eta_\mathrm{QS}\delta\mu_\mathrm{QS}\tilde{\rho}^*_{12,\mathrm{c}} \\ Q_{xy}(y, \rho_{12}, \omega) = \eta q y_1 m_\mathrm{a} \\ M_z(y, \rho_{12}, \omega) = \frac{i\omega\eta q y_1}{c} m_\mathrm{a} \\ \rho_{12,\mathrm{un}} = \frac{i\mu_\mathrm{QS}\tau_2 E^*_{x,1} N_\mathrm{un}}{\hbar R_\rho} \\ N_\mathrm{un} - N_0 = \frac{i\mu_\mathrm{QS}\tau_1\left(E_{x,1}\tilde{\rho}_{12,\mathrm{un}} - E^*_{x,1}\tilde{\rho}^*_{12,\mathrm{un}}\right)}{2\hbar} \\ \tilde{\rho}_{12,\mathrm{c}} = \frac{i\tau_2}{2\hbar R_\rho}\left(\alpha_x\left(m^*_\mathrm{s} + m^*_\mathrm{a}\right)N + 2\mu_\mathrm{QS}E^*_{x,1}N_\mathrm{c}\right) \\ N_\mathrm{c} - N_0 = \frac{i\tau_1}{4\hbar}\left(\alpha_x\left((m_\mathrm{s} + m_\mathrm{a})\tilde{\rho}_{12,\mathrm{c}} - \left(m^*_\mathrm{s} + m^*_\mathrm{a}\right)\tilde{\rho}^*_{12,\mathrm{c}}\right) + 2i\mu_\mathrm{QS}\left(E_{x,1}\tilde{\rho}_{12,\mathrm{c}} - E^*_{x,1}\tilde{\rho}^*_{12,\mathrm{c}}\right)\right) \\ m_\mathrm{s} = \frac{\delta\alpha_\rho\tilde{\rho}^*_{12,\mathrm{c}} + \chi\left(E_{x,1} + E_{x,2}\right)}{R_\mathrm{s}} \\ m_\mathrm{a} = \frac{\delta\alpha_\rho\tilde{\rho}^*_{12,\mathrm{c}} + \chi\left(E_{x,1} - E_{x,2}\right)}{R_\mathrm{a}} \\ R_\rho = \left(1 + i(\omega - \omega_{21})\tau_2\right) \\ R_\mathrm{s} = \left(\omega_0^2 - \omega^2 - 2i\omega\gamma + \sigma\right) \\ R_\mathrm{a} = \left(\omega_0^2 - \omega^2 - 2i\omega\gamma - \sigma\right) \end{cases} \quad (12.10)$$

In contrast to the previous case, with just dipole-like MAs, one is required to take into account both quadrupole and magnetic dipole moments in the propagation equation, which are included in the equation through the first derivatives. Moreover, $E_{x,1}$ and $E_{x,2}$ are the fields driving upper and lower nanowires respectively. Following [13] it is assumed that:

$$E_{x,1} = E_x \quad (12.11)$$

and:

$$E_{x,2} = E_x \exp(iky_1) \quad (12.12)$$

k is the wave vector, and y_1 is the distance between nanowires, see Fig. 4.1. As in [13], the wave vector must be found by consideration of the propagation equation. Substituting (12.11) and (12.12) into (12.10), and taking into account that $ky_1 \ll 1$, system (12.10) becomes:

$$\begin{cases} \frac{\partial^2 E_x}{\partial y^2} + \frac{\omega^2}{c^2}\left(E_x + 4\pi P_x(y,\rho_{12},\omega)\right) + \frac{i4\pi\omega}{c}\frac{\partial M_z(y,\rho_{12},\omega)}{\partial y} = 0 \\ P_x(y,\rho_{12},\omega) = \eta q m_s - \frac{\partial Q_{xy}}{\partial y} + \eta_{QS}(1-\delta)\mu_{QS}\tilde{\rho}^*_{12,un} + \eta_{QS}\delta\mu_{QS}\tilde{\rho}^*_{12,c} \\ Q_{xy}(y,\rho_{12},\omega) = \eta q y_1 m_a \\ M_z(y,\rho_{12},\omega) = \frac{i\omega\eta q y_1}{c} m_a \\ \tilde{\rho}_{12,un} = \frac{i\mu_{QS}\tau_2 E^*_x N_{un}}{\hbar R_\rho} \\ N_{un} - N_0 = \frac{i\mu_{QS}\tau_1\left(E_x\tilde{\rho}_{12,un} - E^*_x\tilde{\rho}^*_{12,un}\right)}{2\hbar} \\ \tilde{\rho}_{12,c} = \frac{i\tau_2}{2\hbar R_\rho}\left(\alpha_x(m^*_s + m^*_a)N + 2\mu_{QS}E^*_x N_c\right) \\ N_c - N_0 = \frac{i\tau_1}{4\hbar}\left(\alpha_x\left((m_s + m_a)\tilde{\rho}_{12,c} - (m^*_s + m^*_a)\tilde{\rho}^*_{12,c}\right) + 2i\mu_{QS}\left(E_x\tilde{\rho}_{12,c} - E^*_x\tilde{\rho}^*_{12,c}\right)\right) \\ m_s = \frac{\delta\alpha_\rho\tilde{\rho}^*_{12,c} + 2\chi E_x\left(1-(ky_1)^2\right)}{R_s} \\ m_a = \frac{\delta\alpha_\rho\tilde{\rho}^*_{12,c} + 2i\chi E_x(ky_1)}{R_a} \\ R_\rho = (1 + i(\omega - \omega_{21})\tau_2) \\ R_s = \left(\omega_0^2 - \omega^2 - 2i\omega\gamma + \sigma\right) \\ R_a = \left(\omega_0^2 - \omega^2 - 2i\omega\gamma - \sigma\right) \end{cases} \tag{12.13}$$

Reduction of (12.13) to the SVA can be done following the same steps as (12.5), (12.6) with the additional assumption that $\frac{\partial}{\partial y}Q_{xy}(y,\rho_{12},\omega) \to ik\eta q y_1 m_a$ and $\frac{\partial}{\partial y}Q_{xy}(y,\rho_{12},\omega) \to ik\eta q y_1 m_a$; wave vector is also set to $\frac{\partial}{\partial y}Q_{xy}(y,\rho_{12},\omega) \to ik\eta q y_1 m_a$. Finally (12.13) becomes:

$$\begin{cases} ik\frac{\partial A_x(y,\omega)}{\partial y} + 2\pi\frac{\omega^2}{c^2}\delta\left[\left(q\eta\alpha_\rho\left(\frac{1}{R_s} - \frac{2iky_1}{R_a}\right) + \eta_\rho\mu_{QS}\right)\rho^*_{12,c}(\omega, A_x)\right. \\ \left. + \eta_\rho\mu_{QS}(1-\delta)\rho^*_{12,un}(\omega, A_x) + 2\chi E_x(ky_1)^2\left(\frac{2}{R_a} - \frac{1}{R_s}\right)\right] = 0 \\ \tilde{\rho}_{12,un} = \frac{i\mu_{QS}\tau_2 E^*_x N_{un}}{\hbar R_\rho} \\ N_{un} - N_0 = \frac{i\mu_{QS}\tau_1\left(E_x\tilde{\rho}_{12,un} - E^*_x\tilde{\rho}^*_{12,un}\right)}{2\hbar} \\ \tilde{\rho}_{12,c} = \frac{i\tau_2}{2\hbar R_\rho}\left(\alpha_x(m^*_s + m^*_a)N + 2\mu_{QS}E^*_x N_c\right) \\ N_c - N_0 = \frac{i\tau_1}{4\hbar}\left(\alpha_x\left((m_s + m_a)\tilde{\rho}_{12,c} - (m^*_s + m^*_a)\tilde{\rho}^*_{12,c}\right) + 2i\mu_{QS}\left(E_x\tilde{\rho}_{12,c} - E^*_x\tilde{\rho}^*_{12,c}\right)\right) \\ m_s = \frac{\delta\alpha_\rho\tilde{\rho}^*_{12,c} + 2\chi E_x\left(1-(ky_1)^2\right)}{R_s} \\ m_a = \frac{\delta\alpha_\rho\tilde{\rho}^*_{12,c} + 2i\chi E_x(ky_1)}{R_a} \\ R_\rho = (1 + i(\omega - \omega_{21})\tau_2) \\ R_s = \left(\omega_0^2 - \omega^2 - 2i\omega\gamma + \sigma\right) \\ R_a = \left(\omega_0^2 - \omega^2 - 2i\omega\gamma - \sigma\right) \end{cases} \tag{12.14}$$

It should be noted that the first (propagation) equation in (12.14) is reduced to (12.7), i.e. to the equation for propagation in a MM with dipole-like MAs.

However, dynamics of $\tilde{\rho}_{12,c}$ in case (12.14) will be affected by the antisymmetric mode and, therefore even in case of very closely spaced nanowires (when retardation can be neglected, which mathematically is expressed by $k \to 0$), the propagation equation will have its own peculiarities.

12.3.2 *Loss Compensation by Completely Uncoupled QS* ($\delta = 0$)

In this chapter, only preliminary data of the solution of (12.14) are presented and a more comprehensive consideration which will be done elsewhere. Figure 12.3 shows propagation of the plane wave in (a) MMs with quadrupoles-like MAs only and (b) quadrupoles-like MAs and uncoupled QS. In the case of the quadrupole-like MAs there are three characteristic frequencies, namely the eigen frequency of a single nanoresonator (which remains the same as previously) $\omega_0 = 314\,\text{THz}$, symmetric and antisymmetric eigen frequencies of the quadrupole-like MAs (coupled nanoresonators) $\omega_s = 324\,\text{THz}$, $\omega_a = 304\,\text{THz}$ respectively.

Stable propagation is achievable for both symmetric and antisymmetric eigenfrequencies, provided parameters for the full compensation (position of the gain maximum, concentration of the QS and pump level) are appropriately chosen. In Fig. 12.3 the stable propagation at the symmetric eigen frequency is shown; the stable propagation at the antisymmetric frequency looks the same and is not presented here. It has to be emphasized that all optical properties of the MM (like dielectric and magnetic response and consequently possibility of the negative refraction) are affected differently: an imaginary part of the effective dielectric constant in the stationary state is zero, while effective magnetic constant remains unaffected. It means in turn, that the loss compensation with the uncoupled QS

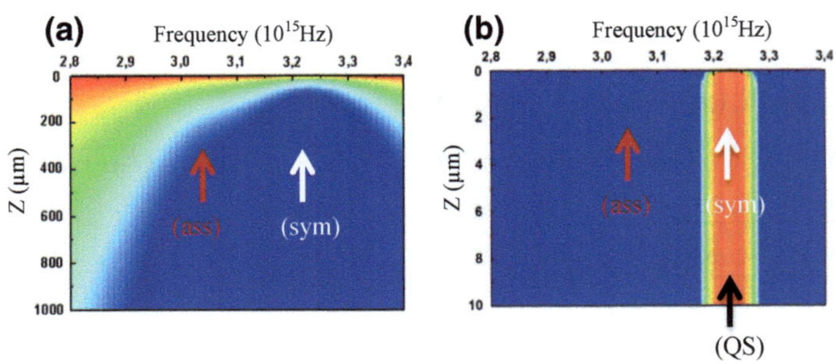

Fig. 12.3 **a** Propagation of the plane wave through the MM (no loss compensation) consisting of quadrupole-like MAs. Two deeps correspond for antisymmetric 3.04 THz and symmetric 3.24 THz modes. Losses near symmetric mode are stronger than near antisymmetric. **b** Loss compensation regime with uncoupled QSs near the symmetric mode 3.24 THz. Positions of the symmetric, antisymmetric, and QS gain peak frequencies are shown by arrows

enhances the effect of negative refraction; from the other side, number of the QSs necessary for the full compensation in the case of uncoupled QSs is pretty high and requires special technological methods.

12.3.3 Loss Compensation by Completely Coupled QS ($\delta = 1$)

For the case of fully coupled QS the physical picture becomes (in compare with 12.2.3, see also Fig. 12.2) even more reach and complicated. Due to the saturation caused nonlinearity the modes become coupled and energy transfer between the modes takes place. Moreover, there are three potential variants of the central gain positioning: one can place the QS gain center coinciding with the eigen nanoresonator frequency 3.14 THz as it was assumed for dipole-like MAs (see Figs. 12.1 and 12.2), and one can match the QS peak gain frequency with the symmetric or antisymmetric eigen frequencies. All three options have been investigated and the results are presented in Fig. 12.4. It has been found, that the resulted frequency pattern (Fig. 12.4) becomes extremely complicated and does not guarantee the stable propagation at the initially found eigen frequencies for symmetric, antisymmetric, and single nanoresonator oscillations. From the other side, stable propagation is possible and the frequency position of the stable propagation depends on the coupling efficiency.

In real experimental realization the coupling efficiency is expected to be spatially inhomogeneous, and the resulted pattern is supposed to be even more complicated. In spite of the fact, that the positions of the stably propagating frequencies seem to be hardly predictable, the coupling (as it was mentioned for the dipole-like MAs) leads to significant reduction of the number of QS necessary for the full loss compensation, and from this point of view remains preferable way of the QS positioning in MM.

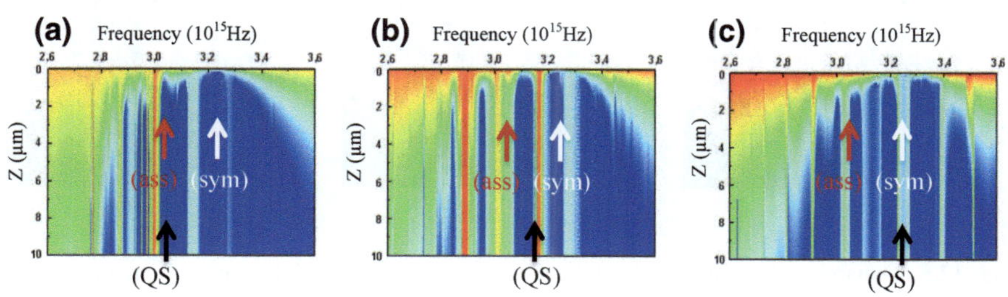

Fig. 12.4 Propagation of the plane wave in the case of totally coupled QS in MM with quadrupole-like MAs. The amount of QS is fixed, ω_{QS} is varied **a** 3.04 × 10^{15} Hz, **b** 3.14 × 10^{15} Hz, and **c** 3.24 × 10^{15} Hz respectively, $N_0 = 0.97$. There are certain frequencies (pattern depends on positioning of the QS peak gain) where losses are totally compensated and stabile propagation is possible. Positions of the symmetric, antisymmetric, and QS gain peak frequencies are shown by arrows

12.3.4 Dynamics of Symmetric and Antisymmetric Modes in MAs at the Propagation in Case of Completely Coupled QS ($\delta = 1$)

The magnetic response of the MM differs MM from any natural material, and it would be extremely interesting to keep this property at the loss compensation scenario as well.

In the frame of the developed in this work approach, the magnetic response depends on the magnitude of the antisymmetric oscillation mode—both magnetic and quadrupole moments are proportional to the magnitude of the antisymmetric mode. Here the magnitudes of the both antisymmetric m_a and symmetric m_s modes and their relation m_a/m_s have been extracted and plotted in Fig. 12.5. It is seen,

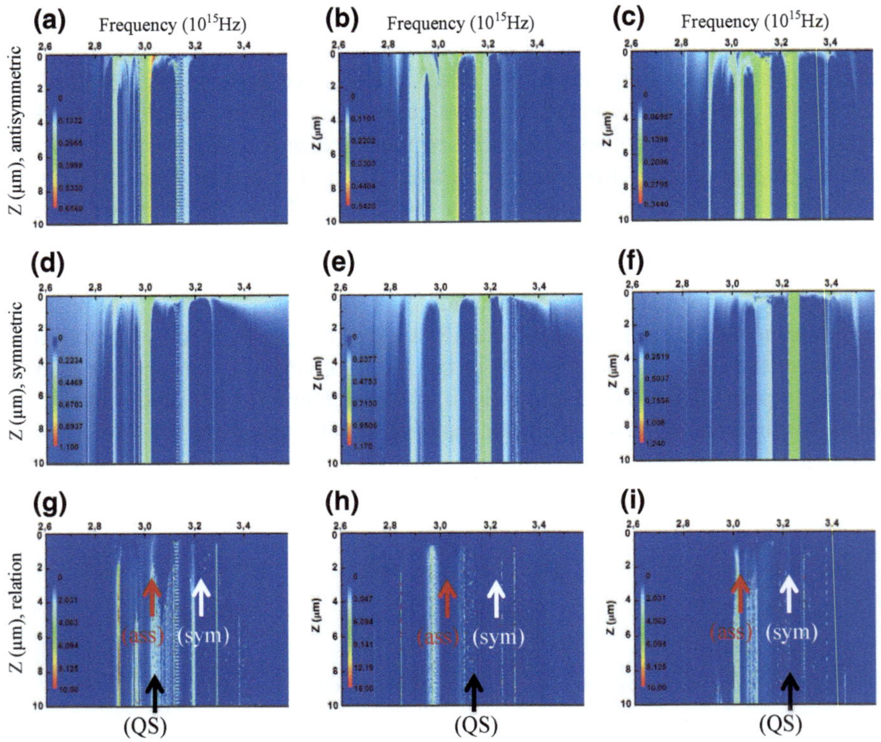

Fig. 12.5 Dynamics of the amplitude of the **a–c** antisymmetric m_a, **d–f** symmetric m_s modes, and (**g–i**) their relation m_a/m_s as a function of frequency for the same parameters as shown in Fig. 12.4: propagation of the plane wave in the case of totally coupled QS in MM with quadrupole-like MAs. The amount of QD is fixed, ω_{QS} is varied **a**, **d**, **g** 3.04×10^{15} Hz (eigen frequency of the symmetric mode of the quadrupole-like MAs), **b**, **e**, **h** 3.14×10^{15} Hz (eigen frequency of the eigen mode of the single dipole-like MAs), and **c**, **f**, **i** 3.24×10^{15} Hz (eigen frequency of the antisymmetric mode of the MAs), $N_0 = 0.97$. Positions of the symmetric, antisymmetric, and QS gain peak frequencies are shown by arrows

that clear domination of the antisymmetric mode (and consequently maximum magnetic response) could be obtained for the QS with the peak gain frequency around eigen frequency of a single nanoresonator. The results presented in Fig. 12.5 indicate that the energy of the inverted QS is transferred to the MAs and is redistributed between the symmetric and antisymmetric modes. Basically, the results reveal complex dynamics of the two nonlinearly coupled modes, where one of the modes is supported by a propagating plane wave. It has to be emphasized, that in this case possible nonlinear interaction between several plane waves was not considered: according to the accepted here SVA approximation (12.5), (12.6) propagation of the external monochromatic wave was assumed. Experimentally it corresponds to the short propagation distances, where generation of new harmonics due to the nonlinear interaction is not significant; this operation mode is expected in the case of single or even multilayer MMs, which can be produced using the state of the art technologies. Nevertheless, in the case of simultaneous propagation of several waves with different carrier frequencies this effect has to be taken into account.

It is expected, that this kind of parametric interaction could be used for loss compensation as well; from the purely theoretical point of view, this parametric nonlinear interaction between modes is also interesting for the investigation of possible generation of the multifrequency stable states (colored stable states) which could probably exist in MMs. Otherwise, according to the results presented in Fig. 12.5 the propagation of a monochromatic plane wave is obviously unstable: the energy of the initial monochromatic mode will be redistributed over many frequencies as the plane wave propagates in the MM.

12.4 Conclusion

In conclusion, the problem of monochromatic plane wave propagation in the MM consisting of quantum MAs has been considered. The system of equations describing the plane wave propagation in the MM with the quantum MAs has been elaborated. Using the results of the previous chapters about the existence of stable and unstable generation modes of the MA, the stable propagation mode has been observed for the case of the fully uncoupled and fully coupled QSs. It has been found, that for the same set of parameters the uncoupled QSs cannot compensate for the optical losses in the MM, which proves that the coupling between the plasmonic nanoresonators and QSs significantly increases the loss compensation efficiency. The case of the plane wave propagation in the MM with the quadrupole-like MAs has been considered only for the monochromatic plane wave propagation and so requires further investigation. The full analysis of stability of the plane wave propagation in MMs with quadrupoles and gain has not been published and is a subject for further investigation.

References

1. A. Boltasseva, H. Atwater, Low-loss plasmonic metamaterials. Science **331**, 290 (2011)
2. P. Tassin, T. Koschny, M. Kafesaki, C. Soukoulis, A comparison of graphene, superconductors and metals as conductors for metamaterials and plasmonics. Nat. Photonics **6**, 259 (2012)
3. S. Wuestner, A. Pusch, K. Tsakmakidis, J. Hamm, O. Hess, Gain and plasmon dynamics in active negative-index metamaterials. Phil. Trans. R. Soc. A **369**, 3525 (2011)
4. O. Hess, J.B. Pendry, S.A. Maier, R.F. Oulton, J.M. Hamm, K.L. Tsakmakidis, Active nanoplasmonic metamaterials. Nat. Mater. **11**, 573 (2012)
5. S. Xiao, V. Drachev, A. Kildishev, X. Ni, U. Chettiar, H.-K. Yuan, V. Shalaev, Loss-free and active optical negative-index metamaterials. Nat. Lett. **466**, 735 (2010)
6. J. Grandidier, G. Colas des Francs, S. Massenot, A. Bouhelier, L. Markey, J.-C. Weeber, C. Finot, A. Dereux, Gain-assisted propagation in a plasmonic waveguide at telecom wavelength. Nano Lett. **9**, 2935 (2009)
7. I. De Leon, P. Berini, Amplification of long-range surface plasmons by a dipolar gain medium. Nat. Photonics **6**, 16 (2012)
8. I. Radko, M. Nielsen, O. Albrektsen, S. Bozhevolnyi, Stimulated emission of surface plasmon polaritons by lead-sulphide quantum dots at near infra-red wavelengths. Opt. Express **18**, 18633 (2010)
9. A. Krasavin, T. Vo, W. Dickson, P. Bolger, A. Zayats, All-plasmonic modulation via stimulated emission of copropagating surface plasmon polaritons on a substrate with gain. Nano Lett. **11**, 2231 (2011)
10. L. Cao, M. Brongersma, Active photonics: ultrafast developments. Nature Photon. **3**, 12 (2009)
11. M. Wegener, J. García-Pomar, C. Soukoulis, M. Nina Meinzer, S. Linden, Toy model for plasmonic metamaterial resonances coupled to two-level system gain. Opt. Express **16**, 19785 (2008)
12. M. Stockman, Spaser action, loss-compensation, and stability in plasmonic systems with gain. PRL **106**, 156802 (2011)
13. J. Petschulat, C. Menzel, A. Chipouline, C. Rockstuhl, A. Tünnermann, F. Lederer, T. Pertsch, Multipole approach to metamaterials. Phys. Rev. B **78**, 043811 (2008)
14. A. Fang, Th Koshny, C. Sokoulis, Self-consistent calculations of loss-compensated fishnet metamaterials. Phys. Rev. B **82**, 121102(R) (2010)
15. E. Andrianov, A. Pukhov, A. Dorofeenko, A. Vinogradov, A. Lisyansky, Forced synchronization of spaser by an external optical wave. Opt. Express **19**, 24849 (2011)
16. E. Desurvire, Erbium-doped fiber amplifiers: principles and applications, Willey-Interscience, ISBN 0-471-26434-2
17. S. Wuestner, J.M. Hamm, A. Pusch, F. Renn, K. Tsakmakidis, O. Hess, Control and dynamic competition of bright and dark lasing states in active nanoplasmonic metamaterials. Phys. Rev. B **85**, 201406(R) (2012)

Chapter 13
Relaxation of Inverted Quantum System Coupled with Metallic Nanoobjects

13.1 Introduction

The problem of the relaxation dynamics of a two-level quantum system (QS) is considered in the frame of the developed in [1] approach and compared with the known and widely used math tools. The commonly accepted in publications approach appears to be questionable from the point of view of basic principles.

The relaxation time modification has been considered in different realizations (including photonic crystals [2] and plasmonic waveguides [3]), but here to be precise only the coupling between the localized nano modes (e.g. localized plasmons) and two-level QSs is considered. Since seminal paper of Purcell [4], and (in more than 30 years) fundamental theoretical works [5, 6], several excellent reviews [7–9] and books about Cavity Quantum Electrodynamics [10, 11] have been published; a full and comprehensive publication reference list can be found there. Among the recent achievements it should be mentioned the experimental realization of the plasmonic patch antenna for control of the spontaneous emission [12] (which has been theoretically considered in [13]) and observation of the ultrafast Rabi oscillation [14], which manifestoes the strong coupling regime [15, 16]. The strong coupling assumes positioning of the quantum system in close proximity to the nanoresonator, which in turns requires inclusion of higher (than dipole) order modes in nanoresonators and dipole-forbidden transitions in the QSs [17–19]. Coupling with realistic shapes of the nanoresonators has been considered in [20] and [21]. The peculiarities of the disordered plasmonic on the radiative and non-radiative relaxation rates have been considered in Cazé et al. [22], and influence of the plasmonic cloak has been investigated in [23]. Accurate calculation of the Green function for complex plasmonic structures allowed estimating the spatially resolved Purcell effect for hyperbolic metamaterials [24] and wire metamaterials [25]. The relaxation time modification due to the presence of a nanoresonator [26], has been taken into account in the modelling of spaser dynamics [27, 28] using basically the same approach as in Novotny and Hecht [29].

A. Chipouline and F. Küppers, *Optical Metamaterials: Qualitative Models*, Springer Series in Optical Sciences 211, https://doi.org/10.1007/978-3-319-77520-3_13

It is worth noting again (see Chap. 9) that the relaxation processes can be described either in the frame of classical or quantum approaches. In the frame of classical approach (ME consisting of the field and charge dynamic equations) there is no steady state solution for the atoms/molecules, because of the electrons in the atoms/molecules lose energy due to the radiation losses at the rotation around the nuclear. From the other side, ME are not compatible with the quantum mechanic approach, i.e. ME, rigorously speaking, cannot be coupled with the density matrix (DM) equations. In spite of that, ME are used with the DM equations to describe optical gain, laser generation etc. (e.g. see Chaps. 9–12). It is stipulated rather by fact that this approach leads to the conclusions supported by the respective experimental confirmations, but not by a rigorous elaboration from basic principles. Nevertheless, in order to describe relaxation processes in the frame of DM, ME will not be used.

13.2 The Accepted Approach for Estimation of Purcell Effect

Analysis of the "state of the art" of the activity in area of the nanoplasmonic assisted relaxation time modification would worth to start from several definitions, namely—what exactly do we mean under "Purcell effect"? From the experimental point of view, the observed in the tests relaxation time modification (without any details about an actual functional form of the relaxation curves and discussion about relaxation channels) is called Purcell effect. In most cases (not only in experimental, but also in vast majority of the theoretical publications) Purcell effect itself is not distinguished from Purcell factor, which is the ratio between the relaxation rates in vacuum and near the nanostructures. This assumes that the relaxation follows exponential time dependence in both cases, namely in free space and in the case of coupling with the nanoresonators. Unfortunately, there is neither theoretical nor experimental evidence that the relaxation in general has an exponential shape; moreover, both experimental [8, 12] and theoretical results [30] show the non-exponential decay law. In addition, it has to be mentioned that the evaluation expressions for the Purcell coefficient, based on the estimated values for the effective mode volume and resonator quality factor, in contrast to the microcavities does not give correct estimation for the plasmonic nanoantenna [31]. Hereafter, it is not apriori assumed that the decay follows an exponential law. It will be shown that the exponent is a good approximation only in the case of relaxation caused by interaction with stochastic environment.

The relaxation dynamics of the QS interacting with a plasmonic nanoresonator can be considered classically and quantum mechanically, we start from the latter one. Due to methodological reasons it is worth to remind basics of the relaxation in a free space. First, it has to be realized that in the frame of Schrödinger equation for the QS in free space there is no relaxation until there is no external fields. The relaxation appears only as a result of consideration of the vacuum fluctuations of the electromagnetic field (second quantization). The spontaneous emission cannot be considered as just influence of the vacuum fluctuations on the QS in the frame of the first perturbation order, but rather as a coupled dynamics of the system "field plus QS". This fundamental question has been perfectly explained in Ginzburg [32], see also different points of view on this problem in Weisskopf [33], Weisskopf [34] and Fermi [35].

A derivation of the spontaneous decay rate in the frame of the quantum electro-dynamic approach can be found in textbooks, for example in Novotny and Hecht [29]. Consideration of the coupled "electric field plus QS" transition rate γ according to the Fermi's Golden Rule gives finally:

$$\begin{aligned} \gamma &= \frac{2\omega}{3\hbar\varepsilon_0}|\mu|^2\rho_\mu(\vec{r}_0,\omega_0) \\ \rho_\mu(\vec{r}_0,\omega_0) &= 3\sum_k\left[\vec{n}_\mu\cdot\left(\vec{u}_k\vec{u}_k^*\right)\cdot\vec{n}_\mu\right]\delta(\omega_k-\omega_0) \end{aligned} \tag{13.1}$$

Here $\vec{\mu}$ is the dipole moment of the transition of the QS, $\vec{\mu}=\mu\vec{n}_\mu$, $\vec{r}_0$ is position of the QS, $\rho_\mu(\vec{r}_0,\omega_0)$ is the partial local density of states (LDOS), $\vec{u}_k$ are the normal modes, $\vec{n}_\mu$ is the unit vector in the direction of $\vec{\mu}$. The electric field is expanded over spatially dependent positive and negative frequency parts:

$$\begin{aligned} \vec{E} &= \sum_k\left[\vec{E}_k^+a_k(t)+\vec{E}_k^-a_k^\dagger(t)\right] \\ \vec{E}_k^+ &= \sqrt{\frac{\hbar\omega_k}{2\varepsilon_0}}\vec{u}_k,\quad \vec{E}_k^- = \sqrt{\frac{\hbar\omega_k}{2\varepsilon_0}}\vec{u}_k^* \end{aligned} \tag{13.2}$$

where:

$$a_k(t)=a_k(0)\exp(-i\omega_kt),\quad a_k^\dagger(t)=a_k^\dagger(0)\exp(i\omega_kt) \tag{13.3}$$

The relaxation dynamics of the probability of the QS to be found in the excited state (population of the excited state) decays exponentially. Decay rate (13.1) describes relaxation into the vacuum field modes $\vec{u}_k$. It is important to underline at this point two facts. First, this decay rate assumes the radiation rate, which means that all the transitions give one photon per transition into the far or near field zones depends on what kind of modes $\vec{u}_k$ are: if the QS is situated in the near/far field zone, then $\vec{u}_k$ are the near/far field zone modes. Second, the interaction with vacuum field has a stochastic character: the vacuum fluctuations occur randomly with some probability. Quantum character of these fluctuations is in the fact, that the virtual photons can not be absorbed and pump a QS, but only stimulate emission of an extra photon in a pumped QS. This important quantum property leads to the appearance of the relaxation in the first order of perturbation expansion—in contrast with a classical case (e.g. interaction with the fluctuations in a thermo bath), where the relaxation appears in the second order of perturbation. We see later, that this differentiates the relaxation due to the vacuum fluctuations and due to the interaction with the environmental classical fluctuations.

The problem now is reduced to the finding of the functional form for the LDOS. There is an elegant way using the methods of Green functions [29], which the vast majority of the recently appeared papers are based on. The eigen functions $\vec{u}_k$ satisfy Helmholz equation, which is for generally inhomogeneous and dispersive (and consequently due to the causality principle lossy) media is:

$$\begin{gathered}\nabla \times \nabla \times \vec{u}_k(\vec{r}, \omega_k) - \varepsilon(\vec{r}, \omega_k)\frac{\omega_k^2}{c^2}\vec{u}_k(\vec{r}, \omega_k) = 0 \\ \nabla \times \nabla \times \overleftrightarrow{G}(\vec{r}_0, \vec{r}, \omega_k) - \varepsilon(\vec{r}, \omega_k)\frac{\omega_k^2}{c^2}\overleftrightarrow{G}(\vec{r}_0, \vec{r}, \omega_k) = \overleftrightarrow{I}\delta(\vec{r}_0 - \vec{r})\end{gathered} \tag{13.4}$$

Here Green function $G(\vec{r}_0, \vec{r}, \omega_k)$ is introduced and $\overleftrightarrow{I}$ is the unit dyad. At the elaboration of the expression for the LDOS through the Green function, it is necessary to assume that the modes $\vec{u}_k$ are orthogonal, which is straightforward for any media without losses, i.e. constant and real permittivity $\mathrm{Re}\{\varepsilon(\vec{r}, \omega_k)\} \neq 0$, $\mathrm{Im}\{\varepsilon\} = 0$. In this case the $\vec{u}_k$ are orthogonal:

$$\int \vec{u}_k(\vec{r}, \omega_k)\vec{u}^*_{k'}(\vec{r}, \omega_{k'})\mathrm{d}^3\vec{r} = \delta_{kk'} \tag{13.5}$$

and expressions for the relaxation rate and LDOS are:

$$\begin{gathered}\gamma = \frac{2\omega_0}{3\hbar\varepsilon_0}|\mu|^2\rho_\mu(\vec{r}_0, \omega_0) \\ \rho_\mu(\vec{r}_0, \omega_0) = \frac{6\omega_0}{3\hbar\varepsilon_0}\left[\vec{n}_\mu \cdot \mathrm{Im}\left\{\overleftrightarrow{G}(\vec{r}_0, \vec{r}_0, \omega_0)\right\} \cdot \vec{n}_\mu\right]\end{gathered} \tag{13.6}$$

Finally, LDOS for the homogeneous isotropic media $\mathrm{Re}\{\varepsilon(\vec{r}, \omega_k)\} = \varepsilon(\omega_k)$, $\mathrm{Im}\{\varepsilon\} = 0$ takes well-known form of the black-body radiation:

$$\rho_\mu(\omega_0) = \varepsilon\frac{\omega_0^2}{\pi^2 c^3} \tag{13.7}$$

And the decay rate is:

$$\gamma = \frac{\varepsilon^{3/2}\omega_0^3|\mu|^2}{3\pi\varepsilon_0\hbar c^3} \tag{13.8}$$

In the case of a dielectric inhomogeneous media ($\varepsilon(\vec{r}, \omega)$, $\mathrm{Im}(\varepsilon(\vec{r}, \omega)) = 0$ which is a good approximation for dielectric in transmission windows) the procedure remains the same excepting more complicated Green function and consequently spatially dependent LDOS and decay rate.

In the case of significant losses (as it takes place for the metallic nanoresonators) orthogonality condition (13.5) fails due to the nonzero imaginary part of the permittivity, which makes (13.4) non-Hermitian. More sophisticated math tools have been suggested for this case [36, 37]; moreover, applicability of the electrodynamic approach was shown to be valid for the case of the very small distances from the metallic surface [38]. Nevertheless, wide application of the Green function in its original formulation (13.6) seems to be stipulated at the moment rather by its relative simplicity than by a full justification of this method for the metallic (dispersive and lossy) nanoresonators [24, 25].

In the frame of the same electrodynamic approach it is rather straightforward to calculate radiative and nonradiative decay rates. Both classical [5] and electrodynamic approaches [6] have been proposed. Taking into account radiative losses in addition to the Joule ones for the nanoresonators, the radiative decay rate can be calculated by energy flux through the surface of the sphere surrounded both the QS and nanoresonator, while the nonradiative decay rate is determined by the calculating of the power dissipated in the nanoresonator due to the Joule heating [30].

13.3 Concerns About the Commonly Accepted Approach

From the other side, the physical picture behind this approach appears to be to some extend controversial.

First, this picture being applied for the near field (zero wave vectors) contradicts to the relativistic principles, which is not surprising and most probably has to be accepted.

Second, all math manipulations with the Green functions and eigen modes assumes that the conditions for the eigen modes to be excited are fulfilled, and that all free electrons in the nanoparticle participate in this mode move coherently. This assumption does not seem to be realistic. In fact, the plasmonic mode relaxation time is about 100 fs, which is comparable with the coherence time of the spontaneous photons, and under this condition the eigen mode just does not have enough time to be formed. This ambiguity is connected with the logic jump between the expansion of the filed over eigen functions (13.4) for the homogeneous and inhomogeneous spaces. In fact, for the homogeneous space microscopic Maxwell equations are used, while introduction of permittivity assumes that we work with macroscopic Maxwell equations. At this transition (homogenization procedure) the electron dynamics is assumed to be coherent, which is questionable for the considered case of a single photon.

Another concern about the use of the eigen mode appearslevel system with pump when we consider an elementary act of the absorption. In fact, one particular photon gets absorbed just by one particular electron, which is a free one, because of the dynamics of the electron does not depend on the nano sizes of the nanoresonator—dynamics of the free electron is not quantized and influences dynamics of the other electrons (model of free electron gas with interaction) after, roughly speaking, the same relaxation time of about 100 fs.

One more concern is connected with the discussed above peculiarities of the vacuum fluctuations. The quantum nature does not possess the process of absorption of the vacuum fluctuation photons by a QS, which in turn leads to the appearance of the relaxation effect in the first order of the perturbation series. The self-influence, described by the Green function at its origin (see [29]), is already the second order of the perturbation theory which evidently exceeds accuracy of the model (remind that the radiative relaxation process is described in the frame of the first perturbation order). It is worth noting here that in the frame of the density matrix approach (will be considered later) the relaxation processes due to the interaction with a thermo bath (classical fields) is described by the second order perturbation terms. In the frame of the Green function approach, the first order perturbation would correspond to the influence of the nanoresonator on the LDOS by taking the Green function not at the origin of the QS, but rather between the nanoresonator and the QS $\overleftrightarrow{G}(\vec{r}_0, \vec{r}_0, \omega_k) \rightarrow \overleftrightarrow{G}(\vec{r}_{\mathrm{NR}}, \vec{r}_0, \omega_k)$, here $\vec{r}_{\mathrm{NP}}$ is the center of the nanoresonator. This qualitative discussion shows that the interpretation of the relaxation rate modification in terms of the classical notations most probably requires further corrections.

13.4 Can Quantum Dynamics Be Described by Harmonic Oscillator Equation?

Now, let us briefly consider the classical approach. In the frame of this approach the quantum dynamics is described by a harmonic oscillator equation. Let us first investigate under which approximation this approach remains valid. We start from the density matrix equation with relaxation in the basis of the eigen function of the unperturbed Hamiltonian for two level system with pump [see Chap. 9 for details, namely (9.4)]:

$$\begin{cases} \frac{d\rho_{12}}{dt} + i\omega_{12}\rho_{12} + \frac{\rho_{12}}{\tau_2} = -\frac{iH_{12}N}{\hbar} \\ \frac{dN}{dt} + \frac{N-N_0}{\tau_1} = -\frac{2iH_{12}\left(\rho_{12}-\rho_{12}^*\right)}{\hbar} \\ \tau_1 = \frac{\tilde{\tau}_1}{W\tilde{\tau}_1+1} \end{cases} \tag{13.9}$$

Introducing $P = \rho_{12} + \rho_{12}^*$, one can finally reduce (13.9) to the following form:

$$\begin{cases} \frac{d^2P}{dt^2} + \frac{2}{\tau_1}\frac{dP}{dt} + \left(\frac{1}{\tau_1^2} + \omega_{21}^2\right)P = \frac{2\omega_{21}H_{12}N}{\hbar} \\ \frac{dN}{dt} + \frac{N-N_0}{\tau_2} = -\frac{2H_{12}}{\hbar\omega_{21}}\left(\frac{dP}{dt} + \frac{P}{\tau_1}\right) \end{cases} \tag{13.10}$$

The first equation in (13.10) could be taken as a trivial harmonic oscillator one in case when these two equations are separable, i.e. N does not depend on P. It could be a good approximation for the unpumped QS without saturation, i.e. under low intensity external Hamiltonian H_{12}; in this case $N \approx N_0 = -1$. In contrast, in case of the relaxation dynamics it is assumed that N evolves from $N = 1$ to $N = -1$, which excludes any reasonable justification for the separation of the first equation as a harmonic oscillator one.

In the case of harmonic oscillator equation the damping parameter is supposed to be responsible for the both line bandwidth and the relaxation time. Let us consider a typical quantum dot with the typical bandwidth of about 50 nm around central wavelength of 1000 nm. Inverse bandwidth spectrum width gives us the approximated relaxation time:

$$\tau \approx \frac{1}{\Delta\nu} = \frac{\lambda^2}{c\Delta\lambda} \sim 100\,\text{fs} \tag{13.11}$$

From the other side, the measured in the tests relaxation time is typically about 1–10 ns. In the frame of the density matrix approach there is no any contradictions, because of there are two different times for the spectrum width (phase relaxation time) and for the energy relaxation time, namely τ_2 and τ_1. The relaxation time, which can be calculated using (13.1) is the energy relaxation time τ_1, because of this time assumes energy transition with photon emission; the τ_2 describes change in the phase of the eigen state without energy transition. Clear, that the harmonic oscillator equation neither can be elaborated within some reasonable approximations, nor can give even qualitative explanation for the different relaxation times and spectrum bandwidth of the QS. It has to be pointed out, that this is again consequence of the quantum character of the QS. For example, in the case of the relaxation of the plasmonic oscillations the relaxation time (again about 100 fs) matches pretty good with the observed bandwidth of the plasmonic resonances which is again comparable with one for the QDs and is typically about 30–50 nm. It means, that the plasmonic dynamics can be safely considered in the frame of classical approach (i.e. harmonic oscillator equation).

13.5 Relaxation in the Frame of Density Matrix Formalism

It is necessary to remind about physical picture behind the elaboration of the density matrix (DM). First, the DM approach and the Schrödinger equation are equivalent and could be transformed one from another [39, 40]. A QS interacting with the external fields is considered. The full Hamiltonian is subdivided by three main parts: the eigen Hamiltonian H_0 which describes the internal energy structure of the QS without any external fields; the Hamiltonian of the interaction between the QS and the stochastic parts of the external fields V_{st}, and the Hamiltonian of the interaction between the QS and the regular part of the external fields V_r. After expansion over the eigen functions of the eigen Hamiltonian H_0 (particles plus field), the interaction with the stochastic and regular Hamiltonians leads to the qualitatively different terms in the final equations. All interactions with the stochastic fields are considered in the frame of the perturbation theory. The interaction with the vacuum fluctuation ("quantum" part of the stochastic interaction) appears in the first order of the perturbations and leads to the relaxation with the time $\tau_{1,\mathrm{r}}$ which stands for the radiative relaxation time; this is exactly the relaxation time found in (13.1). Interaction with the classical part of the stochastic field appeared in the second order of the perturbation, because of in the first order the fluctuation with zero mean value (all stochastic parts are assumed to have zero mean value) does not cause any changes—the probability to excite the QS is exactly the same as the probability for the stimulated transition back to its ground state. This is an important difference between the vacuum (quantum) field fluctuations and the classical ones, which has been already mentioned above. The second order gives a nonzero effect, which appears as an extra nonradiative relaxation time $\tau_{1,\mathrm{nr}}$. The total energy relaxation time is given by:

$$\frac{1}{\tau_1} = \frac{1}{\tau_{1,\mathrm{r}}} + \frac{1}{\tau_{1,\mathrm{nr}}} \tag{13.12}$$

It is important to understand, that the second order of perturbation means self-action, i.e. the QS acts on itself through the in-phase excitation of the external modes. For example, in the case of a thermo bath the QS excites some modes which in turn act back on the QS; due to the phase synchronization the action is not zeroed under the averaging.

The approach with the interaction with the thermo bath also allows us to overcome the fundamental problem of losses. In order to get an orthonormal set of the eigen functions (which is widely used), the Hamiltonian has to be Hermitian. From the other side, the losses have to be somehow incorporated into the consideration. The model with infinite number of eigen modes, weakly coupled with the considered QS allows us to keep the Hamiltonian Hermitian, and at the same time introduce the relaxation processes for the QS. In fact, to this extend the density matrix approach is free from the discussed above problem with the imaginary part of permittivity, which makes eigen functions of (13.5) nonorthogonal and all consequent conclusions (13.6) not fully justified.

The second relaxation time τ_2 appears at the consideration of the phase (not amplitude!) changes of the eigen states. In fact, the probability of the phase changes without the amplitude changes, i.e. without the transitions between the energy levels, is much higher than the transition between the energy levels. Due to this fact, the phase relaxation time τ_2 appears to be shorter (or equal in the limit) than the energy relaxation time τ_1:

$$\frac{1}{\tau_2} = \frac{1}{\tau_1} + \Delta \tag{13.13}$$

It is possible to write down an expression for the extra term Δ using the respective matrix elements of the stochastic Hamiltonian of interaction [41]:

$$\Delta = \frac{1}{\hbar^2} \int w \mathrm{d}\Gamma \int_{-\infty}^{\infty} \mathrm{d}t$$

$$\int_{-\infty}^{t} \mathrm{d}\tau \Big\{ V_{\mathrm{st},1111}(t) V_{\mathrm{st},1111}(\tau) \rho_{11}^{(0)} + V_{\mathrm{st},1112}(t) V_{\mathrm{st},1122}(\tau) \rho_{22}^{(0)} \tag{13.14}$$

$$+ V_{\mathrm{st},2211}(t) V_{\mathrm{st},2211}(\tau) \rho_{11}^{(0)} + V_{\mathrm{st},2222}(t) V_{\mathrm{st},2222}(\tau) \rho_{22}^{(0)} \Big\}$$

Here $V_{\mathrm{st},ijkl}$ are the matrix elements of the stochastic Hamiltonian of interaction, $\rho_{ij}^{(0)}$ are the unperturbed density matrix elements, w and $\mathrm{d}\Gamma$ are the probability of interaction and elementary volume over all parameters of the interaction, which w depends on.

The last Hamiltonian of interaction with the regular fields V_{r} does not require the perturbation approach and is included in (13.9) rigorously as H_{12}. The so called strong coupling regime is one of the phenomena described by this Hamiltonian. From the other side, this interaction can also be treated in the frame of the perturbation approach, but by no means this can be (in general) reduced to the extra relaxation term and packed into the two relaxation times τ_1 and τ_2.

13.6 Physical Picture of Interaction Between QS and Nanoresonator

After having all this said, let us consider the physical picture of interaction between the QS and a nanoresonator in the frame of the density matrix approach. The presence of the nanoresonator modifies the LDOS, which leads to the modification of the radiative relaxation time. The energy of the excitation of the QS can also be transferred to the stochastic and regular modes of the nanoresonators, coupled to the QS. The stochastic modes are just thermal stochastic electron dynamics, while the regular modes are the collective electron dynamics, i.e. the plasmon modes. In order to subdivide these dynamics (i.e. probability of the contribution into stochastic or regular dynamics), the standard approach of the probability distribution function (PDF) of the electrons in the nanoresonator $f_{\mathrm{el}}(t)$ can be used. The kinetic equation is:

$$\frac{\partial f_{\mathrm{el}}}{\partial t} + \left(\vec{v} \cdot \vec{\nabla}_{\mathrm{r}} f_{\mathrm{el}}\right) + \frac{q}{m}\left(\vec{E} \cdot \vec{\nabla}_{\mathrm{v}} f_{\mathrm{el}}\right) + S = 0 \tag{13.15}$$

Here $\nabla_{\mathrm{r}} = i\frac{\partial}{\partial x} + j\frac{\partial}{\partial y} + k\frac{\partial}{\partial z}$, $\nabla_{v} = i\frac{\partial}{\partial v_x} + j\frac{\partial}{\partial v_y} + k\frac{\partial}{\partial v_z}$, S is the impact integral, and $\vec{E}$ is the electric field produced by the QS and is proportional in the near field approximation to the dipole moment of the QS $\vec{E} = \vec{n}_{\mu} \alpha_{\mathrm{nr}}(\rho_{12} + \rho_{21})$, q and m are the charge and mass of electron respectively. Without an external electric field the PDF is just known Maxwell distribution function $f_{\mathrm{el},0} = \left(\frac{m}{2\pi T}\right)^{3/2} \exp\left(-\frac{mv^2}{2T}\right)$, T is the temperature. In the case of an external field the PDF gets modified with typical response time of $\tau_{\mathrm{el}} \sim S^{-1}$, which for typical experimental realizations is about 100 fs, i.e. comparable with τ_2. Remind that the introduction of the permittivity and consequently all the concept of the eigen plasmonic functions of the nanoresonator assumes that the PDF is already stabilized, i.e. for the times much longer than the 100 fs. Taking into account, that the duration of the elementary relaxation process is anyway less or comparable with the 100 fs, we come to the conclusion, that the concept of the eigen functions is rigorously speaking not applicable for the description of the relaxation processes. Instead, we might consider a coupled system of equations which includes density matrix and PDF equations, namely (13.9) and (13.15) and calculate modification of the PDF for the respective excited current in the nanoresonator. This excited current influences back the QS, which actually means second perturbation order or, in other words, self-influence of the QS due to the presence of the nanoresonator. It is also important to include radiative losses into the electron dynamics, i.e. in (13.15), which can be done by an extra term for the impact integral S. This allows us to evaluate the respective contributions to the observed in the tests luminescence enhancement. This approach to the description of the spontaneous relaxation modification has not yet been elaborated, but definitively deserves further consideration and discussion.

13.7 On the Luminescent Measurement

Now it is worth to comment shortly the term of "luminescence enhancement" and discuss the relation between the theoretical picture and observed experiments values. From the point of view of the energy conservation law, any absorbed (pumping) photon can be or reemitted, or relaxed into the nonradiative modes, which corresponds to the final relaxed value of $N = -1$ without pump $W = 0$. The radiated intensity dynamics (which is really measured in the tests) in the case of absence of the nanoresonator follows the exponential relaxation law:

$$I_{\mathrm{rad}}(t) \sim \frac{(1 + N(0)) \exp\left(-\frac{t}{\tau_1}\right) - 1}{\tau_{1,\mathrm{r}}} \tag{13.16}$$

It is important to point out, that the relaxation dynamics follows the total energy relaxation time τ_1 (not only radiative relaxation time $\tau_{1,\mathrm{r}}$!), while an amplitude of the radiated intensity is inversely proportional to the radiative relaxation time $\tau_{1,\mathrm{r}}$ only. In case of CW operation mode at $W\tilde{\tau}_1 > 1$, the inversion is $N = N_0$ and the radiated intensity I_{rad} is:

$$I_{\text{rad}} \sim \frac{N_0(W)}{\tau_{1,\text{r}}} = \frac{W\tilde{\tau}_1 - 1}{\tilde{\tau}_{1,\text{r}}} \sim \begin{cases} W, & \tilde{\tau}_{1,\text{r}} \ll \tilde{\tau}_{1,\text{nr}}, \quad W\tilde{\tau}_{1,\text{r}} \gg 1 \\ W\frac{\tilde{\tau}_{1,\text{nr}}}{\tilde{\tau}_{1,\text{r}}}, & \tilde{\tau}_{1,\text{r}} \gg \tilde{\tau}_{1,\text{nr}}, \quad W\tilde{\tau}_{1,\text{nr}} \gg 1 \end{cases} \tag{13.17}$$

It is hard to separate experimentally an influence of the pump efficiency and the effects caused by relaxation time changes even in the case of a free standing molecule.

13.8 Time Dynamics of Relaxation in Presence of Nanoresonator

Let us now consider changes associated with the coupling of the QS and nanoresonator. The system of equations describing the coupled dynamics has been proposed in [see e.g. 1] and for the case of the relaxation dynamics is written as:

$$\begin{cases} \frac{d\tilde{\rho}_{12}}{dt} + \tilde{\rho}_{12}\left(\frac{1}{\tau_2} + i(\omega - \omega_{21})\right) = \frac{i\alpha_x \tilde{x}^* N}{\hbar} + \xi_\rho \\ \frac{dN}{dt} + \frac{(N - N_0)}{\tau_1} = \frac{i\alpha_x\left(\tilde{x}\tilde{\rho}_{12} - \tilde{x}^*\tilde{\rho}_{12}^*\right)}{2\hbar} \\ 2(\gamma - i\omega)\frac{d\tilde{x}}{dt} + \left(\omega_0^2 - \omega^2 - 2i\omega\gamma\right)\tilde{x} = \alpha_\rho \tilde{\rho}_{12}^* + \xi_x \end{cases} \tag{13.18}$$

Here both relaxation times τ_1 and τ_2 are assumed to be modified due to the presence of the extra stochastic interaction with the non-coherent electrons in the nanoresonator:

$$\begin{cases} \frac{1}{\tau_1} = \frac{P_{\text{r}}}{\tau_{1,\text{r}}} + \frac{P_{\text{nr}}}{\tau_{1,\text{nr}}} \\ \frac{1}{\tau_2} = \frac{1}{\tau_1} + \Delta(P) \end{cases} \tag{13.19}$$

According to (13.14), the extra part differentiating energy and phase relaxation time $\Delta(P)$ is also modified, which is underlined here by the presence on P_{r} and P_{nr} in (13.19). Method of the calculations or estimations for the both Purcell factors P_{r} and P_{nr} is not discussed here, as well as the estimation of $\Delta(P)$. An exact solution for the relaxation dynamics of (13.18) cannot be found analytically, and is supposed in general to be calculated numerically. The results of the typical dynamics (13.18) are presented in Fig. 13.1.

The dynamics is clearly subdivided by two stages: the first one for the time less than the phase relaxation time τ_2 exhibits Rabi oscillations, typical for the strong coupling, and the rest is just trivial exponential relaxation dynamics. This behaviour has been experimentally measured in Vasa et al. [14]. Note that the density matrix approach allows us to consider strong and weak coupling in the frame of the same model, namely in the frame of (13.18): these two regimes correspond to high and low values of the coupling constants α_ρ and α_x; of course, each set of the coupling coefficients corresponds to the respective set of Purcell factors P_{r} and P_{nr}. Both the

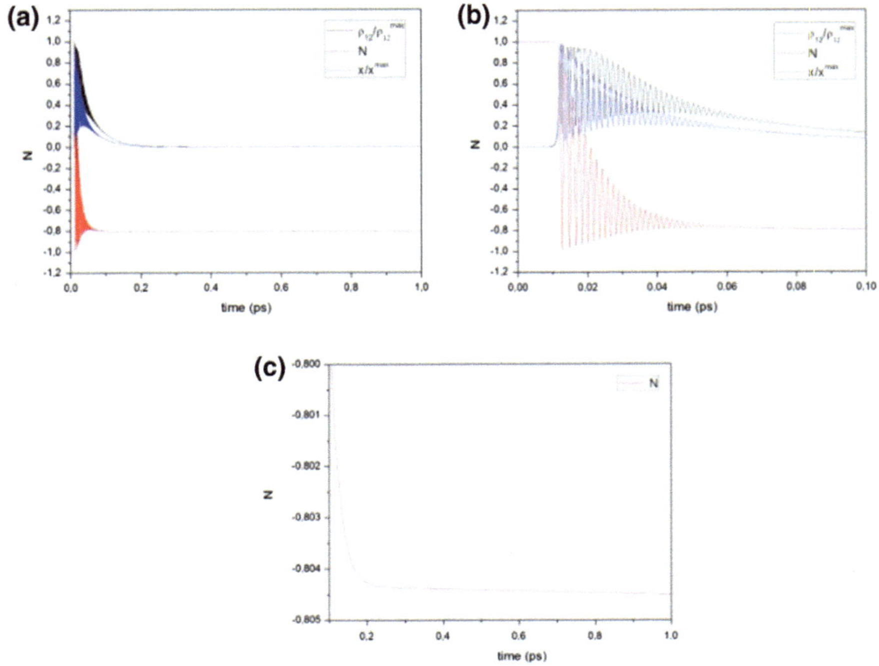

Fig. 13.1 Relaxation dynamics of the system of coupled QS and plasmonic nanoresonator at full resonance $\omega = \omega_0 = \omega_{12}$. The full dynamics (**a**) consists of the fast oscillation at the times less than the phase relaxation time (**b**) and the exponential relaxation for the rest time (**c**). Please, keep in mind the different time scales in (**a–c**)

coupling coefficients and Purcell factor values depend on the distance and mutual positions of the QS and nanoresonator; the method of estimations for these coefficients (for example, used in Stockman [28]) is out of the scopes of this consideration and will be discussed elsewhere.

The dynamics of (13.18) is calculated for the initial conditions $N(t=0) > 0$ (inversion) and $N_0 = -1$ (absence of pump). For the case of the CW operation mode it is assumed that $N_0 > 0$ (permanent pump). N and other two variables ρ_{12} and x are assumed to be found for the stationary case (i.e. $\frac{\mathrm{d}}{\mathrm{d}t} = 0$). As it has been discussed in previous chapters, this system of equations describes the stationary and stochastic (with extra Langevin terms) dynamics of the nanolaser/spaser [26, 28], [Bergman 06]. It has to be realized, that system (13.18) possesses two types of solutions—regular and stochastic. The regular nontrivial solution exists only for the pump rates above the threshold one $W > W_{\mathrm{th}}$; otherwise the regular solution is zero. In contrast, random fluctuations, described by correlators $\langle x(\omega)x^*(\omega')\rangle$, $\langle \rho_{12}(\omega)\rho_{12}^*(\omega')\rangle$ and $\langle x(\omega)\rho_{12}^*(\omega')\rangle$, $\langle x^*(\omega)\rho_{12}(\omega')\rangle$ are nonzero and give extra factors to the estimation for the luminescent enhancement. These calculations are also left for the future works; here it is worth noting that the problem of spontaneous emission enhancement in the frame of the developed approach is methodologically (but not physically!) equivalent to the estimation of the spontaneous emission from any laser below generation threshold, and to this extend is not new (excepting the necessity of the relaxation times modification).

Qualitatively, the luminescent intensity modification can be caused by several factors. First, at the resonance between nanoresonator eigen mode and the pump wavelengths any nanoresonator works as an optical antenna [42], which enhances pump rate efficiency W and therefore increases $N_0(W)$ in (13.17); the coupling between the QS and nanoresonator out of resonance is small and could be neglected to the first approximation. But even in the case of resonance with the emission wavelength (and thus nonresonant for pumping wavelength), the effective pump rate is also enhanced by the presence of the metallic nanoresonator. According to (13.17), both nominator and denominator turn out to be modified and it is extremely hard to separate experimentally these two effects. The enhancement from the coupling with the regular parts of the electron dynamics is proportional to the correlations $\langle x(\omega)\rho_{12}^*(\omega')\rangle, \langle x^*(\omega)\rho_{12}(\omega')\rangle$ [see second equation in (13.18)] and could be roughly estimated by the relation of the phase and the total relaxation times τ_2/τ_1 which is about 10^{-3}. It means that the part of the energy, emitted due to the resonant interaction between the QS and nanoresonator is very small in compare with the total emitted energy and could be safely neglected. The only case when this interaction could place a significant role is the nanoresonator enhanced Dike effect, when all energy due to synchronization between the QS will be emitted in a one short pulse with the duration comparable with the phase relaxation time. Probability of the Dike effect in the considered coupled system is very low at the typical experimental conditions and the pumping rate below the generation threshold. From the other side, above threshold the QS are synchronized due to the nanoresonator (as it takes place in typical "macro" laser), and this state corresponds to the stationary solution of (13.18) (see also [43] about Dike effect for the arrays of the coupled nanolasers/spasers). Neglecting the coherent contribution to the luminescent intensity modification, the only factor (excepting already mentioned pump efficiency modification) is the two Purcell factors P_r and P_{nr}, causing radiative and nonradiative relaxation time modification (13.19) which are supposed to be substituted in (13.17) in order to estimate the effect.

13.9 Conclusion

In conclusion of this chapter:

1. Commonly accepted methods of estimation of the spontaneous relaxation dynamics have been summarized. While the methods are undoubtedly justified for the spontaneous relaxation in a free space, application of the same procedure for the case of the nanoresonators optically coupled to the QS is questionable. The concerns are:

(a) Different orders for the perturbation expansion in the model, which describe principally different physical phenomena, namely coupling with the vacuum fluctuations and self-action at the coupled dynamics with the free electrons in the nanoresonator.
(b) An application of the eigen mode formalism causes a question about necessary time required for the mode formation; otherwise the interaction with stochastic modes of the free electrons appears to be more qualitatively appropriate.
(c) An application of the eigen mode formalism fails for the highly loss materials, i.e. for the metallic nanoresonators due to the non-Hermitian nature of the Helmholtz equation and consequently nonorthogonal eigen functions.
(d) In the frame of the accepted in textbook [29] approach, there is no subdivision between interaction with stochastic and regular fields, which definitively leads to the different relaxation dynamics.

2. The density matrix approach has been analyzed in compare with the commonly accepted ones. It has been shown, that this approach is qualitatively more adequate and free from the mentioned above concerns. It has been suggested to include in consideration one more dynamical equation for the electron distribution function. It is expected, that this approach allows us to estimate the introduced here Purcell factors P_r and P_{nr} for both radiative and nonradiative relaxation times. Further investigations are required.
3. The proposed system of equations allows us to calculate regular and stochastic dynamics above and below the lasing threshold of the nanolaser, and consider both strong and weak coupling regimes in the frame of the same approach.

References

1. A. Chipouline, S. Sugavanam, V.A. Fedotov, A.E. Nikolaenko, Analytical model for active metamaterials with quantum ingredients. J. Opt. **14**, 114005 (2012)
2. M. Maragkou, A.K. Nowak, E. Gallardo, H.P. van der Meulen, I. Prieto, L.J. Martinez, P.A. Postigo, J.M. Calleja, Phys. Rev. B **86**, 085316 (2012)
3. J. Barthes, G. Colas des Francs, A. Bouhelier, J.-C. Weeber, A. Dereux, Phys. Rev. B **84**, 073403 (2011)
4. E.M. Purcell, Prys. Rev. **69**, 681 (1946)
5. R. Ruppin, J. Chem. Phys. **76**(4) (1982)
6. J. Gersten, A. Nitzan, Spectroscopic properties of molecules interacting with small dielectric particles. J. Chem. Phys. **75**(3), 1139 (1981)
7. V. Klimov, M. Ducloy, V. Letokhov, Quantum Electron. **31**(7), 596 (2001)
8. V. Giannini, A. Fernandez-Dominquez, S. Heck, S. Maier, Chem. Phys. **111**, 3888 (2011)
9. D. Dregely, K. Lindfors, J. Dorfmüller, M. Hentschel, M. Becker, J. Wratchup, M. Lippitz, R. Vogelgesang, H. Gissen, Phys. Status Solidi B **249**, 666 (2012)
10. P. Berman (ed.), *Cavity Quantum Electrodynamics* (Academic, New York, 1994)
11. S. Haroche, in Fundamental System in Quantum Optics. ed. by J. Dalibard, J.M. Raimond, J. Zinn-Justin (Les Houches, 1990; Elsevier Science Publishers B V, 1992)

12. C. Belacel, B. Habert, F. Bigourdan, F. Marquier, J.-P. Hugonin, S. Michaelis de Vasconcellos, X. Lafosse, L. Coolen, C. Schwob, C. Javaux, B. Dubertret, J.-J. Greffet, P. Senellart, A. Maitre, Nano Lett. **13**, 1516 (2013)
13. R. Esteban, T.V. Teperik, J.J. Greffet, PRL **104**, 026802 (2010)
14. P. Vasa, W. Wang, R. Pomraenke, M. Lammers, M. Maiuri, C. Manzoni, G. Cerullo, C. Lienau, Nat. Photonics **7**, 128 (2013)
15. C. Van Vlack, P.T. Kristensen, S. Hughes, Phys. Rev B **85**, 075303 (2012)
16. A. Trügler, U. Hohenester, Phys. Rev. B **77**, 115403 (2008)
17. A.M. Kern, O.J.F. Martin, Phys. Rev. A **85**, 022501 (2012)
18. S. Stobbe, P.T. Kristensen, J.E. Mortensen, J.M. Hvam, J. Mork, P. Lodahl, Phys. Rev. B **86**, 085304 (2012)
19. R. Filter, S. Mühlig, T. Eichelkraut, C. Rockstuhl, F. Lederer, Phys. Rev B **86**, 035404 (2012)
20. A.M. Kern, O.J.F. Martin, Nano Lett. **11**, 482 (2011)
21. A.M. Kern, Alfred J. Meixner, O.J.F. Martin, ASCNano **6**(11), 9828 (2012)
22. A. Cazé, R. Pierrat, R. Carminati, Photonics Nanostruct. Fundam. Appl. **10**, 339 (2012)
23. W.J.M. Kort-Kamp, F.S.S. Rosa, F.A. Pinheiro, C. Farina, Phys. Rev. A **87**, 023837 (2013)
24. A. Poddubny, P. Belov, P. Ginzburg, A. Zayats, Y. Kivshar, Phys. Rev. B **86**, 035148 (2012)
25. A. Poddubny, P. Belov, Y. Kivshar, Phys. Rev. B **87**, 035136 (2013)
26. M. Stockman, The spaser as a nanoscale quantum generator and amplifier. J. Opt. **12**, 024004 (2010)
27. D.J. Bergman, M.I. Stockman, PRL **90**, 027402 (2003)
28. M. Stockman, Spaser action, loss-compensation, and stability in plasmonic systems with gain. PRL **106**, 156802 (2011)
29. L. Novotny, B. Hecht, *Principles of Nano-optics* (Cambridge University Press, New York, 2006)
30. H. Mertens, A.F. Koenderink, A. Polman, Phys. Rev. B **76**, 115123 (2007)
31. A.F. Koenderink, On the use of Purcell factors for plasmon antennas. Opt. Lett. **35**, 4208 (2010)
32. V. Ginzburg, UFN **140**, 535 (1983) (in Russian)
33. Weisskopf, Naturwissenschaften **27**, 631 (1935)
34. Weisskopf, Phys. Today **34**(11), 69 (1981)
35. Fermi, Rev. Mod. Phys. **4**, 87 (1932)
36. H. Dung, L. Knoll, D.-G. Welsch, Phys. Rev. A **62**, 3804 (2000)
37. L. Knoll, S. Scheel, D.-G. Welsch, Quant-ph/000621-27 June (2000)
38. M.S. Yeung, T.K. Gustafson, Phys. Rev. A **54**, 6 (1996)
39. U. Fano, Description of states in quantum mechanics by density matrix and operator techniques. Rev. Mod. Phys. **29**, 74 (1957)
40. V.M. Fain, Quantum radio physics, v. 1: photons and nonlinear media. Sovetskoe Radio (1972) (in Russian)
41. V. Akulin, N. Karlov, Intensive resonance interaction in quantum electronics, Nauka (1987) (in Russian)
42. L. Novotny, N. van Hulst, Nat. Photonics **5**, 83 (2011)
43. A. Dorofeenko, A. Zyablovsky, A. Vinogradov, E. Andrianov, A. Pukhov, A. Lisyansky, Opt. Express **21**, 14539 (2013)

Chapter 14
On the Question of Radiative Losses in the Frame of Classic and Quantum Formalism

14.1 Introduction

The classical consideration of the radiative losses for a free dipole (with dynamics described by in the frame of harmonic oscillator (HO) equation) leads to a well-known extra term in the dynamic equation proportional to the third time derivative. Nevertheless, this simple dynamic equation modification is justified for the case of free space only. In the case of the coupled dynamics (e.g. spaser/nanolaser) the radiative losses do not possess this straightforward modelling.

The classic HO model has been widely used in the vast majority of publications devoted to the coupled dynamics of an emitter and plasmonic nanoresonator [1]. This approximation (classic HO instead of quantum DM formalism) actually cannot be used in case of significant inversion variations (see Sect. 13.4). In spite of this trivial textbook fact, the HO model for the quantum emitter dynamics is still surprisingly widely accepted in articles, see recent publications [2–4], and review [5].

Use of DM formalism instead of HO one (e.g. to describe spaser dynamics [6]) lead to a paradox considered in this chapter. The essence of this paradox is rooted in the fact, that the relaxation processes are in fact considered in this case doubly: first by phenomenological energy and phase relaxation times, and second at the consideration of the radiative losses. Appearance of this paradox shows that the problem of dynamics of a quantum emitter coupled with a plasmonic nanoresonator requires fundamentally more accurate treatment.

Particular system, shown in Fig. 14.1 will be kept in mind hereafter. The system consists of nanoresonator (nanoparticle NP), coupled with quantum emitters (Quantum System—QS) as it is shown in Fig. 14.1.

A. Chipouline and F. Küppers, *Optical Metamaterials: Qualitative Models*, Springer Series in Optical Sciences 211, https://doi.org/10.1007/978-3-319-77520-3_14

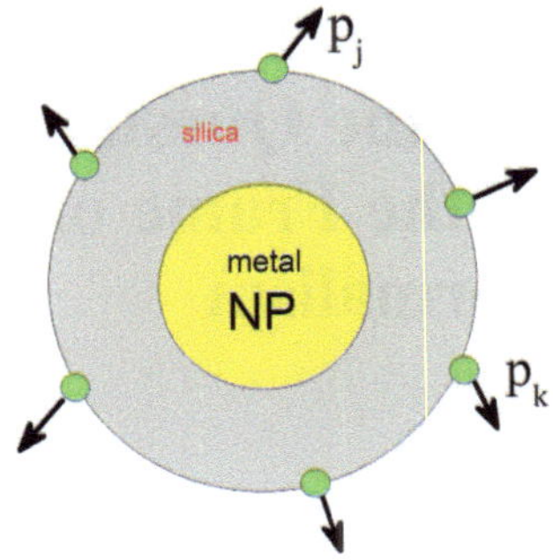

Fig. 14.1 Scheme of the considered object

14.2 Dynamics of Classical and Quantum Dipole

14.2.1 Dynamics of Classical Dipole

The dynamics of a QS interacting with the external field A in the frame of HO equation is:

$$\frac{d^2p}{dt^2} + 2\gamma\frac{dp}{dt} + \omega_0^2 p = \chi A \exp(-i\,\omega\,t) + F \exp(-i\,\omega\,t) \tag{14.1}$$

where χ is some constant. Here function F in classic version (14.1) is the external force (can be an external electric field as well), p—is the dipole moments of the active molecules. External force F brings energy into the system, and is to some extend equivalent to the pump in the case of DM, described by a density matrix.

Energy relaxation is described by γ. This relaxation takes into account the non-radiative relaxation of energy from the HO into a thermo bath. In spite of the fact that (14.1) is probably mostly used equation in physics, the nature of γ and rigorous elaboration of this term is not widely discussed. It is worth to recall it in order to avoid misinterpretation in future discussion. The model described by (14.1) consists of a single oscillator weakly coupled to a lot of similar oscillators, which are in turn weakly coupled with each other. This system of huge number of weakly coupled HOs is called thermo bath or thermostat and appears to be a widely used model in statistical physics. The qualitative picture behind this model is: excitation in HO gets transferred to the other coupled oscillators, which in turn give energy further to the other ones and the initially localized energy (oscillation of a single dipole) gets diffused, or, in other words, thermalized among the other oscillators. Inverse energy transfer is prohibited by the first law of thermodynamics: the dissipated (diffused) energy cannot come back to an origin (i.e. to the single oscillator under consideration). It is also important to realize, that all the oscillators (or, more generally, eigen modes) at the nonzero temperature exhibit stochastic oscillations with arbitrary phase. Hamiltonian of a single HO interacting with thermostat consists of "eigen" part H_0 giving eigen states, regular interaction V_R, and stochastic interaction V_{stoch}:

$$\begin{aligned} &H = H_0 + V_R + V_{stoch}, \\ &\langle V_R \rangle = V_R, \\ &\langle V_{stoch} \rangle = 0 \end{aligned} \tag{14.2}$$

"Eigen" part H_0 leads to dynamic part $\frac{d^2p}{dt^2} + \omega_0^2 p$, regular interaction V_R is responsible for the right side of (14.1) $\chi A \exp(-i\omega t) + F \exp(-i\omega t)$, while stochastic part V_{stoch} results (using iterative approach) in relaxation term $2\gamma \frac{dp}{dt}$ in the second order of approximation of the stochastic interaction: the first order gives evidently zero results. The radiative losses in this case is the consequence of regular interaction, namely $\chi A \exp(-i\omega t)$ in (14.1) and cannot be included in γ.

In Fourie space (14.1) becomes:

$$\begin{cases} (\omega_0^2 - \omega^2 - 2i\gamma\omega)p = \chi A + F \\ p = \frac{\chi A + F}{L_\omega} = \frac{\chi A + F}{\omega_0^2 - \omega^2 - 2i\gamma\omega} = \beta_c(\omega)(\chi A + F) \\ \beta_c(\omega) = \frac{1}{\omega_0^2 - \omega^2 - 2i\gamma\omega} \\ \mathrm{Re}[\beta_c(\omega)] = \frac{\omega_0^2 - \omega^2}{(\omega_0^2 - \omega^2)^2 + 4\gamma^2\omega^2} \\ \mathrm{Im}[\beta_c(\omega)] = \frac{2\gamma\omega}{(\omega_0^2 - \omega^2)^2 + 4\gamma^2\omega^2} \\ |\beta_c(\omega)|^2 = \frac{1}{(\omega_0^2 - \omega^2)^2 + 4\gamma^2\omega^2} \end{cases} \tag{14.3}$$

which describes stationary amplitudes p and A. Here A in the first equation (14.3) is the "self-action" field generated by a dipole at its origin.

14.2.2 Dynamics of Quantum Dipole

Quantum dynamics of the radiating point dipole is considered now in the frame of the DM approach. The dynamics is given by well-known system of equations with artificial pump in SVA (Slowly Varying Approximation) representation (see Chaps. 9 and 13):

$$\begin{cases} \frac{dp}{dt} + p\left(\frac{1}{\tau_2} + i(\omega - \omega_{21})\right) = \frac{i\mu^2 A^* N}{\hbar} \\ \frac{dN}{dt} + \frac{(N - N_0)}{\tau_1} = \frac{i(Ap - A^* p^*)}{2\hbar} \\ p = \mu\rho_{12} \end{cases} \tag{14.4}$$

First of all at the elaboration of the DM approach [and consequently (14.4)] the same subdivision by regular and stochastic interaction (14.2) and all respective math for the elaboration of (2.4) has been accepted [7–9]. Equations (14.4) are written for already averaged values and do not contain stochastic parts—all stochastic processes (interaction with thermo bath and zero field fluctuations) are packed in the two relaxation times τ_1 and τ_2, exactly as the stochastic interaction with the thermo bath has been packed into γ in (14.1). The regular part of the Hamiltonian of interaction, again in full analogy with (14.1) results in right side of (14.4), namely $\frac{i\mu^2 A^* N}{\hbar}$ and $\frac{i(Ap - A^* p^*)}{2\hbar}$.

14.3 Math Formalism for Coupled Dynamics with Radiative Losses

14.3.1 Model Formulation

The problem of the dynamics of an emitter in free space or near some nanoobject is one of the fundamental and has been considered in textbooks [1]. Mathematically, the problem is formulated by a system of coupled equations between dynamics of dipole (in quantum or classical approximations) and Helmholtz equation for the electromagnetic field. We assume here for classical and quantum dipole dynamics HO (14.1) and DM (14.4) equations respectively:

$$\begin{cases} \frac{d^2 p_i}{dt^2} + 2\gamma \frac{p_i}{dt} + \omega_0^2 p_i = \chi A(r_i, t)\exp(-i\omega t) + F(r_i, t)\exp(-i\omega\, t) \\ \Delta E(r,t) - \frac{1}{c^2}\frac{\partial^2 D(r,t)}{\partial t^2} = -\frac{4\pi}{c^2}\frac{\partial^2}{\partial t^2}\sum_i p_i(r_i,t)\delta(r - r_i) \\ E = A\exp(-i\omega t) \end{cases} \tag{14.5}$$

$$\begin{cases} \frac{dp_i}{dt} + p_i\left(\frac{1}{\tau_{2,i}} + i(\omega - \omega_{21})\right) = \frac{i\mu^2 A(r_i,t)^* N}{\hbar} \\ \frac{dN_i}{dt} + \frac{(N_i - N_0)}{\tau_{1,i}} = \frac{i\left(A(r_i,t)p_i - A(r_i,t)^* p_i^*\right)}{2\hbar} \\ \Delta E(r,t) - \frac{1}{c^2}\frac{\partial^2 D(r,t)}{\partial t^2} = -\frac{4\pi}{c^2}\frac{\partial^2}{\partial t^2}\sum_i p_i(r_i,t)\delta(r - r_i) \\ E = A\exp(-i\omega t) \end{cases} \tag{14.6}$$

Here *D*—is the dielectric displacement function, taking into account a possible presence of the nanoobjects. Particular physical picture behind (14.5) and (14.6) is: the coupled equations for the field *E* and polarizability of the emitters p_i. It is assumed that p_i are the dipole moments (in the case of zero dipole moment, p_i is the quadrupole and etc. higher moments, which is rear). In this paper dipole moment is assumed to be nonzero and dominates as a source of radiation for p_i. At the same time, a nanoresonator in the vicinity of the emitter can be taken into account by *D* with all possible eigen modes [10].

It is important to note that the field *E* in Helmholtz equation includes all possible fields: incoming (if any), scattered, and emitted by p_i. The radiative losses are part of this field and are taken into account in (14.5), but NOT included in γ.

As for the radiative losses in (14.6), the consideration in this case is much less trivial. Remind that in the case of HO equation (14.1) the radiative losses are not included due to the fact that the radiative losses are caused not by stochastic part of the interaction Hamiltonian, but rather by the regular one, i.e. by the oscillation driven by an external force F. In the case of quantum consideration, there is a purely quantum contribution to the stochastic part of the interaction, namely interaction with vacuum field fluctuations. They are packed, in addition to the non-radiative relaxation to thermo bath, into the relaxation times τ_1 and τ_2, e.g. for the energy relaxation time τ_1:

$$\frac{1}{\tau_1} = \frac{1}{\tau_{1,\mathrm{r}}} + \frac{1}{\tau_{1,\mathrm{nr}}} \tag{14.7}$$

Here $\tau_{1,\mathrm{r}}$ and $\tau_{1,\mathrm{nr}}$ are the radiative and non-radiative relaxation times respectively. In fact, the radiative relaxation time is given by the well-known expression [1]:

$$\begin{aligned} \frac{1}{\tau_{1,\mathrm{r}}} &= \frac{2\omega}{3\hbar\varepsilon_0}|\mu|^2 \rho_\mu(\vec{r}_0, \omega_0) \\ \rho_\mu(\vec{r}_0, \omega_0) &= 3\sum_k \left[\vec{n}_\mu \cdot \left(\vec{u}_k \vec{u}_k^*\right) \cdot \vec{n}_\mu\right] \delta(\omega_k - \omega_0) \end{aligned} \tag{14.8}$$

Here $\rho_\mu(\vec{r}_0, \omega_0)$ is the local density of states (LDOS), which depends on the presence/absence of any objects in the vicinity of the quantum emitter, e.g. LDOS will be modified in the case of coupling between NP and p_i, see Fig. 14.1.

Now the question is: do we need to consider radiative losses the same way like for HO in (14.5), or there are some differences? In order to answer this question, the solutions for the HO (14.5) and DM (14.6) models are elaborated in parallel.

We are interested in two operation modes, namely stationary states and relaxation dynamics.

14.3.2 Stationary State

The stationary state assumes that the system parameters stationary oscillate at some frequency ω. The slowly varying approximation is:

$$\begin{cases} p_i(r_i, t) = p_i(r_i, \omega)\exp(-i\omega t) \\ E(r, t) = A(r, \omega)\exp(-i\omega t) \\ D(r, t) = D(r, \omega)\exp(-i\omega t) \\ \Delta A(r, \omega) + \frac{\omega^2}{c^2} D(r, \omega) = \frac{4\pi\omega^2}{c^2}\sum_i p_i(r_i, \omega)\delta(r - r_i) \end{cases} \tag{14.9}$$

It has to be pointed out that in principle one can assume Laplace transformation instead of Fourie, but in this case the next step, expression D through the dielectric constant and E, needs to be carefully considered. The dielectric function $\varepsilon(\omega)$ is the

Fourie image of the time dependent dielectric function with all properties implied by causality/passivity principles; Laplace image of the dielectric function is not usually considered.

Assuming:

$$D(r,\omega) = \varepsilon(r,\omega)A(r,\omega) \tag{14.10}$$

we get:

$$\begin{cases} p_i = \frac{\chi A(r_i,\omega)+F(r_i,\omega)}{L_\omega} = \frac{\chi A(r_i,\omega)+F(r_i,\omega)}{\omega_0^2-\omega^2-2i\gamma\omega} = \beta_c(\omega)(\chi A(r_i,\omega)+F(r_i,\omega)) \\ \beta_c(\omega) = \frac{1}{\omega_0^2-\omega^2-2i\gamma\omega} \\ \Delta A(r,\omega) + \frac{\omega^2}{c^2}\varepsilon(r,\omega)A(r,\omega) = \frac{4\pi\omega^2}{c^2}\sum_i p_i(r_i,\omega)\delta(r-r_i) \end{cases} \tag{14.11}$$

Only single QS is assumed in order to demonstrate the paradox, extension on multiple QS is straightforward. The Helmholtz equation allows us to consider any shapes and layouts of NPs, consideration of a single NP and single molecule is stipulated by the fact, that the problem in this case possesses an analytical treatment and at the same time pronouncedly shows the physics behind.

The main step in the transformation of the Helmholtz equation is in the design of the appropriate Green function, taking into account the presence of the NP. It means effectively, that we replace our space with the NP with another space without this NP, but with a more sophisticated Green function. After that, the last equation can be trivially rewritten as:

$$\begin{gathered} \Delta A(r,\omega) + \frac{\omega^2}{c^2}\varepsilon(r,\omega)A(r,\omega) = \frac{4\pi\omega^2}{c^2}p(r_p,\omega)\delta(r-r_p) \\ \Downarrow \\ A(r,\omega) = \frac{4\pi\omega^2}{c^2}G(r,r_p,\omega)p(r_p,\omega) \end{gathered} \tag{14.12}$$

Finally, the problem for stationary states is formulated by the master system of equations for classical case:

$$\begin{cases} p = \beta_c(\omega)(\chi A_{\mathrm{NF}}(r_\mathrm{p},\omega) + F(r_\mathrm{p},\omega)) \\ A_{\mathrm{NF}}(r,\omega) = \frac{4\pi\omega^2}{c^2}G_{\mathrm{NF}}(r,r_\mathrm{p},\omega)p(r_\mathrm{p},\omega) \end{cases} \tag{14.13}$$

and for the quantum case:

$$\begin{cases} p\left(\frac{1}{\tau_2} + i(\omega-\omega_{21})\right) = \frac{i\mu^2 A^*_{\mathrm{NF}}(r_\mathrm{p},\omega)N}{\hbar} \\ \frac{(N-N_0)}{\tau_1} = \frac{i(A_{\mathrm{NF}}(r_\mathrm{p},\omega)p - A^*_{\mathrm{NF}}(r_\mathrm{p},\omega))}{2\hbar} \\ A(r,\omega) = \frac{4\pi\omega^2}{c^2}G(r,r_\mathrm{p},\omega)p(r_\mathrm{p},\omega) \end{cases} \tag{14.14}$$

It is worth noting again, that (14.13) and (14.14) describe the possible stationary states and does not give any information about transient system dynamics.

In addition to the losses in QS itself described by γ (again, radiative losses are not included in γ), there are irreversible losses described by an imaginary part of the dielectric constant. The physical picture of these losses (ohmic losses) is in the interaction of free electron with the solid state lattice resulted in transfer of the kinetic energy of electron into the elementary collective lattice excitations e.g. phonons. Radiative losses are NOT included in the imaginary part of the dielectric constant.

14.3.3 Relaxation Dynamics

Investigation of the relaxation dynamics assumes that we are looking for the solution of (14.5) and (14.6) in time domain. Typical relaxation process assumes that we do not have during the process an extra energy delivery into the system: the system starts at zero time from some nonzero values and relaxes to zero levels of the respective variables. In the case of the linear system, it would be reasonable to assume that the relaxation follows some exponential function. It means that we can use the same substitution (14.9) but the frequency ω in this case becomes a complex value. The imaginary part of this value gives the relaxation time, while real one describes frequency shift; both are supposed to be found from master systems of (14.13) and (14.14) with an extra assumption of complex nature of ω.

14.4 Radiative Losses for Classic and Quantum Dipole in Free Space

14.4.1 Stationary State in Free Space

Results for the p and A_{NF} from (14.13) are:

$$\begin{cases} p = \dfrac{\beta_c(\omega)F}{\left[1-\frac{4\pi\omega^2\chi}{c^2}\beta_c(\omega)G_{\mathrm{NF}}\right]} = \beta_{c,\mathrm{eff}}(\omega)F \\ A_{\mathrm{NF}} = \dfrac{4\pi\omega^2}{c^2}\dfrac{\beta_c(\omega)G_{\mathrm{NF}}F}{\left[1-\frac{4\pi\omega^2\chi}{c^2}\beta_c(\omega)G_{\mathrm{NF}}\right]} = \dfrac{4\pi\omega^2}{c^2}\beta_{c,\mathrm{eff}}(\omega)G_{\mathrm{NF}}F \\ \beta_{c,\mathrm{eff}}(\omega) = \dfrac{\beta_c(\omega)}{\left[1-\frac{4\pi\omega^2\chi}{c^2}\beta_c(\omega)G_{\mathrm{NF}}\right]} \end{cases} \tag{14.15}$$

Amplitude, intensity, and total radiative power in the far field zone can be straightforwardly calculated using the known Green function for far field:

$$\begin{cases} A_{\mathrm{FF}} = \frac{4\pi\omega^2}{c^2} \frac{\beta_c(\omega) G_{\mathrm{FF}} F}{\left[1 - \frac{4\pi\omega^2 \chi}{c^2} \beta_c(\omega) G_{\mathrm{NF}}\right]} = \frac{4\pi\omega^2}{c^2} \beta_{c,\mathrm{eff}}(\omega) G_{\mathrm{FF}} F \\ I_{\mathrm{FF}} = \frac{c}{8\pi} \left(\frac{4\pi\omega^2}{c^2}\right)^2 \left|G_{\mathrm{FF}} p\right|^2 = \frac{2\pi\omega^4}{c^3} \frac{\left|\beta_c(\omega) G_{\mathrm{FF}}\right|^2}{\left|1 - \left(\frac{4\pi\omega^2 \chi}{c^2}\right) \beta_c(\omega) G_{\mathrm{NF}}\right|^2} |F|^2 = \frac{2\pi\omega^4}{c^3} \left|\beta_{c,\mathrm{eff}}(\omega) G_{\mathrm{FF}}\right|^2 |F|^2 \\ P_{\mathrm{FF}} = \frac{c}{8\pi} \left(\frac{4\pi\omega^2}{c^2}\right)^2 \oint \left|G_{\mathrm{FF}} p\right|^2 \mathrm{d}S = \frac{8\pi^2\omega^4}{c^3} \left|\beta_{c,\mathrm{eff}}(\omega)\right|^2 |F|^2 = \frac{8\pi^2\omega^4}{c^3} \frac{\left|\beta_c(\omega)\right|^2 |F|^2}{\left|1 - \frac{4\pi\omega^2 \chi}{c^2} \beta_c(\omega) G_{\mathrm{NF}}\right|^2} \end{cases} \quad (14.16)$$

Systems (14.15), (14.16) are universal. It means that the systems describe stationary state for any possible Green functions, i.e. for any possible nanoobject which can be described by $\varepsilon(r, \omega)$. Let us consider first the problem of the radiative losses in free space. In order to complete (14.16) we have to know Green function for free space in the near field and far field zones, which are [10]:

$$\begin{cases} G_{\mathrm{NF}} = \frac{c^2}{4\pi\omega^2} \frac{2}{3} i k^3 \\ G_{\mathrm{FF}} = \frac{1}{r} \exp(ikr) \end{cases} \quad (14.17)$$

here G_{NF} and G_{FF} are the Green function of the free space in the near field and far field zones respectively. The oscillating dipole is driven by the force F from one side (energy income) and loses energy into thermo bath and radiation. If the energy losses are compensated by the external force F, the system reaches the stationary state. It is important to understand that here Green function for the near field has to be taken, hence $A_{\mathrm{NF}} = \frac{4\pi\omega^2}{c^2} G_{\mathrm{NF}} p$. In this chapter we are interested in free space (no nanoobjects in near field zone). Let us substitute (14.17) into (14.16). The total emitted power becomes:

$$\begin{cases} p = \frac{\beta_c(\omega) F}{\left[1 - \frac{2i\chi}{3} \beta_c(\omega) \left(\frac{\omega}{c}\right)^3\right]} \\ P_{\mathrm{FF,C}}(\omega) = \frac{8\pi^2\omega^4}{c^3} \frac{1}{\left(\left(\omega_0^2 - \omega^2\right)^2 + 4\gamma^2\omega^2\right)} \frac{|F|^2}{\left(1 + \frac{2}{3}\left(\frac{\omega}{c}\right)^3 \frac{\chi\gamma\omega}{\left(\omega_0^2 - \omega^2\right)^2 + 4\gamma^2\omega^2}\right)^2 + \left(\frac{2}{3}\left(\frac{\omega}{c}\right)^3 \frac{\chi\left(\omega_0^2 - \omega^2\right)^2}{\left(\omega_0^2 - \omega^2\right)^2 + 4\gamma^2\omega^2}\right)^2} \\ P_{\mathrm{FF,C}}(\omega_0) = \frac{2\pi^2\omega_0^2}{\gamma^2 c^3} \frac{|F|^2}{\left(1 + \frac{\chi\omega_0^2}{6c^3\gamma}\right)^2} \end{cases} \quad (14.18)$$

which clearly indicates that the emitted power becomes lower if the radiative losses (near field Green function) are taken into account. Effective dipole has been introduced in (14.1), which is characterized by its own spectral response function $\beta_{c,\mathrm{eff}}(\omega)$. The respective dynamic equation for the dipole with the radiative losses is governed by:

$$\beta_{c,\mathrm{eff}}^{-1}(\omega) = \beta_c^{-1}(\omega)\left[1 - \frac{2i}{3}\chi\,\beta_c(\omega)k^3\right] = \beta_c^{-1}(\omega) - \frac{2}{3}\chi k^3 = \beta_c^{-1}(\omega) - \frac{2i}{3}\chi\left(\frac{\omega}{c}\right)^3 \tag{14.19}$$

which corresponds to the extra term proportional to a third derivative in dynamic equation (14.1):

$$\frac{\mathrm{d}^2p}{\mathrm{d}t^2} + 2\gamma\frac{\mathrm{d}p}{\mathrm{d}t} + \omega_0^2p + \frac{2}{3}\frac{\chi}{c^3}\frac{\mathrm{d}^3p}{\mathrm{d}t^3} = F\exp(-i\omega t) \tag{14.20}$$

The radiative losses can be equivalently taken into account by an extra term in the dipole dynamic equation. In this case, the dipole oscillation amplitude p has to be taken from (14.20), not from (14.1).

In text books, the final expressions for the output power (14.16) are usually integrated over the whole spectrum, which is not done here—the final expressions for the integrated power are not observable and are not suited for the analytical consideration. It nevertheless can be done pretty easily using any available numerical package.

Following the same logic as for classic dipole (14.13), consider now a stationary state of quantum oscillator (14.14):

$$\begin{cases} p\left(\frac{1}{\tau 1} + i(\omega - \omega_{21})\right) = \frac{i\mu^2 A_{\mathrm{NF}}^* n}{\hbar} \\ \frac{(n-n_0)}{\tau 1} = \frac{i\left(A_{\mathrm{NF}}p - A_{\mathrm{NF}}^*p^*\right)}{2\hbar} \\ A_{\mathrm{NF}} = \frac{4\pi\omega^2}{c^2}G_{\mathrm{NF}}p^* \end{cases} \tag{14.21}$$

Substituting A_{NF} from the last equation to the other two, we have:

$$\begin{cases} p\left(1 + i\tau_2(\omega - \omega_{21}) - \frac{i\mu^2\tau_2 n}{\hbar}\frac{4\pi\omega^2}{c^2}G_{\mathrm{NF}}^*\right) = 0 \\ n = n_0 - \frac{\tau_1}{\hbar}\frac{4\pi\omega^2}{c^2}\mathrm{Im}[G_{\mathrm{NF}}]|p|^2 \end{cases} \tag{14.22}$$

Along with the trivial solution $p = 0$, (14.22) possesses a nontrivial one given by:

$$\begin{cases} 1 + i\tau_2(\omega - \omega_{21}) = \frac{i\mu^2\tau_2 n}{\hbar}\frac{4\pi\omega^2}{c^2}G_{\mathrm{NF}}^* \\ n = n_0 - \frac{\tau_1}{\hbar}\frac{4\pi\omega^2}{c^2}\mathrm{Im}[G_{\mathrm{NF}}]|p|^2 \end{cases} \tag{14.23}$$

which in turn is reduced to:

$$\begin{cases} \frac{\hbar(1 + i\tau_2(\omega-\omega_{21}))}{i\mu^2\tau_2\frac{4\pi\omega^2}{c^2}G_{\mathrm{NF}}^*} = n \\ n = n_0 - \frac{\tau_1}{\hbar}\frac{4\pi\omega^2}{c^2}\mathrm{Im}[G_{\mathrm{NF}}]|p|^2 \end{cases} \tag{14.24}$$

The first equation gives stationary frequency ω_{eigen} from the condition Im $[n] = 0$, and then consequently $n(\omega_{\text{eigen}})$. From the second equation we get then stationary value of modulus p square:

$$\begin{cases} \operatorname{Im}\left[\frac{\hbar(1+i\tau_2(\omega-\omega_{21}))}{i\mu^2\tau_2\frac{4\pi\omega^2}{c^2}G^*_{\text{NF}}}\right] = 0 \Rightarrow \omega_{\text{eigen}} \\ \operatorname{Re}\left[\frac{\hbar(1+i\tau_2(\omega-\omega_{21}))}{i\mu^2\tau_2\frac{4\pi\omega^2}{c^2}G^*_{\text{NF}}}\right] = n \\ |p|^2 = \frac{\hbar}{\tau_1\frac{4\pi\omega^2}{c^2}\operatorname{Im}[G_{\text{NF}}]}\left(n_0 - \operatorname{Re}\left[\frac{\hbar(1+i\tau_2(\omega-\omega_{21}))}{i\mu^2\tau_2\frac{4\pi\omega^2}{c^2}G^*_{\text{NF}}}\right]\right) \end{cases} \tag{14.25}$$

which is supposed to be an analog of (14.15) for classical dipole. Remember, that the eigen values did not appear at all at the consideration of the classical HO. It reflects the principal difference: a HO model does not assume pumping, but rather driving by an external force. It brings us to the conclusion, that the quantum case is basically closer to a nonlinear oscillator: in both cases the dipole dynamics is nonlinear and possesses an eigen stationary state (will be considered elsewhere).

Stationary state for free space is given by a substitution of the respective near field Green function (14.17) into (14.25):

$$\begin{cases} \operatorname{Im}\left[\frac{\hbar(1+i\tau_2(\omega-\omega_{21}))}{\mu^2\tau_2\frac{2}{3}\left(\frac{\omega}{c}\right)^3}\right] = 0 \Rightarrow \omega_{\text{eigen}} = \omega_{21} \\ \operatorname{Re}\left[\frac{\hbar(1+i\tau_2(\omega-\omega_{21}))}{\mu^2\tau_2\frac{2}{3}\left(\frac{\omega}{c}\right)^3}\right] = n \Rightarrow n = \frac{3\hbar c^3}{2\mu^2\tau_2\omega^3} \\ |p|^2 = \frac{2\hbar c^3}{3\tau_1\omega^3}\left(n_0 - \frac{3\hbar c^3}{2\mu^2\tau_2\omega^3}\right) \end{cases} \tag{14.26}$$

With the following numerical values:

$$\begin{cases} \hbar = 10^{-27}\,(\text{erg} * \text{s}) \\ \omega_0 \sim 10^{15}\,(\text{s}^{-1}) \\ c = 3 * 10^{10}\,(\text{cm/s}) \\ \tau_2 = 100\,\text{fs} = 10^{-10}\,(\text{s}) \\ \mu_{\text{QD}} = 2.5 \times 10^{-17}\,(\text{cgse}) \end{cases} \tag{14.27}$$

rough estimation gives $n \sim 6.5 \times 10^2$, which is impossible because of $-1 < n < 1$. Nevertheless, for order of magnitude higher frequencies the numerical values of n becomes three orders of magnitude lower and requirement $n < 1$ could be satisfied. If it is so, then expression (14.27) assumes that even for a single molecule under some strong enough pump one can get a coherent radiation i.e. set of regularly spaced in time photons. The physical picture behind this expression is following: a generated photon through the self-action produces a stimulated emission causing laser-like auto-oscillation operation mode. This picture is evidently wrong. A spontaneous emission cannot cause at the same time a self-stimulated emission;

in other words, a spontaneously emitted photon cannot cause a stimulated emission of the next photon from the same quantum system. Moreover, the spontaneous emission is already included into consideration by two relaxation times in (14.4). Note, that for the case of linear HO model this problem does not appear—the emitted photon does not produce nonlinear "self-action" but rather produce linear reaction, which results in extra (radiative) losses in final expressions. The demonstrated paradox can be resolved e.g. by removing the part of the Green function, which is responsible for the spontaneous generation, namely $G_{\mathrm{NF}} = \frac{c^2}{4\pi\omega^2}\frac{2}{3}ik^3$. The reason for this modification is in the fact, that the spontaneous emission is already included into the model by two relaxation times and should not be included twice. Hence, as a conclusion: in the frame of the DM formalism the free space part of the Green function has to be artificially excluded from the full expression for the Green function.

It is worth noting, that this in fact fully corresponds to the case of classic oscillator. In both cases, the stochastic interaction is packed into the relaxation time: in the case of classic oscillator there is only one time $\tau = \frac{1}{\gamma}$, and in case of quantum —two relaxation times τ_1 and τ_2. Radiative losses in the case of classic oscillator can be equivalently described by a coupling with Helmholtz equation or by another term in the dynamic equation. In the case of quantum dipole the radiative losses is already included in the relaxation times and consequently the respective field has to be excluded from the right sides of both classics dynamic equation (14.20) and quantum one (14.6). A_{NF} in (14.21) is whatever but NOT the spontaneous photons, which is mathematically manifested by the absence of term $\frac{c^2}{4\pi\omega^2}\frac{2}{3}ik^3$ in the near field Green function. Math prove behind this statement is given in Appendix 14.2.

14.4.2 Relaxation Dynamics in Free Space

The relaxation dynamics is described by the same system of (14.13) for the classic HO but without any external energy sources and assuming that the frequency ω is now complex value:

$$\begin{cases} p = \beta_c(\omega)\chi A_{\mathrm{NF}} \\ A_{\mathrm{NF}} = \frac{4\pi\omega^2}{c^2} G_{\mathrm{NF}} p \end{cases} \tag{14.28}$$

Substituting the second equation in (14.28) into the first one, we get:

$$p\left(1 - \frac{4\pi\omega^2\chi}{c^2}\beta_c(\omega)G_{\mathrm{NF}}(\omega)\right) = 0$$
$$\Downarrow \tag{14.29}$$
$$\frac{4\pi\omega^2\chi}{c^2}\beta_c(\omega)G_{\mathrm{NF}}(\omega) = 1$$

In other words, we become an equation for the eigen values ω_{eigen}. The imaginary part of the eigen frequency describes the relaxation, namely:

$$\tau_{\text{relax}} = \frac{-1}{\text{Im}\left[\omega_{\text{eigen}}\right]} \tag{14.30}$$

The relaxation time, described by (14.30) includes all possible influences in the frame of the elaborated model and is also universal i.e. remains valid in the case of NP in the near field zone. The imaginary part of eigen frequency $\text{Im}[\omega_{\text{eigen}}]$ is always negative and the relaxation time is always positive. The relaxation in free space is obtained by substituting G_{NF} from (14.17) and $\beta_c(\omega)$ from (14.19) into (14.29):

$$\begin{aligned} &\frac{2}{3}i\left(\frac{\omega}{c}\right)^3 \chi \frac{1}{\omega_0^2 - \omega^2 - 2i\gamma\omega} = 1 \\ &\Rightarrow \frac{2i\chi}{3c^3}\omega^3 + \omega^2 + 2i\gamma\omega - \omega_0^2 = 0 \end{aligned} \tag{14.31}$$

Without radiative losses $\chi = 0$ it is reduced to the known quadratic equation for a HO:

$$\omega^2 + 2i\gamma\omega - \omega_0^2 = 0 \tag{14.32}$$

with simple solution:

$$\omega = -i\gamma \pm \sqrt{\omega_0^2 - \gamma^2} \tag{14.33}$$

According to (14.30) the relaxation time is:

$$\tau_{\text{relax}} = \frac{-1}{\text{Im}\left[\omega_{\text{eigen}}\right]} = \frac{1}{\gamma} \tag{14.34}$$

The relaxation processes in quantum emitters require more careful consideration. According to the general rules, relaxation for the quantum oscillator is described by the first equation in (14.23) with $n = -1$ and $\frac{i\mu^2\tau_2}{\hbar}\frac{4\pi\omega^2}{c^2}G_{\text{NF}}^* = 0$, namely:

$$i\tau_2(\omega - \omega_{21}) + 1 = 0 \Rightarrow \omega_{\text{eigen}} = \omega_{21} + \frac{i}{\tau_2} \tag{14.35}$$

and the relaxation time is [in quantum case the sign is opposite in compare with (14.30)]:

$$\tau_{\text{relax}} = \frac{1}{\text{Im}\left[\omega_{\text{eigen}}\right]} = \tau_2 \tag{14.36}$$

Note, that this is so called the phase relaxation time i.e. averaged time between phase jumps in the quantum emitter oscillations. It has to be realized, that this time is NOT the average time between spontaneous photon emissions; the latter is given by τ_1. The relaxation dynamics is described by (14.4) without fields in the right side:

$$\begin{cases} \frac{dp}{dt} + p\left(\frac{1}{\tau_2} + i(\omega - \omega_{21})\right) = 0 \\ \frac{dn}{dt} + \frac{(n-n_0)}{\tau_1} = 0 \end{cases} \tag{14.37}$$

Radiative losses are already included in the relaxation times. The spontaneous emission rate is:

$$P_{\mathrm{FF,Q}}(\omega_{12}) = \hbar\omega_{12}\frac{n_0}{\tau_{1,\mathrm{r}}} \tag{14.38}$$

Note, that the bandwidth of spontaneous emission is not determined in the same mean as for a single photon.

Relaxation rates are usually calculated using both approaches, namely stationary states and by the consideration of the relaxation dynamics. Let us compare these two approaches and consider first stationary case for classic (14.18) and for quantum (14.38) dipoles. Consider a stationary state of the pumped emitter, which is placed near the NP in order to tune the emission rate through the Purcell effect. The emitted power in (14.18) and (14.38) can be presented as photon energy multiplied an effective rate. In both cases (14.18) and (14.38) the measured value is P, which depends on the pump $|F|^2$ and n_0 in (14.18) and (14.38) respectively. The pump in turn depends on the experimental conditions, which means, that the extractable parameter is not just emission rate, but rather emission rate determined by the pump. In many papers the pump in the respective expressions are assumed to be constant and independent on e.g. distance between the emitter and NP, which is evidently wrong. One can conclude that in this kind of experiments the emission rate cannot be in general independently extracted.

Let us now compare two expressions for the emitted power (14.18) and (14.38). Let us also assume that according to the Appendix 14.1 we use a HO equation (14.44) or (14.45), i.e. we substitute $\gamma = \frac{1}{\tau_2}$ in (14.18):

$$P_{\mathrm{FF,C}}(\omega_0) = \frac{2\pi^2\omega_0^2\tau_2^2}{c^3}\frac{|F|^2}{\left(1 + \frac{\chi\omega_0^2\tau_2}{6c^3}\right)^2} \tag{14.39}$$

which is evidently not similar to (14.38). It proves again, that these two approaches (classic and quantum) do not give the same results. The reason is in the fact, that the classic case (14.18) and respectively (14.39) represents a mix of stochastic relaxation process described by γ and regular radiative losses, while the quantum case (14.38) is caused only by the stochastic interactions.

Finally, let us compare relaxation dynamic equations (14.31) and (14.26) with the "gedanken experiment", namely we assume that the pump is turned off abruptly, and we measure the emitted power in the case of classic, or collect the statistics in the quantum cases. The variation of the relaxation time is provided as before by Purcell effect i.e. by an NP near the emitter. First, the solution of (14.16) is evidently not the same as (14.25), which proves again, that these two tests have to be considered as different ones. Second, (14.16) is free from pump values, and hence can be used to measure the respective relaxation: fitting of the relaxation curves for real and imaginary parts of ω gives both γ and χ.

As for the quantum dynamics, the relaxation curve will be proportional to a full (not just radiative or nonradiative) relaxation time, namely:

$$P_{\mathrm{FF,Q}}(t) = \hbar\omega_{12}\frac{n_0(t=0)}{\tau_{1,\mathrm{r}}}\exp\left(-\frac{t}{\tau_1}\right) \tag{14.40}$$

and the extractable parameter is τ_1 ($\frac{1}{\tau_1} = \frac{1}{\tau_{1,r}} + \frac{1}{\tau_{1,\mathrm{nr}}}$), but not $\tau_{1,\mathrm{r}}$. The latter could be also extracted by the fitting of amplitude of (14.38), namely $\hbar\omega_{12}\frac{n_0(t=0)}{\tau_{1,r}}$, but in this case the initial pump values $n_0(t=0)$ will be also different and hardly estimated.

The extracted information thus depends crucially on the model, namely classic or quantum one. In the case of a classic model relaxation parameter γ, pump value $|F|^2$, and coupling parameter χ can be extracted. This takes place due to the fact that we basically have three independent measurements for three mentioned above values, namely measurement of the central frequency shift (real part ofω), decay rate (imaginary part of ω), and amplitude of the emitted power $P_{\mathrm{FF,C}}$. In contrast, in the quantum case there are only two measured values, namely $\frac{n_0}{\tau_{1,r}}$ and $\frac{1}{\tau_1} = \frac{1}{\tau_{1,r}} + \frac{1}{\tau_{1,\mathrm{nr}}}$, and three unknowns, namely n_0, $\tau_{1,\mathrm{r}}$, and $\tau_{1,\mathrm{nr}}$. It means that in the quantum case it is principally impossible to extract all of these three values. In publications, authors often assume that n_0(pump) remains the same and extract numerical values based on this evidently wrong assumption.

14.5 Conclusions

It is shown that the widely accepted HO equation cannot be used to describe dynamics of a quantum emitter. From the other side, a formal substitution of a HO equation (classic dipole) by DM equations leads to unphysical results. This paradox appears in the case of description of e.g. relaxation dynamics (including Purcell effect) and is expected to appear in spaser dynamics in the case of multimode operation. This paradox can be resolved by careful subdivision of the considered fields by regular and stochastic ones, which is actually a prerequisite for the DM equations. It is shown, that the results of relaxation rates obtained from the consideration of the stationary states differ from the results of relaxation dynamics. It is shown, that the independent extraction of the radiative and non-radiative relaxation times in the quantum model (DM) is hardly possible in the frame of commonly accepted tests.

Appendix 1: DM Versus HO Equation

Dynamics of 2-level system is described by the following system of equations:

$$\begin{cases} \frac{d\rho_{12}}{dt} - i\omega_{21}\rho_{12} + \frac{\rho_{12}}{\tau_1} = -\frac{iH_{12}(\rho_{22}-\rho_{11})}{\hbar} \\ \frac{d\rho_{22}}{dt} + \frac{\rho_{22}}{\tau_2} = -\frac{iH_{12}\left(\rho_{12}-\rho_{12}^*\right)}{\hbar} + W \\ \rho_{11} + \rho_{22} = 1 \end{cases} \quad (14.41)$$

Here ρ describes population (diagonal) and polarisation (non diagonal) dynamics, W is a pump rate, and H is a hamiltonian of interaction (for example, for the interaction with an external electric field it becomes $H_{12} = -\mu * E$.

Sometimes it is more convenient to reduce the dynamics to other variables (Bloch equations)

$$\begin{cases} \frac{dQ}{dt} + \frac{Q}{\tau_1} - i\omega_{21}P = \frac{2iH_{12}n}{\hbar} \\ \frac{dP}{dt} + \frac{P}{\tau_1} = i\omega_{21}Q \\ \frac{dn}{dt} + \frac{1+n}{\tau_2} = -\frac{2iH_{12}Q}{\hbar} + 2W \end{cases} \quad (14.42)$$

Here $P = \rho_{12} + \rho_{12}^*, \quad Q = \rho_{12} - \rho_{12}^*, \quad N = \rho_{22} - \rho_{11}$

Taking Q from the first equation and substituting it into the second one, the system in Bloch variables takes the following form

$$\begin{cases} \frac{d^2P}{dt^2} + \frac{2}{\tau_1}\frac{dP}{dt} + \left(\frac{1}{\tau_1^2} + \omega_{21}^2\right)P = \frac{2\omega_{21}H_{12}n}{\hbar} \\ \frac{dn}{dt} + \frac{1+n}{\tau_2} = -\frac{2H_{12}}{\hbar\omega_{21}}\left(\frac{dP}{dt} + \frac{P}{\tau_1}\right) + 2W \end{cases} \quad (14.43)$$

It is clear that the first equation in (14.43) is rather far from a trivial HO one. Nevertheless, if it is assumed that N is not changing too much, then the equations become identical:

$$\frac{d^2P}{dt^2} + \frac{2}{\tau_1}\frac{dP}{dt} + \left(\frac{1}{\tau_1^2} + \omega_{21}^2\right)P = \frac{2\omega_{21}H_{12}n_0}{\hbar} \quad (14.44)$$

For example, in case of interaction with the electric field and low intensity (low intensity means that n can be substituted by just -1 corresponding to the situation with all molecules on the lower level ρ_{11})

$$\frac{d^2P}{dt^2} + \frac{2}{\tau_1}\frac{dP}{dt} + \left(\frac{1}{\tau_1^2} + \omega_{21}^2\right)P = \frac{2\omega_{21}\mu}{\hbar}E \tag{14.45}$$

The second possibility to reduce the system to a HO equation is to assume, that due to the pump W the population difference n is kept constant. In order to prove it mathematically, we have to realize that the first term in right side of the second equation in (14.43) is a multiplication of two fast oscillating functions, while n has much slower dynamics. The math procedure is in substitution both P and H (or, in case of an electric field E) by an anzatz with fast oscillating parts, namely:

$$\begin{aligned} E(t) &= \frac{1}{2}(A(t)^* \exp(i\omega t) + A(t)^* \exp(-i\omega t)) \\ P(t) &= \frac{1}{2}(p(t)^* \exp(i\omega t) + p(t)^* \exp(-i\omega t)) \end{aligned} \tag{14.46}$$

and then reduce the system (14.43) into one equation for slowly varying component p. In this case we have to accept basically Slowly Varying Approximation for the slow amplitudes p and A, which gives us equation for p, but with the first derivative. In this case the system (14.43) anyway cannot be reduced to a HO equation.

But let us just assume that due to some reasons the n is kept on some constant level. In this case (14.44) becomes:

$$\frac{d^2P}{dt^2} + \frac{2}{\tau_1}\frac{dP}{dt} + \left(\frac{1}{\tau_1^2} + \omega_{21}^2\right)P = -\frac{2\omega_{21}\mu n_0}{\hbar}E \tag{14.47}$$

and positive N (inversion) should give us an amplification effect, in this case

$$\frac{d^2P}{dt^2} + \frac{2}{\tau_1}\frac{dP}{dt} + \left(\frac{1}{\tau_1^2} + \omega_{21}^2\right)P = -\frac{2\omega_{21}\mu|n_0|}{\hbar}E \tag{14.48}$$

This equation by no means can be equivalent to the equation of HO with negative absorption. Absorption remains the same (first derivative with some coefficient giving typical relaxation time), and effect of amplification is in a term in right side of (14.48), which is equivalent to a kind of external force.

Conclusion:

1. An absorption coefficient in a HO equation can only have a sign, corresponding to an energy losses.
2. Rigorous quantum description can be reduced to a HO equation only in case of low, unsaturated losses in a passive (unpumped) quantum system.
3. Quantum dynamics, associated with the amplification process (14.43) can NOT be described by a HO equation with an inverted sign of losses anyway.

4. Equation (14.48), which is actually a HO equation with an external force, can be taken to some extend as an equation for polarizability, but even this equation can NOT be obtained through a rigorous math from the DM approach.

There are basically no reasons to use a HO equation for amplification, the system (14.41) is not too complicated and can be analyzed analytically.

Appendix 2: Maxwell Equations and Density Matrix Formalism

A starting point for the consideration is the Helmhoz equation:

$$\begin{aligned}
&\Delta A + k^2 A = \frac{4\pi\omega^2}{c^2} P \\
&P = \mu(\langle\psi_2|\mu A|\psi_1\rangle + \langle\psi_2|\mu A|\psi_1\rangle^*) \\
&H\psi = E\psi \\
&H = H_0 + \mu A; \quad H_0\psi_k = E_k\psi_k \\
&A = A_{\text{regular}} + A_{\text{zero fluctuatins}} + A_{\text{stochastic}} \Rightarrow \\
&\Rightarrow H = H_0 + \mu\left(A_{\text{regular}} + A_{\text{zero fluctuatins}} + A_{\text{stochastic}}\right) \\
&\qquad + V_{\text{stochastic}} = H_0 + \mu A_{\text{regular}} + \mu(A_{\text{zero fluctuatins}} + A_{\text{stochastic}})
\end{aligned} \tag{14.49}$$

$A_{\text{zero fluctuations}}$ are the zero (vacuum) fluctuations of the electric field.

The last scopes in the last equation for A are the combination of the interaction with the vacuum fluctuations and thermo bath. Both are considered as stochastic functions with zero averaged values. As for the Helmholz equation:

$$\begin{aligned}
&\Delta A + k^2 A = \frac{4\pi\omega^2}{c^2} P \\
&A = A_{\text{regular}} + A_{\text{zero fluctuatins}} + A_{\text{stochastic}} \Rightarrow \left(\Delta + k^2\right) A_{\text{regular}} \\
&\qquad + \left(\Delta + k^2\right)(A_{\text{zero fluctuatins}} + A_{\text{stochastic}}) = \frac{4\pi\omega^2}{c^2} P \Rightarrow \\
&\left(\Delta + k^2\right) A_{\text{regular}} = \frac{4\pi\omega^2}{c^2} P\left(A_{\text{regular}}\right) \\
&\left(\Delta + k^2\right)(A_{\text{zero fluctuatins}} + A_{\text{stochastic}}) = 0
\end{aligned} \tag{14.50}$$

Substituting A into the equation for polarization we see, that the stochastic part gets canceled:

$$
\begin{aligned}
P &= \mu(\langle\psi_2|\mu|\psi_1\rangle + \langle\psi_2|\mu A|\psi_1\rangle^*) \\
A &= A_{\text{regular}} + A_{\text{zero fluctuatins}} + A_{\text{stochastic}} \Rightarrow P = \mu\left(\langle\psi_2|A_{\text{regular}}|\psi_1\rangle + \langle\psi_2|A_{\text{regular}}|\psi_1\rangle^*\right) \\
&\quad + \mu(\langle\psi_2|(A_{\text{zero fluctuatins}} + A_{\text{stochastic}})|\psi_1\rangle + \langle\psi_2|(A_{\text{zero fluctuatins}} + A_{\text{stochastic}})|\psi_1\rangle^*) \\
&= \mu\left(\langle\psi_2|A_{\text{regular}}|\psi_1\rangle + \langle\psi_2|A_{\text{regular}}|\psi_1\rangle^*\right)(\Delta + k^2)A_{\text{regular}} = \frac{4\pi\omega^2}{c^2}P(A_{\text{regular}})
\end{aligned}
\tag{14.51}
$$

As for the Schrödinger equation, it gets transferred into the DM equations, where both $A_{\text{zero fluctuatins}}$ and $A_{\text{stochastic}}$ (i.e. interaction with vacuum fluctuations and with thermo bath) are packed into the two relaxation times τ_1 and τ_2.

$$
\begin{aligned}
H\psi &= E\psi \\
H_0\psi_k &= E_k\psi_k \\
H &= H_0 + \mu\left(A_{\text{regular}} + A_{\text{zero fluctuatins}} + A_{\text{stochastic}}\right) \\
&= H_0 + \mu_{\text{regular}} + \mu(A_{\text{zero fluctuatins}} + A_{\text{stochastic}})
\end{aligned}
$$

$$
\begin{cases}
\frac{dp}{dt} + p\left(\frac{1}{\tau_2} + i(\omega - \omega_{21})\right) = \frac{i\mu^2 A^*_{\text{regular}} n}{\hbar} \\
\frac{dn}{dt} + \frac{(n-n_0)}{\tau_1} = \frac{i\left(A_{\text{regular}} p - A^*_{\text{regular}} p^*\right)}{2\hbar} \\
n = \rho_{22} - \rho_{11} \\
p = \mu\rho_{12} \\
n_0 = \frac{(W\bar{\tau}_1 - 1)}{(W\bar{\tau}_1 + 1)}
\end{cases}
\tag{14.52}
$$

In DM equations the both relaxation times depend on the environment according to the local density of state which in turn depends on imaginary part of the respective Green function.

References

1. L. Novotny, B. Hecht, *Principles of Nano-optics* (Cambridge University Press, New York, 2006)
2. A. Poddubny, P. Belov, Y. Kivshar, Phys. Rev. B **87**, 035136 (2013)
3. M. Decker, I. Staude, I. Shishkin, K. Samusev, P. Parkinson, V. Sreenivasan, A. Minovich, A. Miroshnichenko, A. Zvyagin, C. Jagadish, D. Neshev, Y. Kivshar, Dual-channel spontaneous emission of quantum dots in magnetic metamaterials. Nat. Commun. **4**, 2949 (2013)
4. C. Sauvan, J.P. Hugonin, I.S. Maksymov, P. Lalanne, Theory of the spontaneous optical emission of nanosize photonic and plasmon resonators. PRL **110**, 237401 (2013)
5. M. Agio, D. Cano, The Purcell factor of nanoresonators. Nat. Photon. **7** (2013)
6. M. Stockman, The spaser as a nanoscale quantum generator and amplifier. J. Opt. **12**, 024004 (2010)
7. U. Fano, Description of states in quantum mechanics by density matrix and operator techniques. Rev. Modern Phys. **29**, 74 (1957)

8. V.M. Fain, *Quantum Radio Physics, Vol. 1: Photons and Nonlinear Media* (Sovetskoe Radio, Moscow, 1972) (in Russian)
9. V. Akulin, N. Karlov, *Intensive Resonance Interaction in Quantum Electronics* (Nauka, Moscow, 1987) (in Russian)
10. V. Pustovit, T. Shahbazyan, Phys. Rev. B **82**, 075429 (2010)

Correction to: Optical Metamaterials: Qualitative Models

Correction to:
A. Chipouline and F. Küppers, *Optical Metamaterials: Qualitative Models*, Springer Series in Optical Sciences 211, https://doi.org/10.1007/978-3-319-77520-3

In the original version of the book, department name of the volume editors was corrected from Technical University of Darmstadt to Darmstadt University of Technology, and the primary affiliation "Skolkovo Institute of Science and Technology, Moscow, Russia" was added to both the volume editors Prof. Arkadi Chipouline and Prof. Franko Küppers. The book has been updated with the changes.

The updated version of the book can be found at
https://doi.org/10.1007/978-3-319-77520-3

A. Chipouline and F. Küppers, *Optical Metamaterials: Qualitative Models*, Springer Series in Optical Sciences 211, https://doi.org/10.1007/978-3-319-77520-3_15

Results and Conclusions

Main Achieved Results

In conclusion, a new and self-consistent approach for the qualitative consideration of various properties of the optical MMs has been presented. The developed approach has been incorporated in more general homogenization model for the compound materials with magnetic response. The main results could be presented in form of the "Multipole tree of knowledge" (in analogy with Metamaterials tree of knowledge introduced by Prof. N. Zheludev in [50]), which emphasizes central role of the multipole expansion in the elaborated approach.

The red apples in Fig. 1 show already developed applications, while the green ones present the applications which are planned to be further investigated.

The summarized conclusions of the work are:

1. Basics of homogenization procedure for Maxwell equations has been developed in maximum general form and is supposed to fill the gap between microscopic Maxwell equations and various forms of averaging procedures, suggested for MMs. The results have been published as a big review in form of "Feature Article" for journal "Metamaterials" [4] and presented and discussed at the international conferences.
2. The homogenization model for MMs has been developed based on multipole expansion approach and relations between this model and the elaborated basics of the homogenization have been found. The results have been published in [2, 5, 10] and presented at the international conferences.
3. The model has been applied to the optical MMs in order to consider linear and nonlinear effects in the MMs in the frame of the developed unified approach. The results have been published in [7, 9, 11, 12, 13, 14, 15] and presented at the international conferences.
4. The developed approach has been extended on the case of quantum MMs, i.e. MMs with MAs consisting of coupled plasmonic nano resonators and quantum

A. Chipouline and F. Küppers, *Optical Metamaterials: Qualitative Models*, Springer Series in Optical Sciences 211, https://doi.org/10.1007/978-3-319-77520-3

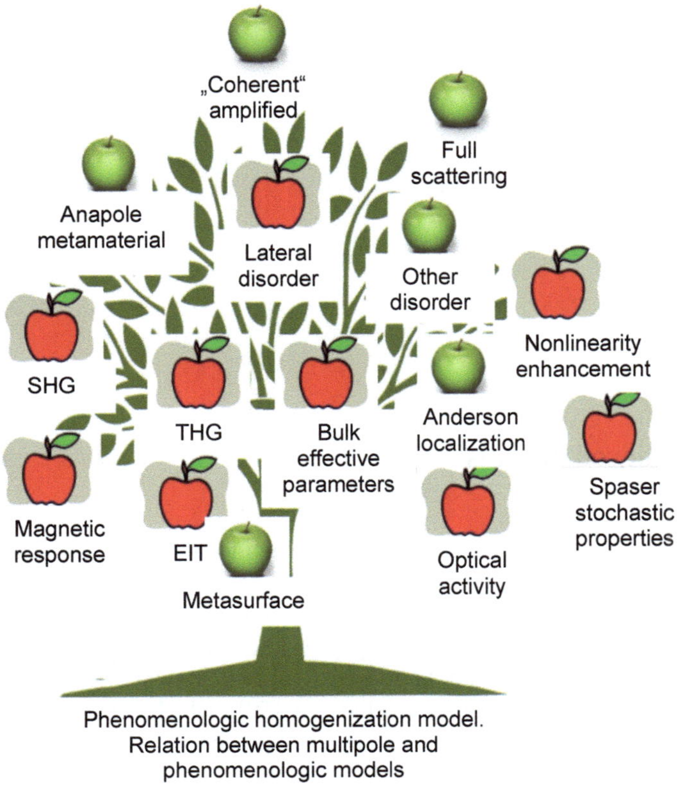

Fig. 1 "Multipole tree of knowledge". Red apples show the fully developed applications, while green ones present the problems which are planned to be solved

ingredients. The results have been published in [3, 6] and presented at the international conferences.

5. The developed model for quantum MMs has been applied to the cases of nonlinear MMs, dynamics of spaser, and MMs with gain. The results have been published in [8, 9], and presented at the international conferences.

Prepared Publications and Presentations

The list of the prepared publications with the related to the presented work results is summarized below:

Book chapter:

1. J. Petschulat, C. Rockstuhl, C. Menzel, A. Chipouline, A. Tünnermann, F. Lederer, and T. Pertsch, Multipole metamaterials. in *Plasmonics and*

Plasmonic Metamaterials, ed. by G. Shvets, I. Tsukerman (World Scientific Publishing Co. Pte. Ltd., Singapore, 2012)

Journal and ArXiV publications:

2. A. Chipouline, J. Petschulat, A. Tuennermann, T. Pertsch, C. Menzel, C. Rockstuhl, F. Lederer, Multipole approach in electrodynamics of MM. Appl. Phys. A **103**, 899–904 (2011)
3. A. Chipouline, V.A. Fedotov and A.E. Nikolaenko, Analytical Model for MM with Quantum Ingredients. ArXiv **1104.0110** (2011)
4. A. Chipouline, C. Simovski, S. Tretyakov, Basics of averaging of the Maxwell equations for bulk materials. Metamaterials **6**, 77 (2012)
5. A. Chipouline, S. Sugavanam, J. Petschulat, T. Pertsch, Metamaterials with interacting Metaatoms. ArXiv (2012). http://arxiv.org/abs/1205.6839.
6. A. Chipouline, S. Sugavanam, V.A. Fedotov, A.E. Nikolaenko, Analytical model for active metamaterials with quantum ingredients. J. Opt. **14,** 114005 (2012)
7. A. Chipouline, S. Sugavanam, J. Petschulat, T. Pertsch, Extension of the multipole approach to random metamaterials. Adv. Optoelectron. **2012 (2012)**, Article ID 161402 (2012)
8. A.S. Chirkin, A.V. Chipouline, Generalized expression for the natural width of the radiation spectrum of quantum oscillators. JETP Lett. **93**, 114 (2011)
9. A. Nikolaenko, N. Papasimakis, A. Chipouline, F. De Angelis, E. Di Fabrizio, N. Zheludev, THz bandwidth optical switching with carbon nanotube metamaterial. Opt. Express **20**, 6, 6068 (2012)
10. J. Petschulat, C. Menzel, A. Chipouline, C. Rockstuhl, A. Tünnermann, F. Lederer, T. Pertsch, Multipole approach to metamaterials. Phys. Rev. B **78**, 043811 (2008)
11. J. Petschulat, A. Chipouline, A. Tüunnermann, T. Pertsch, C. Menzel, C. Rockstuhl, F. Lederer, Phys. Rev. A **80**, 063828 (2009)
12. J. Petschulat, J. Yang, C. Menzel, C. Rockstuhl, A. Chipouline, P. Lalanne, A. Tüennermann, F. Lederer, T. Pertsch, Opt. Express **18**(14), 14454 (2010)
13. J. Petschulat, A. Chipouline, A. Tünnermann, T. Pertsch, C. Menzel, C. Rockstuhl, T. Paul, F. Lederer, Simple and versatile analytical approach for planar metamaterials. Phys. Rev. B **82**, 075102 (2010)
14. E. Pshenay-Severin, U. Hübner, C. Menzel, C. Helgert, A. Chipouline, C. Rockstuhl, A. Tünnermann, F. Lederer, T. Pertsch, Double-element metamaterial with negative index at near-infrared wavelengths. Opt. Lett. **34**, 1678 (2009)
15. E. Pshenay-Severin, A. Chipouline, J. Petschulat, U. Huebner, A. Tuennermann, and T. Pertsch, Optical properties of metamaterials based on asymmetric double-wire structures. Opt. Express **19**, 6269 (2011)
16. J. Reinhold, M.R. Shcherbakov, A. Chipouline, V.I. Panov, C. Helgert, T. Paul, C. Rockstuhl, F. Lederer, E.-B. Kley, A. Tünnermann, A.A. Fedyanin, T. Pertsch, The contribution of the magnetic resonance to the third harmonic generation from a fishnet metamaterial. Phys. Rev. B **86**, 115401 (2012)

Conferences:

17. J. Petschulat, A. Chipouline, T. Pertsch, A. Tünnermann, C. Rockstuhl, C. Menzel, F. Lederer, Analytic description of plane wave propagation in metamaterials. SPIE photonics Europe 2008, Strasbourg, France, 2008
18. C. Rockstuhl, C. Menzel, T. Paul, J. Petschulat, E. Pshenay-Severin, C. Helgert, A. Chipouline, T. Pertsch, F. Lederer, Properties of bulk metamaterials. Invited, SPIE photonics Europe 2008, Strasbourg, France, 2008
19. J. Petschulat, A. Chipouline, C. Menzel, C. Rockstuhl, T. Pertsch, F. Lederer, Light propagation and effective parameters of nanowire based metamaterials: An analytical approach. JTuA123, CLEO 2008, San Jose, CA, USA, 4–9 May 2008
20. J. Petschulat, C. Menzel, C. Rockstuhl, A. Chipouline, T. Pertsch, F. Lederer, The dispersion relation of light in metamaterials: An analytical approach. META'08 conference, Marrakesh-Morocco, 7-10 May 2008.
21. A. Chipouline, J. Petschulat, C. Menzel, C. Rockstuhl, T. Pertsch, F. Lederer, "The dispersion relation of light in metamaterials: An analytical approach", "On the way to nanotechnological revolution". Workshop of Russian and former Soviet Union scientists, France, Porquerolles, 5–11 Oct 2008.
22. J. Petschulat, C. Menzel, C. Rockstuhl, A. Chipouline, T. Pertsch, F. Lederer, A. Tünnermann, Electrodynamic multipole properties of metamaterials. NFO'08, Argentina, Buenos-Aires, 2008.
23. J. Petschulat, A. Chipouline, T. Pertsch, A. Tünnermann, C. Rockstuhl, C. Menzel, F. Lederer, Analytic description of plane wave propagation in metamaterials. SIAM 2008, Rome, Italy.
24. J. Petschulat, A. Chipouline, E. Pshenay-Severin, A. Tünnermann, T. Pertsch, C. Menzel, C. Rockstuhl, T. Paul, F. Lederer, Linear and nonlinear properties of metamaterials: analytical modelling based on multipole expansion. 2nd European Topical Meeting on Nanophotonics and Metamarials, Seefeld, Austria, 5–8 Jan 2009
25. J. Petschulat, A. Chipouline, E. Pshenay-Severin, A. Tünnermann, T. Pertsch, C. Menzel, C. Rockstuhl, T. Paul, F. Lederer, Analytical modelling of linear and nonlinear properties of metamaterials based on multipole expansion. SPIE Optics + Optoelectronics meeting, Prague Congress Centre, Prague, Czech Republic, 20–24 Apr 2009
26. A. Chipouline, J. Petschulat, C. Menzel, C. Rockstuhl, A. Tuennermann, F. Lederer, T. Pertsch, Multipole approach in electrodynamics of metamaterials. Invited, META'10 conference, Cairo, 2010.
27. A. Chipouline, Special session organization "Analytical modeling of metamaterials". META'10 conference, Cairo, 2010
28. A. Chipouline, Homogenization of metamaterials based on multipole approach. Metamaterials 10, Karlsruhe, 2010
29. A. Chipouline, A. Chirkin, E. Pshenay-Severin, M. Mundus, T. Pertsch, On the question of semiclassical nano laser description. Metamaterials 10, Karlsruhe, 2010

30. A. Chipouline, S. Sugavanam, C. Helgert, J. Petschulat, T. Pertsch, Application of multipole model for analytical description of disorder in metamaterials. Metamaterials 10, Karlsruhe, 2010
31. A. Chipouline, Homogenization of metamaterials based on multipole approach. Invited, CAOL'2010, Sevastopol, 2010
32. A. Chipouline, M. Mundus, A. Chirkin, Generalized Schawlow-Townes formula for natural bandwidth of nano laser. Poster, CLEO'2011, München, 2011
33. A. Chipouline, Analytical modeling of optical properties of metamaterials coupled with quantum systems. Invited, Bio-plasmonic conference, Jena, 2010
34. A. Chipouline, V. Fedotov, On the question of quantum metmaterials, Nanometa 2011, oral, nanophotonics and metamaterials, Seefeld, Tirol, Austria, 2011
35. A. Chipouline, S. Sugavanam, C. Helgert, J. Petschulat, A. Tünnermann, T. Pertsch, A new route toward the description of disordered metamaterials: merging statistical methods and the multipole expansion. Poster, nanophotonics and metamaterials, Seefeld, Tirol, Austria, 2011
36. A. Chipouline, S. Sugavanam, Multipole expansion and spatial dispersion in metamaterials. Poster, Nanophotonics and Metamaterials, Seefeld, Tirol, Austria, 2011
37. A. Chipouline, M. Mundus, T. Pertsch, Semiclassical model of nano laser, Poster, nanophotonics and metamaterials, Seefeld", Tirol, Austria, 2011
38. A. Chipouline, J. Petschulat, A. Tünnermann, T. Pertsch, A generic model for analysis metamaterals with gain. Poster, nanophotonics and metamaterials, Seefeld", Tirol, Austria, 2011
39. A. Chipouline, V.A. Fedotov, A.E. Nikolaenko, Coupling classic and quantum objects: from nano-laser to quantum metamaterials. Invited, days of diffraction, St. Petersburg, 2011
40. A. Chipouline, Multipole model for metamaterials with gain: from nano laser to quantum metamaterials. Invited, SPIE'2011, Prague, 2011
41. A. Chipouline, Multipole approach for analytical modeling of passive and active metamaterials. Invited, ICMAT 2011, Singapore, 2011
42. A. Chipouline, Multipole approach for analytical modeling of metamaterials. Invited, Photonica 2011, Belgrade, 2011
43. A. Chipouline, V. Fedotov, Towards quantum magnetic metmaterials. Proceedings Nanometa, 2011, THU4s.3 (96), 2011
44. A. Chipouline, S. Sugavanam, V.A. Fedotov, A.E. Nikolaenko, Simple Analytical description of CNT nonlinear response enhanced by metamaterials. Metamaterials'12 conference, Sankt-Petersburg, 2012
45. A.V. Chipouline, Course of nanophotonics in view of modernization of courses of electrodynamics: 4 years of experience. Metamaterials'12 conference, Sankt-Petersburg, 2012
46. A. Chipouline, C. Simovski, Basics of homogenization of Maxwell equations. Metamaterials'12 conference, Sankt-Petersburg, 2012

47. M.Yu. Saygin, A.V. Chipouline, A.S. Chirkin, Analysis of the nanolaser linewidth using semiclassical laser model. Metamaterials'12 conference, Sankt-Petersburg, 2012
48. A.V. Chipouline, Homogenization of metamaterials based on multipole approach. Novosibirsk, Nano-photonics Workshop, 2012
49. A.V. Chipouline, V. Fedotov, Special session "Active and Quantum Metamaterials". META 2012, Paris, 2012
50. N. Zheludev, The road ahead for metamaterials, Science **328**, 582 (2010)

Possible Future Developments and Applications of the Presented Approach

The presented in this work model is planned to be further developed. The proposed activity can be summarized in the shown in Fig. 2 scheme. Two possible ways of expansions (multipole and anapole ones) in combination with the quantum ingredient are supposed to become the "elementary bricks" for various applications in area of bio oriented research, optical components for telecommunication and signal processing, and superconducting-based MMs.

The main points of the plan are:

1. Systematic investigation of the elementary artificial building blocks (MA), consisting of coupled plasmonic nanoresonators and quantum objects of different types.

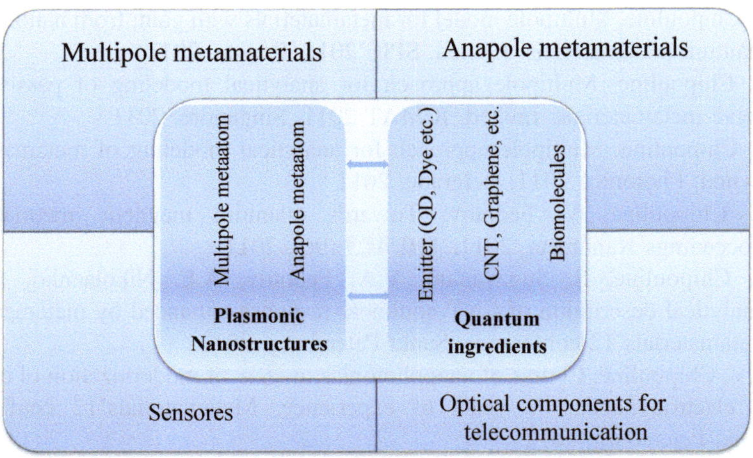

Fig. 2 Schematic representation of the future development of the multipole model and its possible applications

2. Investigation of electro-magnetic properties of MMs, constructed/engineered from the aforementioned elementary building blocks.
3. Investigation of new optical components for application in areas of sensors and telecommunication designed and built using the described above elementary building blocks.

Acknowledgements Acknowledgments start from greatest appreciation of the invaluable support of my family, my beautiful wife Marina and my son Fedor for understanding and tolerance during the preparation of this work and for continuous support me during all our life; to my parents, whose interest to my work really inspired me from my student time and who up to now remain a great example of positive emotions and enthusiasm in various situations.

Author is extremely thankful for the tolerance and inestimable long term assistance and supervising of Prof. Falk Lederer, and organizational support and attention of Prof. Andreas Tünnermann. Author specially acknowledges superior scientific environment and financial support organized by Prof. Thomas Pertsch in his group of nanophotonics, IAP/FSU.

Author gratefully acknowledges valuable scientific discussions with Prof. Falk Lederer, Prof. Thomas Pertsch, Prof. Andreas Tünnermann, Prof. Carsten Rockstuhl, and all members of the nanophotonics group at IAP/FSU and theoretical optics group in Institute of Solid State Physics and Optics, FSU Jena. Special acknowledgments have to be given to Dr. Jörg Petschulat, who has made crucial contributions to the development of this approach and to Dr. Christoph Menzel, who has done multiple numerical calculations in support of the analytical constructions, to Dr. Ekaterina Pshenay-Severin for the theoretical and experimental work on asymmetric structures, and to Dr. Christian Helgert for the technological support.

Acknowledgments go to the member of technological group headed by Dr. Bernhardt Kley for the multiple discussions and sample preparations.

Author would like specially acknowledge technical and organizational support of Dr. Bodo Martin.

Author is extremely thankful for the organizational support received every day from the secretaries of the IAP, namely from Frau Christin Weber and Frau Julia Vetter, Frau Sabina Rockstroh, Frau Carolla Steinberg, Frau Anja Kluge, and Frau Silvia Abbe; from secretariat of ASP from Frau Szilvia Mammel, Frau Ricarda Knetsch, and Frau Dorit Schmidt.

Author also specially acknowledges inestimable editing support received from one of the ASP student Mr. Alex Brown.

Author acknowledges with great pleasure multiple scientific discussion and joint work with Prof. E. Ilichev (IPHT, Jena), Prof. A. Ustinov (KIT, Karlsruhe), Prof. N. Zheludev, Dr. V. Fedotov, Dr. A. Nikolaenko, Dr. E. Plum (ORC, University of Southampton, UK), Prof. Sergey Turitsyn, Mr. Srikanth Sugavanam (Aston University, UK), Prof. S. Tretyakov, Prof. C. Simovsky (University Aalto, Helsinki, Finland), Prof. M. Stockman (Georgia University, USA), Prof. Ildar Gabitov (Arizona State University, USA), Prof. Z. Vardeni (University of Utah, USA), Prof. Kikuo Cho (University Osaka, Japan), Dr. Ashod Azaryan (University Boudreaux, France), Prof. A. S. Chirkin and Prof. Fedyanin (Moscow State University, Russian Federation), and Prof. A. Vinogradov (MIPT, Russian Federation).

Author would like to give special thanks to the great teacher whose unforgettable lectures turned out to be decisive in formation of author's mind in his student's time: Academician Prof. Evgenii Dianov, Prof. Victor Veselago, Prof. Nikolai Karlov, Academician Prof. Vitaliy Ginzburg, Academician Prof. Peter Kapiza, Academician Prof. Sergey Kapiza, and Academician Prof. Alexander Prokhorov.

Bibliography

1. R. Adair, L.L. Chase, S.A. Payne, J. Opt. Soc. Am. B **4**, 875 (1987)
2. R. Adair, L.L. Chase, S.A. Payne, Phys. Rev. B **39**, 3337 (1989)
3. G. Bachelier, I. Russier-Antoine, E. Benichou, C. Jonin, P.-F. Brevet, JOSA B **25**, 955 (2008)
4. D. Bethune, Opt. Lett. **6**, 287 (1981)
5. N. Bloembergen, *Nonlinear Optics* (Benjamin Press, New York, 1965)
6. N. Bloembergen, W.K. Burns, M. Matsuoka, Opt. Commun. **1**, 195 (1969)
7. P. Butcher, D. Cotter, *The Elements of Nonlinear Optics* (Cambridge University Press, Cambridge, 1990)
8. D.J. Cho, F. Wang, X. Zhang, Y.R. Shen, Phys. Rev. B **78**, 121101(R) (2008)
9. K. Cho, *Reconstruction of Macroscopic Maxwell Equations: A Single Susceptibility Theory*, vol. 237, Springer Tracts in Modern Physics (Springer, Berlin, Heidelberg, 2010). https://doi.org/10.1007/978-3-642-12791-5
10. M. Choy, R. Byer, Phys. Rev. B **14**, 1693 (1976)
11. N. Feth et al., Opt. Lett. **33**, 1975 (2008)
12. A.A. Golubkov, V.A. Makarov, Boundary conditions for electromagnetic field on the surface of media with weak spatial dispersion. Phys. Usp. **38**, 325 (1995)
13. J. Jayabalan, P. Manoranjan, A. Banerjee, K.C. Rustagi, Phys. Rev. B **77**, 045421 (2008)
14. P.B. Johnson, R.W. Christy, Phys. Rev. B **6**, 4370 (1972)
15. E. Kim, F. Wang, W. Wu, Z. Yu, Y.R. Shen, Phys. Rev. B **78**, 113102 (2008)
16. M.W. Klein, C. Enkrich, M. Wegener, S. Linden, Science **313**, 502 (2006)
17. M.W. Klein, M. Wegener, N. Feth, S. Linden, Opt. Express **15**, 5238 (2007)
18. T. Koshny, P. Markos, D. Smith, C. Sokoulis, Resonant and antiresonant frequency dependence of the effective parameters of metamaterials. Phys. Rev. E. **68**, 065602(R) (2003)
19. S. Kujala, B.K. Canfield, M. Kauranen, Y. Svirko, J. Turunen, PRL **98**, 167403 (2007)
20. J. Park, *Fourier modal method and its applications in computational nanophotonics*. Crc Pr Inc, ISBN-10: 1420088386, ISBN-13: 978-1420088380, 2012.
21. A. Maluckov, L. Hadzievski, N. Lazarides, G. Tsironis, Phys. Rev. E **77**, 046607 (2008)
22. A. Mary, S. Rodrigo, F. Garcia-Vidal, L. Martin-Moreno, PRL **101**, 103902 (2008)
23. R. Miller, APL **5**, 17 (1964)
24. P. Mühlschlegel, H.-J. Eisler, O. Martin, B. Hecht, D. Pohl, Science **308**, 1607 (2005)
25. T. Paul, C. Rockstuhl, F. Lederer, JOSA B **27**, 1118 (2010)
26. J. Petschulat, J. Yang, C. Menzel, C. Rockstuhl, A. Chipouline, P. Lalanne, A. Tüennermann, F. Lederer, T. Pertsch, Opt. Express **18**(14), 14454 (2010)
27. V. Pustovit, J. Sotelo, G. Niklasson, JOSA A **19**, 513 (2002)

A. Chipouline and F. Küppers, *Optical Metamaterials: Qualitative Models*, Springer Series in Optical Sciences 211, https://doi.org/10.1007/978-3-319-77520-3

28. J. Reinhold, M. Shcherbakov, A. Chipouline, V. Panov, C. Helgert, T. Paul, C. Rockstuhl, F. Lederer, E.-B. Kley, A. Tünnermann, A. Fedyanin, T. Pertsch, The contribution of the magnetic resonance to the third harmonic generation from a fishnet metamaterial. Phys. Rev. B **86**, 115401 (2012)
29. A. Sarychev, V. Shalaev, *Electrodynamics of Metamaterials* (World Sci, Singapore, 2007)
30. I. Shadrivov, A. Kozyrev, D. Weide, Y. Kivshar, Appl. Phys. Lett. **93**, 161903 (2008)
31. I. Shadrivov, A. Kozyrev, D. Weide, Y. Kivshar, Opt. Express **16**, 20266 (2008)
32. N. Sharma, PRL **98**, 217402 (2007)
33. Y. Shin, A. Chavez-Pirson, Y. Lee, Opt. Lett. **25**, 171 (2000)
34. D. Smith, W. Padilla, D. Vier, S. Nemat-Nasser, S. Schultz, PRL **84**, 4184 (2000)
35. M. Strandberg, Phys. Rev. **106**, 617 (1957)
36. M. Thiel, G. von Freymann, S. Linden, M. Wegener, Opt. Lett. **34**, 19 (2009)
37. Y. Zeng, W. Hoyer, J. Liu, S. Koch, J. Moloney, Phys. Rev. B **79**, 235109 (2009)
38. A. Zharov, I. Shadrivov, Y. Kivshar, PRL **91**, 037401 (2003)

Index

A. Chipouline and F. Küppers, *Optical Metamaterials: Qualitative Models*, Springer Series in Optical Sciences 211, https://doi.org/10.1007/978-3-319-77520-3

MIX
Papier aus verantwortungsvollen Quellen
Paper from responsible sources
FSC® C105338

If you have any concerns about our products,
you can contact us on
ProductSafety@springernature.com

In case Publisher is established outside the EU,
the EU authorized representative is:
Springer Nature Customer Service Center GmbH
Europaplatz 3, 69115 Heidelberg, Germany

Printed by Libri Plureos GmbH
in Hamburg, Germany